Multi-Agent Systems for Concurrent Intelligent Design and Manufacturing

Multi-Agent Systems for Concurrent Intelligent Design and Manufacturing

Weiming Shen, Douglas H. Norrie
and Jean-Paul A. Barthès

London and New York

First published 2001
by Taylor & Francis
11 New Fetter Lane, London EC4P 4EE

Simultaneously published in the USA and Canada
by Taylor & Francis Inc,
29 West 35th Street, New York, NY 10001

Taylor & Francis is an imprint of the Taylor & Francis Group

© 2001 Weiming Shen, Douglas H. Norrie and Jean-Paul Barthès

Printed and bound in Great Britain by
Biddles Ltd, Guildford and King's Lynn

Publisher's Note
This book has been prepared from camera-ready copy provided by the
authors.

British Library Cataloguing in Publication Data
A catalogue record for this book is available from the British Library

Library of Congress Cataloging in Publication Data
Shen, Weiming
   Multi-agent systems for concurrent intelligent design and manufacturing /
Weiming Shen, Douglas H. Norrie, and Jean-Paul Barthès.
      p. cm.
   Includes bibliographical references and index.
   1. Design, Industrial - Computer-aided design 2. Intelligent agents
(Computer software) 3. Concurrent engineering. I. Norrie D.H. II. Barthès,
Jean-Paul, III. Title.

TS171.4.S543 2000
658.5'752'028563-dc21

                                                                                                00-033764

ISBN 0-7484-0882-7

Table of Contents

Part II Important Issues in System Implementation 53

Chapter 4 Knowledge Representation in Agent-Based Concurrent Design and Manufacturing Systems.. 55

Part I: Introduction

Part Three

1

General Introduction

1.1 MOTIVATION

Agent technology is widely recognized as a promising paradigm for the next generation of design and manufacturing systems. Researchers have already applied agent technology to concurrent engineering, collaborative engineering design, manufacturing enterprise integration, supply chain management, manufacturing planning, scheduling and control, material handling, and holonic manufacturing systems. The technology is sufficiently mature that some universities have created graduate courses in this area in departments of Computer Engineering, Electrical Engineering, Mechanical Engineering, Manufacturing Engineering, and Industrial Engineering. Outside of these departments, an increasing number of engineers and researchers wish to learn more about this emerging technology. For background reference, there are relevant texts on AI or DAI such as "Artificial Intelligence: A Modern Approach" (Russell and Norvig, 1995), "Foundations of Distributed Artificial Intelligence" (O'Hare and Jennings, 1996), "Readings in Distributed Artificial Intelligence" (Bond and Gasser, 1988). For agent technology, there are also recent texts such as "Multi-Agent Systems" (Ferber, 1999, the translation of his 1995 work "Les Systemes Multi-Agents"), "Software Agents" (Bradshaw, 1997), "Readings in Agents" (Huhns and Singh, 1997), "Agent Technology: Foundations, Applications and Markets" (Jennings and Wooldridge, 1998), "Multiagent Systems: A Modern Approach to Distributed Artificial Intelligence" (Weiss, 1999), as well as more specialized works in the series "Lecture Notes in Artificial Intelligence" and in numerous conference and workshop proceedings. However, there are no reference books on *agent technology for concurrent engineering design and manufacturing* describing systematically and synthetically the relevant theories and methods and giving practical information and examples for developing multi-agent systems in this domain. The objective of this book is to fill this gap.

1.2 BOOK ORGANIZATION

The book is divided into four parts:

- The first part describes concurrent engineering design and manufacturing and its requirements and also gives an introduction to Distributed Artificial Intelligence and to agent technology.

- The second part discusses key issues in developing agent-based concurrent engineering design and manufacturing systems: knowledge representation, agent architectures, system organization, learning, communication, cooperation, coordination, negotiation, conflict resolution, and ontology. Most of these issues are common to agent-based systems, but some are specific to agent-based concurrent design and manufacturing systems. These issues are discussed using examples from engineering design and manufacturing.
- The third part reviews significant and relevant agent-based systems. All selected projects are described in a unified format, to facilitate understanding and comparison of the different approaches.
- The fourth part introduces methodology, standards, frameworks, tools, languages, and finally a general procedure for building agent-based concurrent engineering design and manufacturing systems.

A note about the title: Concurrent Engineering is now evolving into Collaborative Engineering (see for example, Mills 1999) and the technology described in this book will provide support for this new area. However, we use the widely understood term Concurrent Engineering in this book because of its familiarity to most practitioners.

1.3 HOW TO USE THIS BOOK

This is the first book systematically describing the principles, key issues, and applications of agent technology in relation to concurrent engineering design and manufacturing. It can be used as a reference book by professional engineers and researchers in this area. It can also be used as a university course text or as a supplementary text for graduate students.

It is not strictly necessary to read through this book sequentially. However, readers with little knowledge in the area should read Chapter 2 first to understand the requirements for next generation concurrent design and manufacturing systems as well as concepts and terms used throughout the book, and then read Chapter 3 for an introduction to distributed artificial intelligence and agent technology.

Chapters 4 to 12 are on key issues that are important for implementing agent-based concurrent design and manufacturing systems. However, Chapter 6 (Agent Architectures) and Chapter 7 (System Architecture) are perhaps more important than the other chapters in this sequence. Thus, it is recommended that Chapters 6 and 7 be read next, then Chapters 8, 9, 10 and so on. Chapters 4 and 5 may be returned to later although readers who have some knowledge in this area may start the second part of the book with Chapters 4 and 5.

Chapters in the third part dealing with application examples are optional. Readers may wish only to read the chapters related to their interests or projects, in this section. Many cited references and web sites are provided for those who need more detailed information. Readers, such as graduate students, who want to compare the different approaches used in various projects may need to work through all three application chapters, or through the sections in each chapter relevant to the application domain of interest.

The two chapters in the last part of the book are written primarily for practitioners in this area and provide some direction on selecting an appropriate methodology, framework, tools and programming languages when implementing a prototype. It is strongly suggested that the cited web sites be visited to get updated information, since commercial software especially can be upgraded within a short time period. The final chapter provides some guidelines for practitioners to follow during prototype design and implementation.

REFERENCES

Bond, A.H. and Gasser, L. (eds) (1988) *Readings in Distributed Artificial Intelligence*, Morgan Kaufmann, San Mateo, CA.

Bradshaw, J. (eds.) (1997) *Software Agents*, MIT Press.

Ferber, J. (1995) *Les Systèmes Multi-Agents*, InterEditions, Paris.

Ferber, J. (1999) *Multi-Agent Systems*, Addison-Wesley.

Huhns, M. and Singh, M. (eds.) (1997) *Readings in Agents*, Morgan Kaufmann.

Jennings, N. and Wooldridge, M. (eds.) (1998) *Agent Technology: Foundations, Applications and Markets*, Springer.

Mills, C.A. (1999) *Collaborative Engineering as an Element of the Integrated Manufacturing Enterprise*, CASA/SME Blue Book Series, Society of Manufacturing Engineers.

O'Hare G.M.P. and Jennings, N. (eds.) (1996) *Foundations of Distributed Artificial Intelligence*, Wiley-Interscience.

Russell, S. and Norvig, P. (1995) *Artificial Intelligence: A Modern Approach*, Prentice Hall.

Weiss, G. (ed.) (1999) *Multiagent Systems: A Modern Approach to Distributed Artificial Intelligence*, MIT Press.

2

Concurrent Engineering Design and Manufacturing

2.1 INTRODUCTION

Engineering Design and Manufacturing is often abstracted as a series of stages as shown on Figure 2.1. Although this is intended to apply to the many different situations encountered in the domain, it is too general to account for the actual organization of the engineering process.

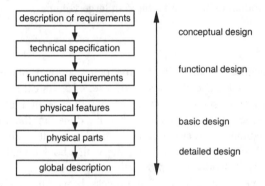

Figure 2.1 General description of design stages.

As mentioned previously, the design and manufacture of large complex artifacts requiring the cooperation of a number of specialists is of particular interest. The three following examples focus respectively on the very different domains of harbor design, automobile design, and plant design.

Harbor Design

The example is taken from Monceyron and Barthès (1992). Designing a harbor is a long process which may last up to 10 years. The basic question to be answered is how much of the harbor must be laid inland and how much will be off shore. The resulting decision is a compromise between the cost of building expensive structures like breakwaters and the cost of continually removing silt. The first step in harbor design is to define the geometry, then iterations take place which include the design of breakwaters, the computation of the residual wave motion inside the harbor, and discussions with many different specialists, like firemen, pilots, dockers, etc. Because many data are lacking when starting the design, the first phase may last up to two years. During this preliminary design phase, the designers study different possible -choices (sometimes up to 40 variants), compute the financial implications, and rank them. At the end of the preliminary phase only one or two solutions are kept. They will be used as the basis for detailed computations, after some additional measurements are acquired. The core team of designers for this activity involves merely a dozen of specialists working at the same time.

Automobile Design

Designing a new model of an automobile (e.g. a hybrid vehicle (with both electrical and gasoline engines) involves a number of people, working together during a preliminary phase in order to produce an operational prototype demonstrating the feasibility of some choices. The prototype will then serve as the basis for the design of the industrial product to be manufactured in large volume.

Plant Design

Design in chemical engineering (e.g., for petrochemical plants), starts from a chemical process, which basically consists of a set of chemical reactions. A first step consists in transforming such equations into a sequence of unit operations which determine what kind of transformation will be done on the raw materials and intermediate compounds in order to obtain the final product(s). The result is usually a single blue sheet (Process Flow Sheet) together with a book containing the result of a number of computations done to validate the early design (e.g., mass balance, and heat balance, safety computations). The Process Flow Sheet is then given to the engineering bureau, where specialists do the detail design, i.e., produce a (large) number of drawings (and may build a mock-up), called Process and Instrumentation Diagrams (P&ID). The latter are used on the construction site by field engineers to actually build the plant.

The three previous examples are good candidates for the kind of engineering activities that we have in mind. More generally, design may proceed from fairly abstract requirements obtained from the marketing department or specified in a contract, through a conceptual design step, then through preliminary design. The result from preliminary design can be a set of specifications or a real operational prototype (e.g., automobile, pilot plant). Then follows a step which is a mixture of redesign (possibly including scale up) and detailed design, then construction or manufacturing, assembly, testing, and start up. The important point here is that currently the manufacturing or construction team is different from the design team, which itself is different from the preliminary design team. The steps follow a clearly identified waterfall style as shown on Figure 2.2. The general approach in Concurrent

Engineering is to overlap the stages of this linear progression in design, but unless the stages are coordinated carefully this approach can cause considerable difficulty.

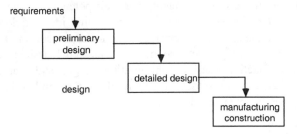

Figure 2.2 Global design steps.

In an actual situation, the Preliminary Design phase can be quite complex: for harbor design it consists of computing and selecting among dozens of different solutions to keep only one or two, the whole process lasting two years; for an automobile, it consists of gathering a design team, putting them together (the design platform), and having them build a prototype, which may also last well over a year; in chemical engineering, the preliminary design consists mainly of qualifying the process by doing computations involving material and heat balance, on many possible variations of the target process, then storing these for transfer to the engineering bureau or to an external customer (in some cases this stage includes building a small scale pilot plant).

The Detailed Design phase consists of starting from the result of the Preliminary Design phase and detailing the design according to the manufacturing or construction constraints, taking into account data that were ignored during the preceding phase.

The Manufacturing or Construction phase is also quite different, since it requires the set up of the manufacturing process or the construction site, and it is organized and implemented by people other than those worked in the two previous steps, and usually with a quite different culture.

Concurrent engineering will require appropriate sharing of data and knowledge among all those who contribute to the various stages and sub-stages of designing, building, or manufacturing of the product.

Now if we turn to research, the picture is quite different. Researchers are more interested in the conceptual aspects of the design activities, and preferably in activities that can be performed by a single person or a small group of persons working closely together. Many papers have been published on the theory of design and how to improve design tools or methods in specific restricted sub-domains. Most of the papers address the preliminary design phase, mainly because problems at this stage remain tractable, while at a later stage their nature changes to become mostly organizational and of a different nature. Does this mean that we should restrict our study of the engineering process to preliminary design? This is clearly not the aim of those who seek to introduce agent technology into the field.

2.2 ENGINEERING DESIGN

Before introducing the concepts of agents and their potential for the design domain, we review a few (traditional and new) concepts found in the field of engineering design in this section and in manufacturing in the following section. Let us start with these concepts: basic design, detailed design, conceptual design, creative design, innovative design, functional design, configuration design, catalog design, parametric design, feature-based design, redesign, routine design, compiled design, intelligent design, and knowledge intensive design.

2.2.1 Design and Design Process

Before examining each concept or pair of concepts, let us recall the purpose of design. Among the many definitions proposed for design, Takala (1989) gives the following one: "Designing is the process of finding or constructing a product or its explicit representation, or model of it, which will implicitly satisfy specified functional and other requirements expected by its users." This definition summarizes well the modern meaning of the word design.

In Europe the word "design" (or "designer") often has a different connotation, more artistic and less technical, perhaps inherited from its origins. According to Orel (1991) the concept of design appeared during the Renaissance, when artists were talking of *disegno* to describe the quality or the aesthetics of a picture or of a model.

Design also implies a process which can be described as such. According to Treur (1991): "In general a design process starts by establishing the need for a certain object and by expressing a number of properties that the object should possess; we call these properties the (initial) design requirements. The aim of a design process is to construct an explicit description of the structure of an object satisfying the design requirements."

2.2.2 Basic Design vs. Detailed Design

Basic design is concerned with the development of a rough feasible solution to the problem to be solved. In our examples basic design is equated with the preliminary design phase, i.e., building an automobile prototype, computing a process flow sheet, or selecting an approximate harbor layout. Detailed design will have to specify the missing information needed to build the final product. In our examples, basic design and detailed design are performed by different teams.

2.2.3 Conceptual Design

At the beginning of a design process (during preliminary design), one usually distinguishes a special phase called conceptual design. Many authors have given definitions of conceptual design. Among them:

> The earliest phase of the design process is distinguishable from the later phases in that it is concerned more with the understanding of the problem and making general rather than specific decisions about the solution. [...] Conceptual design commences with high level descriptions of requirements and proceeds with a high level description of a solution.
>
> McNeil et al., 1998

> Conceptual design [...] is the formulation of abstract ideas with approximate concrete [...] representations.
>
> Takala, 1989

> The early or conceptual stage of the [design] process is dominated by the generation of ideas which are subsequently evaluated against general requirements' criteria. There follows a process whereby additional data are incorporated allowing decisions to be made between competing alternatives as more tangible evidence of function is derived.
>
> Baker et al., 1991

Although the conceptual design phase is undertaken with limited information, it will condition the resulting steps of the design. Poor choices will lead to poor designs, and poor designs will lead to poor products.

As is evident, at the conceptual design stage, information is incomplete, which renders the problem of representing the designed product and the design process quite difficult. Several representations have been proposed for this stage such as bond graphs (Finger and Rinderle, 1989), or the sketching of abstractions (Mukherjee and Liu, 1995). Furthermore, there is a lack of available tools for supporting the conceptual design stage, which is quite understandable regarding the abstract nature of this step. Authors like Karni and Arciszewski (1997) insist that such tools should act as assistants rather than being prescriptive.

Welch and Dixon (1994) further partition conceptual design into two stages: phenomenological design (transforming a functional requirement to a behavioral description); and embodiment design (matching the behavioral description to initial physical systems).

2.2.4 Creative and Innovative Design

Terms like creative or innovative design imply that something new is introduced into the product. Creative design could apply to inventions, e.g., a new painting style.

Innovative design applies to the redesign of existing equipment by using different materials or technologies, e.g., replacing steel parts with polymers in automobile bodies, replacing hydraulic with electrical control in airplanes (Airbus 320), etc. A specific approach to innovative design is proposed by Otto and Wood (1998) who suggest doing reverse engineering followed by a redesign phase, and illustrate this by redesigning a wok. Other authors are interested in combining artificial intelligence techniques for trying to automate the creative and innovative phases of the design process (Sushkov et al., 1995).

2.2.5 Functional Design

Functional design refers to the design step intermediate between technical specifications and physical features (embodiment) as shown in Figure 2.1. Originally, the important role of functional design was to express the requirements as a set of consistent functional constraints to be later satisfied by actual physical mechanisms. Although this design step still exists, it has been refined over the years. Moreover, it has been found that functional approaches have a much wider scope in the product life cycle, as argued by Chittaro and Kumar (1998) who stress the importance of functional representation and reasoning as applied to design, diagnosis, or automation. They make the distinction between function as a relation between input and output of some quantity (energy, matter, information), and the relation between the goal of a human user and the behavior of a component, mentioning however that the distinction is sometimes difficult to enforce.

Functions and functional representation are used as a general representation mechanism in design, for example:
- to capture design rationale (Chandrasekaran et al., 1993),
- to support conceptual design (Chakrabarti and Bligh, 1994), or for stamped metal parts (Mukherjee and Liu, 1995),
- to perform design analysis (Price, 1998),
- to organize knowledge about a device and its environmental interactions, supporting adaptive design as an alternative to routine design (Prabhakar and Goel, 1998).

Several specialized workshops and conferences are devoted each year to functional approaches: workshops at AAAI, IJCAI, or organized by the International Functional Modeling Applications Association (IFMAA).

2.2.6 Configuration Design, Catalog Design

Configuration Design is defined by Carlson-Skalak and coworkers (1998) as "the engineering task in which a configuration is created by assembling off-the-shelf components into a functional system." The process consists mainly of exploring a large solution space for locating feasible combinations of components. The search can be helped by using different techniques like:

-   reasoning about the function of a component eventually abstracted at different levels, which was promoted by McDermott (1982) or Mittal and coworkers (1986),
-   applying neural network approaches,
-   using grammars as done by Schmidt and Cazan (1997),
-   using genetic algorithms, as proposed by Brown and coworkers (1993), Carlson-Skalak et al. (1998), Gero and Kazakov (1998), Jo and Gero (1998).

2.2.7 Parametric Design

Parametric design is mainly based upon geometric transforms and was found in the commercial CAD systems as early as the mid eighties. At the time, it allowed representation of part families by setting a number of dimensions as variables. Because the part was actually programmed as a macro, it could accommodate simple structural changes (e.g., changing the number of holes in a plate). A more complex approach in research was to use variational geometry to determine the final value of dimensions by optimization techniques (Gossard et al., 1988). Another approach consists of using geometrical constraints as in the SOAR system (Arbab, 1989), or more recently, coupled with incremental specification as in the ICAD™ system. However, the most sophisticated approach to parametric design is included in the adaptation step of case-based reasoning approaches as used by Hua and coworkers (1996) who offer to design new houses from a collection of existing ones.

Parametric design (also called constraint-based design) can be applied in conjunction with feature-based design as shown by Fu and de Pennington (1993).

2.2.8 Feature-Based Design

The idea of feature-based design is to reduce the complexity of part design by a partial discretization of the design space. It is a kind of intermediate step to reduce the design to configuration design. At the same time the approach tries to bridge the gap with manufacturing (Shah and Rogers, 1993). As stated by Rosen (1993) "features are meaningful abstractions of geometry that engineers use to reason about components, products, and processes. For design activities, features are design parameters and can incorporate information relevant to life-cycle activities such as manufacturing." Nielsen and coworkers (1991) talk about the possibility of capturing and using the design intent. Work has progressed in two directions: developing design systems based on features (Cutkosky et al., 1992), and developing libraries of features as proposed by Gene and coworkers (1998) for snap-fit assemblies.

Two other lines of work are worth mentioning in the domain of feature-based design: the first one being feature extraction (related to the problem of recognizing features in previous designs or in geometric models); the second focusing on defining a generic representation that could be used at a later stage in manufacturing.

2.2.9 Redesign and Routine Design

Redesign consists of starting from an existing product and trying to improve it. A good example is given by Otto and Wood (1998) when they first use reverse engineering on a kitchen appliance and then try to systematically improve the old design by testing alternative solutions. The approach is offered as a possibility to produce innovative designs.

Routine design is usually related to the repetitive design of the same family of product. A new product is produced by modifying previous ones. Most of our products follow this process from appliances to computers, cars, airplanes, factories. Obtaining innovative products is highly desirable and researchers have offered new approaches which often introduce techniques from artificial intelligence. An example is the use of grammars in architecture (Oksala, 1993) or the use of history to explore the design space in furniture design (Takala, 1989).

2.2.10 Compiled Design

Compiled design is a dream pursued by many industrialists. The main idea is to automatically produce manufacturing specifications from product requirements. In the domain of VLSI design, after the introduction of new techniques (e.g., Mead and Conway (1980)) a high degree of automation was achieved, leading to the concept of the silicon compiler. In the field of mechanical engineering, the situation is more complex because the designer deals with continuous rather than discrete variables. However, an approach like feature-based design introduces some level of discretization bringing the design process one step closer to automation. Researchers like Whitney (1996) are quite pessimistic about a possible transfer of VLSI design and manufacturing techniques to the world of mechatronic systems. This view however is not shared by everybody, and Antonsson (1997) estimates that "significant potential exists for the development of approaches to compilation of mechanical systems (in analogy to compilation of digital VLSI)."

2.2.11 Intelligent Design

During the second part of the eighties, intelligent design was based on introducing and using knowledge within the design tools. Bridge (1989) wrote: "The term *intelligent CAD* is understood as an environment to support the designer's intellectual activities which contains tools with built-in design knowledge." At the same time, Rogier and Tolman (1989) wrote: "the word 'intelligent' in intelligent design must be interpreted as: 'providing the possibility to manipulate and extend the amount of knowledge stored in the system'. This amount of knowledge stored in the system may be divided into three categories: (1) domain specific knowledge, (2) procedural knowledge, (3) meta-knowledge."

Some time later, Tomiyama and coworkers (1995) distinguished several generations of intelligent CAD systems: "the first generation was based on the expert system technology [...]; the second generation intelligent CAD further incorporated abilities for constraint management and solving [...]; the third generation incorporates a representation of both design process and design objects."

In practice, intelligent design was developed along three axes:

(1) introducing mechanisms from artificial intelligence into CAD environments;
(2) coupling CAD tools with expert systems;
(3) developing new design tools from artificial intelligence bases.

Among the large number of projects developed in the late eighties and early nineties we select a few to illustrate these different approaches. For the first approach which consists of augmenting CAD tools with mechanisms for knowledge, one can mention:

- the introduction of constraints and reasoning about constraints in the geometric world (Arbab, 1989; Bridge, 1989; Veltkamp, 1991), or more recently Guan and MacCallum (1995);
- the use of constraints and TMS (Faltings et al., 1991; Baykan and Fox, 1991);
- the mixing of numeric and symbolic approaches as proposed by Murtagh and Shimura (1989);
- the use of grammars to generate families of designs (Aksoy and Saglamer, 1993; Oksala, 1993);
- the development of new representation languages for adding knowledge to traditional design tools using objects or frames, allowing to add rules (Krause et al., 1989; Bijl, 1989; Barthès, 1993);
- the use of learning techniques to improve the design process (Persidis and Duffy, 1989);
- the use of object-oriented approaches for new CAD tools, e.g. IIICAD (Xue et al., 1989);
- the use of neural networks and genetic programming mechanisms to limit search in large design spaces (Brown et al., 1993; Carlson-Skalak et al., 1998; Gero and Kazakov, 1998; Jo and Gero, 1998).

The second approach was more favored by industry, since it could solve legacy problems by interfacing traditional CAD environments with intelligent systems. Some interesting projects were proposed along this line:

- using CAD environments as slave systems driven from an intelligent external system, as in the ARCHIX project used to design front-wheel drive trains for automobiles (Thoraval, 1991), or in the ANAXAGORE project where the goal was to optimize satellite configurations (Trousse, 1993);

- using multi-expert environments to integrate intelligent as well as legacy systems, as was done by Monceyron and Barthès (1992) for harbor design.

The third approach considered the design process from the point of view of artificial intelligence and developed new tools using artificial intelligence concepts, techniques, or languages (such as LISP or PROLOG). Activities ranged from modeling the early design steps to producing actual design tools:

- logic was considered for providing a backbone to the design approach by Tweed and Bijl (1989), and Treur (1991);
- logical approaches were also used to develop specialized systems, as was done by Kishi (1989) for creative conceptual design;
- frame languages provided a flexible representation allowing the modeling of product and knowledge with the same formalism;
- case-based reasoning was proposed as a mechanism for reusing previous designs, as was mentioned previously;
- among the full scale products developed, one could mention NuDraw and ICAD™. NuDraw was developed by the MIT AI Lab and LMI and was the first system to integrate simulation in the design of printed circuit boards. ICAD™ was developed as a design system for mechanical engineering and introduced incremental compiling of constrained specifications to generate and maintain complex designs;
- researchers also proposed using intelligent agents, which will be discussed further in this book.

Clearly artificial intelligence approaches fare well in two situations: (1) in the early design steps when there are not enough data to feed the traditional CAD systems, and (2) when the design space is discrete and the search is highly combinatorial, since some artificial intelligence techniques used are similar to those in combinatorial optimization, which leads to an efficient reduction of the search space.

A number of international meetings have been held on the subject of intelligent design and the corresponding proceedings contain numerous references (e.g., IFIP WG 5.2, or Eurographics ICAD Group).

2.2.12 Knowledge Intensive Design

Knowledge intensive design results from a change of perspective when considering the design process in the product life cycle. While preceding approaches focused on the design process itself, more recently the trend is to consider the entire product life cycle and the knowledge involved with this. Thus Mäntylä (1995) states that "knowledge capitalization and knowledge deployment are essential characteristics to a Knowledge Intensive CAD system." Tomiyama and coworkers (1995) introduced the concept of Knowledge Intensive Engineering as follows: "Knowledge Intensive

Engineering [is] a new way of engineering activities in various product life cycle stages conducted with more knowledge in a flexible manner to create more added value. It is crucial, therefore, to share and reuse various kinds of knowledge involved in the product life cycle." They then defined knowledge intensive design in the following way: "A knowledge intensive design environment should allow generation and flexible exchange of design knowledge and information that will be further used in various stages of the life cycle. It is not just another 'knowledge-based design' system that assists designers in an intelligent manner." Thus, the emphasis is put on the organizational aspects of the design process rather than on the technical aspects. One question remains, namely what kind of technology can be used for implementing knowledge intensive environments. Barthès (1995), Brown and coworkers (1995), Tomiyama and coworkers (1995), among other researchers, offer the concept of agents as a possible enabling technology.

2.2.13 Concurrent Design

Concurrent Design (also called Cooperative Design, Collaborative Design and Interdisciplinary Design) is the process of designing a product through cooperation among those associated with the life cycle of the product. This would include those from functions as disparate as design, manufacturing, assembly, test, quality and purchasing as well as those from suppliers and customers (Sprow, 1992). The objectives of such a Concurrent Design team might include optimizing the mechanical function of the product, minimizing the production or assembly costs, or ensuring that the product can be easily and economically serviced and maintained (Hartley, 1992). A number of emerging technologies including distributed objects, agents and the Internet technology have been proposed to implement concurrent design systems. This book is primarily focused on the applications of agent technology to concurrent design and manufacturing.

2.3 ADVANCED MANUFACTURING SYSTEMS

This section introduces four advanced manufacturing systems: Computer-Integrated Manufacturing Systems (CIMS); Flexible Manufacturing Systems (FMS); Intelligent Manufacturing Systems (IMS); and Holonic Manufacturing Systems (HMS). The description of traditional manufacturing systems is outside the scope of this book. However, there are numerous texts and references in this area (e.g. the one by Miltenburg (1995)).

2.3.1 Computer Integrated Manufacturing Systems (CIMS)

Computer Integrated Manufacturing (CIM) is not a new concept, but it is still evolving with new features being added. Computer Integrated Manufacturing

Systems (CIMS) emerged as the result of developments in both manufacturing and information technology (computer technology). The computer plays an important role in integrating the following areas of a CIMS, as shown in Figure 2.3 (Kusiak and Heragu, 1988):

- Part (component) and product design
- Tool and fixture design
- Process planning
- Programming of numerically controlled (NC) machines, material handling systems (MHS), etc.
- Production planning
- Machining
- Assembly
- Maintenance
- Quality control
- Inspection
- Storage and retrieval

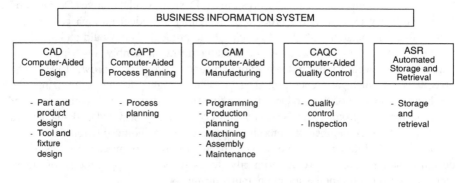

Figure 2.3 Structure and functional areas of a CIMS.

To emphasize the computer aspect, the terms computer-aided design (CAD), computer-aided process planning (CAPP), computer-aided manufacturing (CAM), computer-aided quality control (CAQC), and automated storage and retrieval (ASR) have been commonly used. Each refers to one or more of the listed functional areas. CAD refers to part and product design and tool and fixture design; CAPP refers to process planning; CAM refers to the programming of manufacturing hardware, and to production planning, machining, assembly, and maintenance; CAQC refers to quality control and inspection; and ASP involves storage and retrieval of raw materials, finished products, and in-process inventory.

The computer plays a leading role in the automation and integration of the hardware components, that is the machines, material handling carriers, as well as the software components (tools) of the manufacturing system. A detailed discussion of these functional areas can be found in (Kusiak and Heragu, 1988). A research literature review of CIMS is outside the scope of this book but many texts on Computer-Integrated Manufacturing Systems (CIMS) exist and numerous proceedings of international conferences on CMIS are also widely available.

Until recently, the CIM concept has been promoted as being the concept that will deal with the challenges of modern manufacturing systems. But, CIM often wrongly assumes (or at least those who implement CIMS do) that organizational structures and procedures can be automatically improved by incorporating computer technology (Warnecke, 1993). Unfortunately this is more likely to lead to the cementing of the existing procedures and structures of the company in question than to their improvement.

New problems have also been identified during the implementation of CIM systems in different companies. CIM aims to totally integrate and optimize the systems, and to systematize the activities of a company. Consequently, its use can result in large investments, long lead times, and rigid systems, due to the large size of the integrated system and its centralization. Furthermore, this rigidity and the emphasis on automation may lead to a loss of both flexibility and adaptability in changing environments. This may stagnate future progress unless the creativity of the worker is intensively used for continuous improvement and innovation. To overcome these problems, new concepts and new types of advanced manufacturing systems have been proposed and developed, including Flexible Manufacturing Systems (Section 2.3.2) to address the lack of flexibility; and Intelligent Manufacturing Systems (Section 2.3.3) to address the problems of system reconfiguration, coordination, and adaptation. Holonic Manufacturing Systems (Section 2.3.4) and Agent-Based Manufacturing Systems (which are the main focus of this book) are two types of recently proposed and developed Intelligent Manufacturing Systems.

2.3.2 Flexible Manufacturing Systems (FMS)

A flexible manufacturing system is one that is capable of producing a variety of products (or parts) with minimal time lost in changeover from one product to the next. A flexible manufacturing system typically consists of a set of cells, a material handling system that connects those cells, and service centers (e.g., material warehouse, tool room, or repair equipment). The cell is an autonomous unit that performs a specific manufacturing function (e.g., a machining center, inspection machine, or a load-unload robot) (Jha 1991).

FMS is also defined as "a set of machines in which parts are automatically transported under computer control from one machine to another for processing." A FMS must have the ability to process a wide variety of parts or assemblies without intervention from outside to change the system.

We may note from the definitions of the CIMS and FMS that CIMS is a much more advanced version of computer automated factory and includes all FMS functions as well as the business of the company.

Flexible manufacturing systems are designed for small batch (low volume) and high variety conditions (many different products). They are designed to fill the gap between high-production transfer lines and low-production NC machines. However, it must be noted that only under certain circumstances are FMSs clearly justified. From the R&D literature, it can be easily be established that FMSs may be appropriate in the following situations:

-    Production of families of parts;
-    Random launching of parts into the system;
-    Reduced manufacturing lead times;
-    Reduced in-process inventory;
-    Increased machine utilization;
-    Reduced direct and indirect labor;
-    Better management control.

However, designing a good FMS that will work in a particular factory with a fixed product mix is a difficult task. Most FMSs are designed with the help of simulation. Many different simulation tools and languages are available to help in the investigation of the cells' layouts, sizes, types of equipment, and scheduling policies prior to purchase and installation.

The detailed introduction of FMS concepts and a research literature review of FMS are also outside the scope of this book. Interested readers may refers to related handbooks, textbooks and conference proceedings.

2.3.3 Intelligent Manufacturing Systems (IMS)

Developments in Artificial Intelligence have already had a significant impact on manufacturing systems. Almost all areas of Computer-Integrated Manufacturing (CIM) shown in Figure 2.3 have been affected by artificial intelligence. Researchers and engineers are currently developing intelligent manufacturing systems for applications ranging from material handling and assembly to long-term planning.

An international IMS (Intelligent Manufacturing Systems) Consortium was officially established in 1995. It is an industry-led, international Research & Development Program whose objective is to develop the next generation of manufacturing and processing technologies through global cooperation (Hayashi, 1993). Its members include companies, universities and research institutions from Australia, Canada, the European Union, Japan, Switzerland, and the United States. More information about the IMS consortium and its research projects is available through its web site: http://www.ims.org/.

In order to show the coverage of the IMS research, we list here some of the research themes of the IMS consortium (http://www.ims.org/):

Total product life cycle issues

- Future general models of manufacturing systems, e.g., "agile manufacturing" (see Section 2.4), "fractal factory[1]", "bionic manufacturing[2]" and "holonic manufacturing systems" (Section 2.3.4).
- Intelligent communication network systems for information processes in manufacturing.
- Environment protection, minimum use of energy and materials.
- Recyclability and refurbishment.
- Economic justification methods.

Process issues

In order to meet the needs for rapid response to changing requirements, for saving human and material resources and to improving working conditions for employees, the following themes are identified:

- Clean manufacturing processes that can minimize effects on environment, e.g., Process emission minimized systems; Process disposal minimized systems; Factory (process) life-cycle pre-assessed systems, etc.
- Energy efficient processes that can meet manufacturing requirements with minimum consumption of energy, e.g., Minimum consumption of energy; Integrated cycled process for less energy consumption; Modules of energy conservation type; Production management technology of energy conservation type.
- Technology innovation in manufacturing processes, including Methods that can quickly produce different products through "Rapid Prototyping Methods"; Manufacturing processes that can flexibly respond to changes in labor conditions, changes of products or materials.
- Improvement in the flexibility and autonomy of processing modules that compose manufacturing systems, e.g., open distributed systems and their modules that can match both unmanned, man-machine mixed and labor intensive systems, and can metamorphologically architect system components in correspondence with changes of products.
- Improvement in interaction or harmony among various components and functions of manufacturing, e.g., open infrastructure for manufacturing; inter-connected information systems such as "remote ID" among respective modules.

---

[1] The term comes from fractal geometry in describing and analyzing objects in multi-dimensional spaces, especially focusing on the fractional dimension where Euclidean geometry is not suitable. The main characteristic of fractals is self-similarity, implying recursion, and the pattern-inside-of-pattern (Warnecke 1993; Strauss and Hummel, 1995).
[2] The Bionic System draws parallels with biological systems which exhibit many features including autonomous and spontaneous behaviors, and social harmony, within hierarchically ordered relationships (Okino 1992; 1993).

Strategy/Planning/Design tools

Manufacturing takes place in a global economy. How and where raw materials are transformed is a strategic decision. The decision is complicated in terms of what to make and where to make or buy it, in what is becoming a single global economy.

Many of today's manufacturing organizations are designed using vertical and hierarchical structures. The move towards heterarchical structures is and will continue to require major changes in organizations, systems and work practices. Methodologies and tools are needed to help us to define appropriate manufacturing strategies and to design appropriate organizations and business/work processes. Such methodologies and tools includes:

- Methods and tools to support business process re-engineering.
- Modeling tools to support the analyses and development of manufacturing strategies.
- Design support tools to support planning in an extended enterprise or virtual enterprise environment.

Human/Organization/Social issues

- Promotion and development projects for improved image of manufacturing.
- Improved capability of manufacturing workforce/education, training.
- Autonomous offshore plants (integration of supplementary business functions in subsidiaries)
- Corporate Technical Memory – keeping, developing, accessing.
- Appropriate performance measures for new paradigms.

Virtual/Extended Enterprise issues

The extended enterprise is an expression of the market driven requirement to embrace external resources in the enterprise without owning them. Core business focus is the route to excellence but product/service delivery requires the amalgam of multiple world class capabilities. Changing markets require a fluctuating mix of resources. The extended enterprise, which can be likened to the ultimate in customizable, reconfigurable manufacturing resource, is the goal. The process is applicable even within large organizations as they increasingly metamorphoses into umbrellas for smaller units/focused factories.

The operation of the extended enterprise requires take up of communications and database technologies which are near to the current state of the art.

Research and Development opportunities in this area are:

- Methodologies to determine and support information processes and logistics across the value chain in the extended enterprise.
- Architecture (business, functional and technical) to support engineering co-operation across the value chain, e.g., concurrent engineering across the extended enterprise.

-   Methods and approaches to assign cost/liability/risk and reward to elements of the extended enterprise.
-   Team-working across individual units within the extended enterprise.

The IMS European Working Group has organized two international workshops on IMS (IMS98, 1998; IMS99, 1999). The AAAI Special Interest Group on Manufacturing (SIGMAN) has been very active recently. Its workshops (1995, 1996, 1998) have highlighted some of the significant recent advances in academic research and industrial practice. Interested readers may refer to (Luger, 1998) for the Proceedings of the 1998 AI & Manufacturing Workshop. Other information is available through SIGMAN web site at: http://sigman.cs.umn.edu/.

An IMS webliography with URL links to the web sites of related research groups, programs/projects, international journals and conferences, or other related resources has been created by the first author and is available through the Internet at: http://www.ucalgary.ca/~wshen/weblio.htm.

2.3.4 Holonic Manufacturing Systems (HMS)

The HMS concept was proposed in 1994 by the HMS consortium as a test case under the international Intelligent Manufacturing Systems (IMS) Research Program (Hayashi 1993). "Holon" is a word coined by combining 'holos' (the whole) and 'on' (a particle) following Koestler (1967). Key holonic concepts (Christensen, 1994) already developed by the HMS Consortium include:

-   *Holon*: An autonomous and cooperative building block of a manufacturing system for transforming, transporting, storing and/or validating information and physical objects. The holon consists of an information processing part and often a physical processing part. A holon can form part of another holon.
-   *Holarchy*: A system of holons which can cooperate to achieve a goal or objective. The holarchy defines the basic rules for cooperation of the holons and thereby limits their autonomy.
-   *Holonic Manufacturing System* (HMS): A holarchy which integrates the entire range of manufacturing activities from order booking through design, production and marketing to realize the agile manufacturing enterprise.

A holon has interfaces through which it exchanges information, material, or resources (which may themselves be holons) with other holons or its environment. The generic activity model for a holon can therefore be represented as in Figure 2.4.

An HMS is therefore a manufacturing system where key elements, such as raw materials, machines, products, parts, AGVs, etc., have autonomous and cooperative properties (Christensen, 1994; Deen, 1994). In an HMS, each holon's activities are determined through cooperation with other holons, as opposed to being determined by a centralized mechanism. In this type of systems, intelligent entities called 'holons' have a physical part as well as a software part. Detailed description of holonic concepts and holonic manufacturing systems can be found in (Christensen, 1994; Deen, 1994; Van Leeuwen and Norrie, 1997)

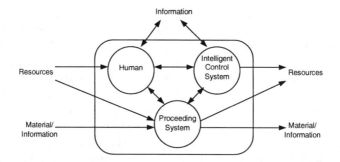

Figure 2.4 Generic Activity Model of a Holon (Christensen, 1994).

With the same concepts and similar system architecture, the partners of HMS consortium have developed their own testbeds using their existing software and hardware environments. Most of the research results on HMS are reported only internally in the HMS consortium. However, some results have been published, on material handling (Christensen et al., 1994), manufacturing planning and scheduling (Hasegawa et al., 1994; Biswas et al., 1995; Sousa and Ramos, 1997), and intelligent manufacturing control (Brennan et al., 1997; Wang et al., 1998).

2.4 NEXT-GENERATION OF CONCURRENT DESIGN AND MANUFACTURING SYSTEMS

The manufacturing enterprises of the 21st century will be in an environment where markets are frequently shifting, new technologies are continuously emerging, and competitors are multiplying globally. Manufacturing strategies should therefore shift to support global competitiveness, new product innovation and introduction, and rapid market responsiveness. The next generation of concurrent design and manufacturing systems will thus be more strongly time-oriented, while still focusing on cost and quality. Such systems will need to satisfy the following fundamental requirements:

- *Enterprise Integration and cooperation:* In order to support global competitiveness and rapid market responsiveness, an individual or collective manufacturing enterprise will have to be integrated (connected) with its related management systems (e.g., purchasing, orders, design, production, planning & scheduling, control, transport, resources, personnel, materials, quality, etc.) as well as with its customers, suppliers and other partners via networks. Thus there is a credible need for cooperation among various systems or subsystems. Such cooperation should be in an efficient and quick-response manner.

- *Heterogeneous Environments:* Such concurrent design and manufacturing systems will need to accommodate heterogeneous software and hardware as well as human beings in both their manufacturing and information environments. Humans (e.g., project managers, product designers, production engineers, etc.) and computers need to be integrated to work collectively at various stages of the product development and even through the whole product life cycle, with rapid access to required knowledge and information. Bi-directional communication environments are required to allow effective, quick communication between human and computers.

- *Interoperability:* Heterogeneous information environments may use different programming languages, represent data with different representation languages and models, and operate in different computing platforms. The sub-systems and components in such heterogeneous environments should interoperate in an efficient manner. Translation and other capabilities will be needed to enable such interoperation or interaction.

- *Distributed Concurrent Engineering Issues:* The usability of concurrent design and manufacturing systems should be enhanced by interfacing with design evaluation systems such as design for manufacturability (DFM), design for assembly (DFA), design for capacity and capability (DFCC), design for performance (DFP), and design for environment (DFE) systems. A large (or even huge) volume of design and manufacturing data and knowledge will need to be shared by distributed manufacturing enterprises using the Internet, intranets or local networks.

- *Agility:* Considerable attention must be given to reducing product cycle time to be able to respond to customer desires more quickly. Agile manufacturing is the ability to adapt quickly in a manufacturing environment of continuous and unanticipated change and thus is a key component in manufacturing strategies for global competition. To achieve agility, manufacturing facilities must be able to rapidly reconfigure and interact with heterogeneous systems and partners. Ideally, partners are contracted with "on the fly" only for the time required to complete specific tasks.

- *Scalability:* Scalability means that additional resources (software and hardware) can be incorporated into the organization as required. This capability should be available at any working node in the system and at any level within the nodes. Expansion of resources should be possible without disrupting organizational links previously established.

- *Fault Tolerance:* The system should be fault tolerant both at the system level and at the subsystem level so as to detect and recover from system failures at any level and minimize their impacts on the working environment.

In subsequent chapters, the potential of intelligent agents for such next-generation systems will be further explored.

REFERENCES

Aksoy, M. and Saglamer, G. (1993) A Shape-Grammar for Courtyard and Houses, In Beheshti, M.R. and Zreik, K. eds., *Advanced Technologies*, Elsevier Science Publishers, pp. 251-258.

Antonsson, E.K. (1997) The Potential for Mechanical Design Compilation, *Research in Engineering Design*, 9:191-194.

Arbab, F. (1989) Examples of Geometric Reasoning in OAR. In Akman, V., ten Hagen, P.J.W., Veerkamp, P.J. eds., *Intelligent Systems III - Implementation Issues*, Springer Verlag, pp. 32-57.

Baker, K., Ball, L., Culverhouse, P., Dennis, I., Evans, J., Jajodzinski, P., Pearce, P., Scoththern, D. and Venner, G. (1991) A Psychologically Based Intelligent Design Aid, In ten Hagen, P.J.W., Veerkamp, P.J. eds., *Intelligent CAD Systems III*, Springer Verlag, pp. 21-39.

Barthès, J-P. A. (1993) Object-Oriented Environments for Design, In Beheshti, M.R. and Zreik, K. eds., *Advanced Technologies*, Elsevier Science Publishers, pp. 389-398.

Barthès, J-P. A. (1995) Do we have the Technology for Supporting Knowledge Intensive CAD in Large Design Projects? In Tomiyama, T., Mäntylä, M., Finger, S. eds., *Knowledge Intensive CAD*, Volume 1, pp. 17-27.

Baykan, C.A. and Fox, M.S. (1991) Constraint Satisfaction Techniques for Spatial Planning, In ten Hagen, P.J.W., Veerkamp, P.J. eds., *Intelligent CAD Systems III*, Springer Verlag, pp. 187-204.

Bijl, A. (1989) Relations, Functions & Constraints without Prescription. In *Preprints of the 3rd IFIP W5.2 Workshop on Intelligent CAD*, Osaka, Japan, pp. 36-54.

Biswas, G., Bagchi, S. and Saad, A. (1995) Holonic Planning and Scheduling for Assembly Tasks, TR CIS-95-01, Center for Intelligent Systems, Vanderbilt University, USA.

Brennan, R.W., Balasubramanian, S., and Norrie, D.H. (1997) A dynamic control architecture for metamorphic control of advanced manufacturing systems, In *Proceedings of the International Symposium on Intelligent Systems and Advanced Manufacturing,* Pittsburgh, PA, pp. 213-223.

Bridge, S.F. (1989) Intelligent Representation of Geometric Knowledge, In Akman, V., ten Hagen, P.J.W., Veerkamp, P.J. eds., *Intelligent Systems III - Implementation Issues*, Springer Verlag, pp. 275-290.

Brown, D.R. and Hwang, K.Y. (1993) Solving Fixed Configuration Problems with Genetic Search, *Research in Engineering Design*, 5:80-87.

Brown, D.C., Dunskus, B., Grecu, D.L. and Berker, I. (1995) SINE: Support For Single Function Agents, In *Proceedings of Applications of AI in Engineering*, Udine, Italy.

Carlson-Skalak, S., White, M.D. and Teng, Y. (1998) Using an Evolutionary Algorithm for Catalog Design, *Research in Engineering Design*, 10:63-83.

Chakrabarti, A. and Bligh, T.P. (1994) An Approach to Functional Synthesis of Solutions in Mechanical Conceptual Design. Part I: Introduction and Knowledge Representation, *Research in Engineering Design;* 6:127-141.

Chandrasekaran, B., Goel, A.K. and Iwasaki Y. (1993) Functional Representation as Design Rationale, *IEEE Computer*, 26:48-56.

Chittaro, L. and Kumar, A.N. (1998) Reasoning about Function and its Application to Engineering, *Artificial Intelligence in Engineering*, 12:331-336.

Christensen, J.H. (1994) Holonic manufacturing systems: initial architecture and standards directions, In *Proceedings of First European Conference on Holonic Manufacturing Systems*, Hanover, Germany.

Christensen, J.H., Struger, P.J., Norrie, D.H. and Schaeffer, C. (1994) Material handling requirements in holonic manufacturing systems. In *Proceedings of International Symposium on Material Handling Research*, Grand Rapids, MI.

Cutkosky, M.R., Tenenbaum, J.M. and Brown, D.R. (1992) Working with Multiple Representations in a Concurrent Design System, *ASME Journal of Mechanical Design*, 114(3):515-524.

Deen, S.M. (1994) A cooperation framework for holonic interactions in manufacturing, In *Proceedings of the Second International Working Conference on Cooperating Knowledge Based Systems (CKBS'94)*, DAKE Centre, Keele University.

Faltings, B., Haroud, D., Smith, I. (1991) Dynamic Constraint Satisfaction with Continuous Variables, In *Preprints of IFIP WG 5.2 Working Conference on Intelligent CAD*, Colombus, Ohio, pp. 399-415.

Finger, S. and Rinderle, J.R. (1989) A Transformational Approach to Mechanical Design using a Bond Graph Grammar, *ASME Design Theory and Methodology*, Montreal, Canada.

Fu, Z. and de Pennington, A. (1993) Constraint-Based Design Using an Operational Approach, *Research in Engineering Design*, 5:202-217.

Gene, S., Messler, R.W.Jr. and Gabriele, C.A. (1998) A Hierarchical Classification Scheme to Define and Order the Design Space for Integral Snap-Fit Assembly, *Research in Engineering Design*, 10:94-106.

Gero, J.S., and Kazakov, V.A. (1998) Evolving Design Genes in Space Layout Planning Problems, *Research in Engineering Design*, 10:163-176.

Gossard, D.C., Zuffante, R.P. and Sakurai, H. (1988) Representing Dimensions, Tolerances, and Features in MCAE Systems, *IEEE Computer Graphics and Applications*, 8(2):51-59.

Guan X. and MacCallum, K.J. (1995) Modelling of Vague and Precise Geometric Information for Supporting the Entire Design Process, In Tomiyama, T., Mäntylä, M., Finger, S. eds., *Knowledge Intensive CAD*, Volume 1, pp. 301-319.

Hartley, J. (1992) *Concurrent Engineering*, Cambridge, Mass.: Productivity Press.

Hasegawa, T., Gou, L., Tamura, S., Luh, P.B., and Oblak, J.M. (1994) Holonic Planning and Scheduling Architecture for Manufacturing. In *Proceedings of the 2nd International Working Conference on Cooperating Knowledge-based Systems*, University of Keele.

Hayashi, H. (1993) The IMS International Collaborative Program. In *Proceedings of 24th ISIR*, Japan Industrial Robot Association.

Hua, K., Faltings, B. and Smith, I. (1996) CADRE: Case-Based Geometric Design, *Artificial Intelligence in Engineering*, 10:171-183.

IMS98 (1998) First International Workshop on Intelligent Manufacturing Systems, Lausanne, Switzerland. (http://www.mech.kuleuven.ac.be/pma/project/imswg/wshop3/workshop.htm)

IMS99 (1999) Second International Workshop on Intelligent Manufacturing Systems, Leuven, Belgium. (http://www.mech.kuleuven.ac.be/pma/project/imswg/wshop6/workshop.htm)

Jha, N.K. (1991) *Handbook of Flexible Manufacturing Systems*, Academic Press, Inc., San Diego, CA.

Jo, J.H. and Gero, J. (1998) Space Layout Planning Using an Evolutionary Approach, *Research in Engineering Design*, 10:149-162.

Karni, R. and Arciszewski, T. (1997) A Tool for Conceptual Design of Production and Operations Systems, *Research in Engineering Design*, 9:146-167.

Kishi, Y. (1989) An Architecture of Design Intelligence Modeling for Creative Conceptual Design Processing, In *Preprints of the 3rd IFIP W5.2 Workshop on Intelligent CAD*, Supplements version, Osaka, Japan, pp. 48-72.

Koestler, A. (1967) *The Ghost in the Machine*, Arkana Books, London.

Krause, F.L., Vosgenau, F.M. and Yaramanoglu, N. (1989) Implementation of Technical Rules in a Feature-Based Modeller, In Akman, V., ten Hagen, P.J.W., Veerkamp, P.J. eds., *Intelligent Systems III - Implementation Issues*, Springer Verlag, pp. 195-208.

Kusiak, A. and Heragu, S.S. (1988) Computer-integrated manufacturing: a structural perspective, *IEEE Network*, 2(3):14-22.

Luger, G.F. (1998) *Artificial Intelligence and Manufacturing - Research Planning Workshop – State of the Art & State of the Practice*. The AAAI Press, Menlo Park, CA.

Mäntylä, M. (1995) Product Modelling for Knowledge-Intensive CAD: Towards a Research Agenda, In Tomiyama, T., Mäntylä, M., Finger, S. eds., *Knowledge Intensive CAD -1*, Preprints of the IFIP WG 5.2 Workshop, Espoo, Finland, pp. 105-122.

McDermott, J. (1982) R1: A Rule-Based Configurer of Computer Systems, *Artificial Intelligence*, 19:39-58.

McNeil T., Gero J.S. and Warren (1998) Understanding Conceptual Electronic Design Using Protocol Analysis, *Research in Engineering Design*, 10:129-140.

Mead, C. and Conway, L. (1980) *Introduction to VLSI Design*, Addison-Wesley.

Miltenburg, J. (1995) *Manufacturing Strategy: How to Formulate and Implement a Winning Plan*, Productivity Press, Portland, OR.

Mittal, S., Dym, C.L. and Morjaria, M. (1986) Pride: An Expert System for the design of Paper Handling Systems, *Computer*, July, 1986.

Monceyron, E. and Barthès, J-P. A. (1992) Architecture for ICAD Systems: An Example from Harbor Design, *Revue Science et Techniques de la Conception,*1:49-68.

Mukherjee, A. and Liu, C.R. (1995) Representation of Function-Form Relationship for the Conceptual Design of Stamped Metal Parts, *Research in Engineering Design*, 7:253-269.

Murtagh, N. and Shimura, M. (1989) A Constrained-Based Hybrid Engineering Design System, In *Preprints of the 3rd IFIP W5.2 Workshop on Intelligent CAD*, Supplements version, Osaka, Japan, pp. 287-301.

Nielsen, E.H., Dixon, J.R. and Zinsmeister, G.E. (1991) Capturing and Using Design Intent in a Design with Features System, In *Proceedings of ASME Design Theory and Methodological Conference*, pp. 95-102.

Okino, N. (1992) A Prototype of Bionic Manufacturing System, In *Proceedings of the International Conference on Object-Oriented Manufacturing Systems*, Calgary, Alberta, Canada, pp. 485-492.

Okino, N. (1993) Bionic Manufacturing Systems, In Peklenik, J. ed., *Flexible Manufacturing Systems: Past-Present-Future*, CIRP, pp. 73-95.

Oksala, T. (1993) Typology versus Change: Studies in the Generation of Apartment Types in Housing Production, In Beheshti, M.R. and Zreik, K. eds., *Advanced Technologies*, Elsevier Science Publishers, pp. 435-442.

Orel, T. (1991) Le rôle des technologies d'information dans le processus de design: une approche du design management, In *Sur la modélisation des processus de conception créative*, EUROPIA. Paris, France, pp. 7-27.

Otto, K.N. and Wood K.L. (1998) Product Evolution: A Revere Engineering and Redesign Methodology, *Research in Engineering Design*, 10:226-243.

Persidis, A. and Duffy, A. (1989) Learning in Engineering Design, In *Preprints of the 3rd IFIP W5.2 Workshop on Intelligent CAD*, Supplements version, Osaka, Japan, pp. 91-110.

Prabhakar, S. and Goel, A.K. (1998) Functional Modeling for Enabling Adaptive Design of Devices for New Environments, *Artificial Intelligence in Engineering*, 12:417-444.

Price, C.J. (1998) Function-Directed Electrical Design Analysis, *Artificial Intelligence in Engineering*, 12:331-336.

Rogier, J.L.H. and Tolman, F.P. (1989) The BiCad System, In Akman, V., ten Hagen, P.J.W., Veerkamp, P.J. eds., *Intelligent Systems III - Implementation Issues*, Springer Verlag, pp. 281-310.

Rosen, D.W. (1993) Feature-Based Design: Four Hypotheses for Future CAD Systems, *Research in Engineering Design*, 5:125-139.

Schmidt, L.C. and Cagan, J. (1997) GGREADA: A Graph Grammar-Based Machine Design Algorithm, *Research in Engineering Design*, 9:195-213.

Shah, J.J. and Rogers, M.T. (1993) Assembly Modeling as an extension to Feature-Based Design, *Research in Engineering Design*, 5:218-237.

Sousa, P. and Ramos, C. (1997) A dynamic scheduling holon for manufacturing orders, *Journal of Intelligent manufacturing*, 9(2):107-112.

Sprow, E. (1992) Chrysler's Concurrent Engineering Challenge, *Manufacturing Engineering*, 108(4):35-42.

Strauss, R.E. and Hummel, T. (1995) The new industrial engineering revisited – information technology, business process reengineering, and lean management in the self-organizing "fractal company", In *Proceedings of 1995 IEEE Annual International Engineering Management Conference*, pp. 287-292.

Sushkov, V.V., Mars, N.J.I, and Wogum, P.M. (1995) Introduction to TIPS: A Theory for Creative Design, *Artificial Intelligence in Engineering*, 9:177-189.

Takala, T. (1989) Design Transactions and Retrospective Planning. Tools for Conceptual Design, In Akman, V., ten Hagen, P.J.W., Veerkamp, P.J. eds., *Intelligent CAD Systems II*, Springer Verlag, pp. 262-272.

Thoraval, P. (1991) *Modélisation des systèmes d'aide à la conception*, PhD Thesis, Université de Technologie de Compiègne, France.

Tomiyama, T., Umeda, Y., Ishii, M., Yoshioka, M. and Kiriyama, T. (1995) Knowledge Systematization for a Knowledge Intensive Engineering Framework, In Tomiyama, T., Mäntylä, M., Finger, S. eds., *Knowledge Intensive CAD*, Volume 1, pp. 55-80.

Treur, J. (1991) A Logical Framework for Design Processes. In ten Hagen, P.J.W., Veerkamp, P.J. eds., *Intelligent CAD Systems III*, Springer Verlag, pp. 3-19.

Trousse, B. (1993) Towards a Multi-Agent Approach for Cooperative Distributed Design Assistants, In Beheshti, M.R. and Zreik, K. eds., *Advanced Technologies*, Elsevier Science Publishers, pp. 451-460.

Tweed, C, Bijl, A. (1989) MOLE: A Reasonable Logic for Design, In Akman, V., ten Hagen, P.J.W., Veerkamp, P.J. eds., *Intelligent Systems III - Implementation Issues*, Springer Verlag, pp. 146-167.

Van Leeuwen, E.H. and Norrie, D.H. (1997) Intelligent manufacturing: holons and holarchies. *Manufacturing Engineer*, 76(2):86-88.

Veltkamp (1991) Geometric Constraint Management with Quanta. In *Preprints of the IFIP WG 5.2 Working Conference on Intelligent CAD*, Colombus, Ohio, pp. 381-397.

Wang L., Balasubramanian S. and Norrie D. (1998) Agent-based Intelligent Control System Design for Real-time Distributed Manufacturing Environments, In *Working Notes of the Agent-Based Manufacturing Workshop*, Minneapolis, MN, pp. 152-159.

Warnecke, H.J. (1993) *The Fractal Company*, Springer-Verlag.

Welsh, R.V. and Dixon, J.R. (1994) Guiding Conceptual Design Through Behavioral Reasoning, *Research in Engineering Design*, 6:169-188.

Whitney, D.E. (1996) Why Mechanical Design cannot be like VLSI Design, *Research in Engineering Design*, 8:125-138.

Xue, D., Takeda, H., Kiriyama, T., Tomiyama, T. and Yoshikawa, H. (1989) An Intelligent Integrated Interactive CAD, In *Preprints of the 3rd IFIP W5.2 Workshop on Intelligent CAD*, Osaka, Japan, pp. 478-497.

3

Distributed Artificial Intelligence and Agents

3.1 CLASSIC AI AND DAI

Artificial intelligence was born in the late fifties. It had two main goals: (1) a scientific goal which consisted in trying to understand how humans reason; (2) an engineering goal which consisted in designing and building "intelligent" machines. Which goal was promoted depended on whom you were talking to. In 1963 Feigenbaum and Feldman (1963) wrote: "The purpose of this volume is to inform the non specialist about current research on intelligent behavior by computer." They added in the introduction, three main objectives for artificial intelligence: (1) learning heuristic methods and rules; (2) inductive inference (modeling the world and pattern recognition); and (3) understanding natural language. Thirty years later, Davis and coworkers (1993) have classified the various approaches used in artificial intelligence in broad categories according to their intellectual origins: (1) psychology; (2) biology; (3) statistics; and (4) economics. In between, artificial intelligence experienced successes and failures, was praised and despised, but developed numerous original hardware and software techniques which today are widely used in everyday computing.

Starting in 1965 with the DENDRAL project, artificial intelligence developed the field of expert systems. The approach initially consisted of modeling experts' knowledge as condition-action rules acting on a representation of the external world. What was new with respect to traditional computer programs was that the knowledge was explicitly described and coded in the programs as symbolic structures. Thus, a computer could both act as mimic and expert by utilizing this recorded knowledge. The next step forward came when the tasks to be solved happened to be complex and required diverse skills. Thus, in the mid seventies, the HEARSAY speech understanding system was developed to record data and partial conclusions in a single location and then use several expert systems to work on different aspects of the data simultaneously (Erman and Lesser, 1975). The different sets of rules were called *knowledge sources* and the overall architecture a *blackboard*. This amounted to doing parallel distributed problem solving and introduced the new field of distributed artificial intelligence (DAI). The various sources could be considered to be separate experts working on the same problem. If these experts were distributed on different machines, new concepts like cooperation became important.

A little later at MIT, Hewitt was interested in modeling intelligence by decomposing inter-communicating knowledge-based problem solving experts into a society of primitive actors. This led to further research in communication mechanisms (Hewitt, 1979). At about the same time, Smith (1979) introduced the concept of the *contract-net* for distributing tasks dynamically among the nodes of a computing network, which in fact was the first high-level protocol for task sharing.

Later on, distributed artificial intelligence introduced more formal concepts, leading to the research themes discussed in the next section.

3.2 RESEARCH THEMES IN DAI

Distributed artificial intelligence involves study of the distribution of intelligent processes among independent entities. Artificial intelligence was initially focused on the behavior of a single entity, whereas distributed artificial intelligence is interested in the behavior of multiple interacting entities that we will call *agents*. If we restrict ourselves to the engineering point of view of designing and building systems of agents, the main problems are:

- how to construct a specific agent;
- how will a system of agents interact;
- how can we control societies of agents?

We will look at different types of agents in Section 3.5. However, there exist two extreme approaches to an agent's internal architecture. The first consists of building very simple agents using purely reactive mechanisms, and to let them interact freely to reach a common goal. The second is to design complex cognitive agents, which are capable of reasoning and of structured interaction with other agents. The two approaches are the extremes of a continuum of agents capable of displaying both a reactive and a cognitive behavior.

The main research themes at the level of designing a particular agent are:

- how do we organize each processing agent locally;
- how do we model its expertise;
- what do we model of the external world;
- what are the basic mechanisms which will allow an agent to interact with other agents? (This also involves communication, see below.)

Interaction among agents occurs at different levels. The first and basic level is for an agent to be capable of exchanging information physically, i.e., sending and receiving messages. Issues at this level are addressed by much research in the domain of networks and telecommunications. Then, assuming that the transportation problems are solved, the second level can be seen along three dimensions:

- dispatching strategy;
- message structure;
- inner message complexity.

The dispatching strategy concerns the mode of interaction among agents. Indeed, communication can be point to point, multicast, or broadcast. The type of communication may vary in time according to needs. More sophisticated protocols like the contract-net (Smith, 1979) can be seen as implementing a distribution strategy at this level. Message structure concerns the sequences of messages. A message can be isolated, e.g., a spontaneous message containing a single piece of information, or else, a message can be part of a dialog implementing a complex strategy of cooperation or of negotiation. Research in this area involves communication languages, models of interaction, behavior patterns. These are further considered in Chapter 8. Inner message complexity concerns the content of a message. The complexity may range from very simple, e.g., object-oriented types of message (action parameters) to very complex as in natural language. This is addressed in Section 3.4.

One of the main problem arising with distributed processing is how to make sure that decisions taken locally with incomplete information contribute to the global goal. Various approaches are possible to global control including:

- do nothing;
- implement hierarchical control;
- use optimization techniques;
- use a market approach;
- use a collaborative approach.

Doing nothing consists of letting each agent solve subtasks corresponding to subproblems and return the result to the calling agent or to the agent specified as a possible continuation. Hierarchical control is used in complex blackboard architectures for controlling the sequencing of each agent. It is also used in the master-slave type of hierarchical architectures. Optimization techniques like simulated annealing can be used to find additional solutions to a global problem. The market approach uses a monetary or equivalent approach to reach equilibrium between supply and demand. In the collaborative approach agents are continually interchanging information, and collaborate in static or dynamically formed groups to harmonize their aggregate actions with global goals.

Other issues to be addressed are:

- how to formulate and distribute problems and gather results;
- how to solve conflicts;
- how to make sure that local decisions are consistent with an overall goal;
- how to allow agents to transfer knowledge;
- how to represent what is going on within an agent.

3.3 MODELS OF DAI SYSTEMS

In this section we briefly review some of the research on the blackboard architecture, the contract net, actors, and then introduce agents.

3.3.1 Blackboard Architecture

As mentioned in Section 3.1, the blackboard architecture was proposed in the HEARSAY project as a means to organize and control large AI systems. The HEARSAY project was developed at CMU as a speech understanding system. Its first version HEARSAY I (Reddy et al., 1976) was based on the idea of cooperating independent knowledge sources separating acoustics knowledge from syntax and from semantics. The simplest organization of a blackboard approach gathers facts and hypotheses in a single location called the blackboard and lets the various independent sets of rules access the information as needed, i.e., be data-driven (Figure 3.1a). In theory, any knowledge source can be activated asynchronously and independently. In practice, some constraints have to be placed on the order of activation. This can be done by giving scores to hypotheses written on the blackboard, or by adding a knowledge source in charge of sequencing the activation of the others, as shown on Figure 3.1b.

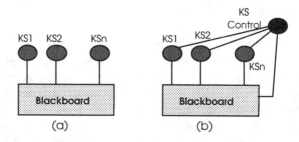

Figure 3.1 Blackboard architecture: (a) simple, (b) with a control KS.

The main advantages of a blackboard architecture are: (1) separate, independent knowledge sources; (2) shared memory; and (3) possibilities of parallel asynchronous processing. Gathering knowledge into separate, independent knowledge sources allows for replacement of knowledge sources or extension by simple addition of new knowledge sources. The shared memory contains all the data, hypotheses, and contextual information used for solving the problem at hand. Inspecting the memory yields an overview of the state of the computation. Finally, knowledge sources can be distributed over several independent processors sharing a common memory, allowing for a simple form of parallel processing which may increase the efficiency of the overall process.

Following HEARSAY I, many blackboard systems have been described ranging from early systems like CRYSALIS (Engelmore and Nii, 1977) to more recent approaches. In particular, Monceyron and Barthès (1992) showed that a blackboard architecture could be used to integrate legacy systems in the domain of harbor design. An interesting application was proposed by Hayes-Roth to design the internal structure of an agent as a blackboard system in the GUARDIAN project (Hayes-Roth et al., 1989). Other noted applications are the DVMT project, an application for monitoring distributed vehicles (Corkill and Lesser, 1983), and the MINDS project, an application for retrieving documents (Huhns et al., 1987).

3.3.2 Contract Net

The introduction of the contract-net is a milestone in the history of distributed artificial intelligence. The contract-net mechanism was developed by Smith (1979) and demonstrated on a distributed sensing system (buoys and computing nodes). As in the blackboard approach, knowledge is distributed among knowledge sources. Each knowledge source is a node in a network. Contrary to the blackboard approach, there is no shared memory where data and partial results are made available to the various knowledge sources. The contract-net implements the following approach for distributing tasks dynamically to nodes in the network. When a node has to solve a problem for which it does not have expertise, it broadcasts a task-announcement message which describes the task to be solved. The task announcement has a time duration after which it is no longer valid. Nodes that receive the message and that can solve the problem can return a bid message. The node which issued the task-announcement message, called the *manager*, waits for bids and selects one (or more) bidder for executing the task, who is called the contractor. Thus, the choice of the *contractor* is done after selection by the manager and by mutual agreement. To be able to function correctly, the system must include a high-level protocol capable of defining several types of messages with a structured content. The main advantages of the approach are stated by Smith as:

- avoiding bottlenecks since tasks are not always sent to the same processing nodes;
- avoiding channel saturation, by using a carefully crafted set of messages;
- increasing reliability, by distributing the control; and
- favoring extensibility, since the nodes are anonymous in the network.

The basic contract-net protocol remains widely used today. Other protocols based upon the contract-net have been proposed: the multi-phase negotiation protocol by Conry et al. (1991); the interactive protocol by Sandholm (1993); the Market-Driven Contract Net by Baker (1991); the Extended Contract Net Protocol (ECNP) replacing *grant* and *reject* by *temporal grant, temporal reject, definitive grant and definitive reject* by Fischer et al. (1995); and the B-Contract introducing two levels of priority by Scalabrin (1996).

3.3.3 Actors

The contract-net protocol offered an early practical means for dealing with open systems from a software engineering point of view. The actor model proposed by Hewitt offers a model of computation for open systems at a finer grain than the contract-net approach. Blackboards have independent knowledge sources sharing a common memory, the contract-net approach has knowledge sources attached to distinct nodes of a computation network and both approaches use a limited number of moderately complex knowledge sources. In the actor approach, Hewitt pursued two goals: (1) preparing to solve really large problems; (2) defining a theoretical framework for this.

Hewitt (1979) argued that available hardware and software technology are insufficient for solving large AI problems due mainly to the fuzziness of the objectives in AI context and to the unpredictable nature of intermediate computations. He proposed using a new paradigm: *actors*. At that time, the previous approaches such as the blackboard and the contract-net were more experimental in nature. Hewitt insisted on developing a rigorous theoretical framework independent of the actual hardware and software technologies.

In the actor approach, problem solving is viewed as the result of the interaction of the activities of knowledge sources working independently and locally (with limited expertise and limited knowledge) each one communicating with a limited number of other knowledge sources.

An actor is a computational entity which is described by its behavior. An actor has a *script* which specifies the actions it should take, and a list of *acquaintances* which is a finite collection of actors it knows about. An actor system has two more components: (1) a mail delivery system; (2) tasks. The mail delivery system ensures that messages are delivered to an actor in an orderly fashion and serializes messages in a large enough mailbox. A task is an identified unique structure which contains a message and is processed by the mail delivery system.

The computation mechanism of an actor is defined as follows. When an actor receives a message, it can do three things according to its script:

- create new tasks;
- create new actors;
- create a replacement actor with a new behavior.

Each new task has a unique identity and is a message sent to another actor. When a new actor is created, its mailbox is created at the same time. The replacement actor is a new actor similar to the previous one, which may have a modified script or a modified list of acquaintances, leading to a modified behavior. Once an actor becomes active, it produces actions according to its script and then dies when it finishes its actions. A new incoming message is processed by its replacement actor in the same fashion. Dead actors and old tasks (messages) are garbage collected. In order for a computation to converge, there are some predefined actors, e.g., for doing a multiplication on two numbers and returning the result, which are called rock bottom actors (Theriault, 1983).

Special programming languages which have been defined for actor systems include PLASMA, SAL, ACT, as well as specific computer architectures like APIARY (Hewitt, 1980).

In addition to the theoretical framework, an advantage of the actor approach is the ability to create new actors as needed, adjusting the number of processing elements to the size and complexity of the problem to be solved. In addition, actors can migrate freely from machine to machine, helping to improve load balancing.

The actor approach had been considered for a long time to be part of the domain of object-oriented concurrent programming (Yonesawa and Tokoro, 1987) which itself evolved as a new programming paradigm. Subsequently, actors have often been considered to be a form of agent. Recently, Jamali et al. (1999) proposed using actors as building blocks for developing systems of mobile agents using Java and XML.

3.3.4 Agents

In artificial intelligence, the concept of the agent grew from the early work on blackboards, contract-net, and actors. Separately, in applied fields such as manufacturing, object-oriented systems were being developed with increasing intelligence being incorporated into the objects. What began as passive objects became 'active objects' or 'rule-based objects' or 'intelligent objects' and finally 'intelligent agent objects' as this stream of evolution merged with that of DAI (Norrie and Kwok, 1991). The many agent systems which have been proposed and developed can be classified along a continuum ranging from *reactive systems* to *cognitive systems* as indicated on Figure 3.2.

Researchers interested in reactive systems, whose agents often resemble actors, use agents which only react to external stimuli and do not communicate directly or at all (except for control signals). Such agents have also a limited internal state. One can distinguish two groups of researchers in the domain of reactive agents:

(1) those like Brooks (1990) who think that it is too difficult to build models of the external world and that it is better to use the world itself as a model;
(2) others like Ferber (1996) who are interested in building large populations of reactive agents and seeing if from the agent's local behavior, a global social structure can emerge (in a way similar to ant colonies, for example).

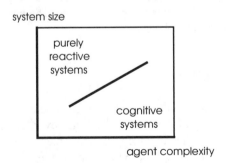

Figure 3.2 Different types of agents in artificial intelligence.

The first area of research led to various types of agent architecture organized as layers responsible for different types of reaction (see Chapter 6 for details), and led to different sorts of robots interacting directly with the world. This line of work also relates to a thesis by Minsky (1985) that thinking processes result from the interaction of non-intelligent elementary processes. In the second area, problems like the hunting of prey by groups have been studied, as well as different fashions of communication through the (usually simulated) environment.

The cognitive approach to agents considers often quite complex agents, having internal symbolic models of the world and their expertise, and capable of sophisticated reasoning and communication. The goal has been to study the behavior of such agents singly and in groups, to implement various communication and negotiation protocols, and to experiment with different internal structures for the agents. Since each agent can be quite complex, their number in a multi-agent system may not only be limited by the current hardware and software capabilities, but also by the complexity of the resulting systems.

In between the two extremes, many hybrid agent systems are being built for applications in a variety of domains. Quite often they possess limited reasoning capacity and are specialized for solving particular problems. Examples ranging from air traffic control to portfolio management can be found in Jennings and Wooldridge (1998) and more recent surveys of agent systems.

The concept of agent is today very popular and the term has been applied quite widely, even to simple code modules that do not embody any artificial intelligence technology. There are other types of agents whose heritage is not distributed artificial intelligence, but object-oriented programming, distributed systems, human/ machine interfaces, or other domains. Few really agree on a definition for a software agent despite recent efforts to provide one. Franklin and Graesser (1997) discuss several definitions before giving their own point of view with a "natural kinds classification of autonomous agents." Nwana (1996) gives another point of view; the first chapter of Bradshaw (1997) has a good discussion on the subject; and the first part of Intelligent Agents III (Müller et al., 1997) contains several papers on the subject of "what is an agent." Finally, Henkel (1999) in a special issue of IEEE Intelligent Systems has given a good analysis of the confusion that reigns in the domain. We will examine in Section 3.5 different types of agents. The next section examines the difference between agents and objects.

3.4 OBJECTS VS. AGENTS

Quite often agents are confused with objects or considered as a natural extension of objects. The exact meaning of the term 'object', however, depends on whom you are talking to. Among the domains in which the term object is used, four are of interest to us: (1) artificial intelligence; (2) programming languages; (3) databases; and (4) operating systems. Note that in this section, we do not consider domains such as philosophy, logic, computer graphics, design or manufacturing in which the word object may have slightly or radically different meaning.

In artificial intelligence, objects are used for representing the world; the main quality required of objects is that they allow modeling of the complex reality of the external world. This is developed in Section 3.4.1.

In (object-oriented) programming languages, objects are used as the basis for a programming paradigm with a goal of efficiency achieved through a structure factorizing representation (class/subclasses) and code (methods). The anticipated result is modular reusable code.

In the domain of object-oriented databases objects are used to organize (usually heterogeneous) data for persistent storage, and to permit organized access. A characteristic of this approach is that a large number of heterogeneous objects can be accessed by many users simultaneously. For more information on the subject, refer to Zdonik and Maier (1990).

In the domain of operating systems, objects are used for their modularity and for the capability of distributing computation over several machines. The main characteristics of the approach are flexibility and efficiency.

In this section, we will distinguish two main usages for objects: firstly for representation as discussed in section 3.4.1, and secondly for programming as discussed in Sections 3.4.2 to 3.4.4. Many of the current approaches do not make a clear distinction between objects and agents, and we will state our position in Section 3.4.5. One of the obvious differences between objects (developed using the object-oriented programming paradigm) and agents is in their communication. In object-oriented programming, an object can only interpret and act upon a message defined in its class (usually the message is the name of a method in this class). Agents in a multi-agent system, however, will have a common Agent Communication Language (ACL) in which they communicate with each other.

3.4.1 AI and Objects

For reasoning, AI programs need to represent the external world. To do so they build on philosophical foundations. For objects, these foundations were created by Democritus, Plato, and then Thomas Aquinas, who state that the world is composed of objects which can be idealized and described with properties. (Thomas Aquinas then adds that there is no reason why properties could not be themselves objects.) This philosophical background, coupled with the possibility of manipulating lists of symbols, made object representation quite natural in artificial intelligence. In 1975 Minsky proposed the concept of frame, mixing object representation and computation in a mechanism for modeling the thinking process. A number of frame representation languages were then developed and refined, e.g., FRL (Goldstein and Roberts, 1977), KRL (Bobrow and Winograd, 1977), RLL (Greiner and Lenat, 1980), KEE (Fikes and Keller, 1985), SRL (Fox et al., 1986).

The frame model survived many fashions and is used today, for example, in systems for representing ontologies like Cyc (Lenat and Guha, 1990) and Ontolingua (Gruber, 1992). As mentioned previously, the goal of using objects (or frames) in artificial intelligence is that of being able to model reality while focusing on expressivity, and knowing that, although our information about reality is incomplete and sometimes inconsistent, we still can reason and act. To give a single example, AI representations must be able to express beliefs about persons they need to model. Thus, taking an example from RLL, one must be able to state simultaneously that Mary is 24, but that John believes that she is only 22, leading to multiple values for the single (simple) age attribute. Of course the reasoning mechanism must be organized to deal with this sort of data, which makes artificial intelligence systems quite different and much more complex than their typical business counterparts. At the same time, one must note that traditional applications in artificial intelligence either have been quite small, or did not pay much attention to efficiency in the presence of large amounts of information.

Another approach worth mentioning in artificial intelligence is the use of classless representations, wherein a class is replaced with the concept of *prototype*. Classless objects are useful when dealing with recognition situations where properties are acquired progressively through sensors, leading to possible changes in the classification of the object being recognized. For example, a round orange object can be considered in turn as an orange, a basketball, or a pumpkin, according to the information gained progressively about size or weight. If we reason with classes, we will have to change class as we change the interpretation to a different type of object. Most of the differences between AI representation structures and data structures found in more traditional computing come from the fact that AI tries to model the real world rather than some controlled synthetic model (as in accounting).

Agent systems, if they are connected to and interacting with the external world, will need to have a (usually complex) representation of the world and of situations in the world.

3.4.2 Objects and CORBA

The CORBA approach addresses the issue of interoperability among objects in the domain of object-oriented programming. Object-oriented languages emphasize programming efficiency by stressing modularity of data structures and code sharing. However, a major problem lies with the number of different languages, and platforms, reducing the reusability of the developed products. Thus, a desirable goal is that any object written in any language and running on any type of platform should be accessible from anywhere else. The Object Management Group (OMG) has developed standards to assist in achieving this goal. One result is the Common Object Request Broker Architecture, known as CORBA. Both the Distributed Common Object Model (DCOM) developed by Microsoft and the CORBA standard allows the development of component-based software. In CORBA, the Interface Definition Language allows each object to present a standardized interface and to make services available externally. In addition the architecture offers additional services like naming, security, or licensing.

A good short discussion of CORBA can be found in (Brenner et al., 1998). Detailed and recent information about CORBA is available through the web site at http://www.corba.org/, and about COM/DCOM at http://www.microsoft.com/COM/.

3.4.3 Distributed Objects

Objects can be distributed on different machines as is demonstrated every day by numerous applications. Objects can be Java programs or system resources in object-based operating systems. A single object can also be distributed on several machines, as in the case of distributed databases. An application involving discrete simulation environments, the OOSE project (Whittington, 1993) demonstrated how objects could be partially extracted from an OO database and sent to various machines for specific processing (simulation, displaying, storage).

3.4.4 Passive Objects and Active Objects

In Object-Oriented Analysis and Design, often a distinction is made between *active* and *passive* objects. For instance, Colbert (1989) defines an object as active if it "displays independent motive power," while a passive object "acts only under the motivation of an active object."

In fact, active objects are an extension of the concept of demon (daemon). They qualify objects as able to react to some specified conditions and, for example, to notify the user. Active objects are classified as being reactive.

3.4.5 From Objects to Agents

Agents are often portrayed as being a special brand of active objects. As such, they are mainly used for efficient modular programming with reusability of the code as a major objective. Although agents are modular and can display a behavior which resembles that of objects, they are in essence quite different. Agents have their own execution thread (like actors), they can have internal state modeling mental attitudes, which may lead to an agent refusing to answer a message. Moreover, as noted in Section 3.4, they will use a common language for communication. It is true, however, that simple reactive agents are not very different in some respects from active objects.

The main difference from the designer's point of view between an agent and an object is the following: in object programming, the object sending the message is effectively responsible for ensuring that the address is accurate, the method corresponding to the message exists, and that the parameters are adequate. In the agent approach, interpretation of the message lies with the receiving agent which might simply ignore the message if it cannot understand it or if the message is not relevant to its current goals, or might interpret it in a quite different fashion than the sender intended. This change of perspective is one of the major differences between objects and agents.

3.5 DIFFERENT TYPES OF AGENTS

In this section we discuss some of the different models that have been proposed for agents. Software agents can obviously be distinguished from robots involving hardware in contact with the external world. Mobile agents are software agents that can migrate from a machine to another; reactive agents are simple software (and /or hardware) agents equipped with a simple mechanism for reacting to the changes in the environment; cognitive agents are capable of complex reasoning; autonomous agents pursue their own goals; interface agents are used as intermediates between a human user and the automatic system; and intermediate agents (also called middle agents) are agents that play a specific role in complex agent architectures. The following sections give further details on these.

3.5.1 Software Agents

For authors like Nwana (1996): "Software agents have evolved from multi-agent systems (MAS), which in turn form one of the three broad areas which fall under DAI, the other two being Distributed Problem Solving and parallel AI. Hence, as with multi-agent systems, they inherit many of DAI's motivations, goals and potential benefits." For Brenner et al. (1998) agents can be classified as follows: "At the highest level, three categories of agents can be distinguished: human agents, hardware agents and software agents. [...] Intelligent software agents are defined as being a software program that can perform specific tasks for a user and possesses a degree of intelligence that permits it to perform parts of its tasks autonomously and to interact with its environment in a useful manner." For Jennings and Wooldridge (1998), "an intelligent agent is a computer system that is capable of *flexible* autonomous action in order to meet its design objectives. By flexible we mean that the system must be responsive [...] proactive [...], and social [...]."

In fact, there is little agreement on what is a software agent and it would be necessary to develop an ontology of agents to classify the various types of software agents. Franklin and Graesser (1997) proposed a (limited) taxonomy of agents, and

Nwana (1996) proposed a classification of software agents distinguishing: collaborative agents; interface agents; mobile agents; information agents; reactive agents; hybrid agents; heterogeneous agent systems; and smart agents. In this section, we review briefly some of these different types of agents, although this review is by no means complete.

3.5.2 Mobile Agents

Mobile agents are software agents capable of moving from one machine to another automatically. One advantage compared to a static agent residing on a particular machine is that this can decrease the network communication load. Indeed, if we consider for example an agent involved in E-commerce trying to buy videos on specific topics, then the agent will first have to select which sites to visit, then for each interesting site it will request samples of videos in order to select those of interest, and after having repeated the process for a number of sites, it will be able to place an order. The whole process will generate considerable traffic on the network. An alternative approach consists in sending the agent from the index site (a yellow pages service) to each provider in turn and doing the selection process locally. If the agent is small enough this will generate much less traffic on the net. The feasibility of this approach requires that (1) the agent be small enough to be shipped efficiently; and (2) the agent be able to execute on various platforms. Several approaches have been developed allowing code to migrate from machine to machine, to accomplish this.

The basic requirement is that the executing agent code together with its environment be serialized, shipped over the network, and then reconstructed on the target machine. A useful trick is to use an intermediate code which will be executed by a local interpreter. Languages such as Smalltalk, Lisp, Python, safe-Tcl, Telescript, or Java, are good candidates for implementing this approach. Each site receiving a roaming agent must provide a local environment containing the interpreter, and also the necessary software for the incoming agent to access the local services. In addition, the environment must solve all specified security problems (identification, authorization to access resources, etc.). This approach is based on the concept of remote programming introduced by White (1976). Figure 3.3 (from White (1997)) shows the local organization for Telescript mobile agents (see also Nwana (1996), Brenner et al. (1998), and Bradshaw et al. (1999) for more detailed discussions).

The text by Wayner (1995) shows how such agents can be programmed in practice using XLisp and safe-Tcl. Among systems developed to date are: Odyssey from General Magic; Concordia from Mitsubishi; Aglets from IBM (Lange and Oshima, 1998); and Voyager from ObjectSpace. A brief description of these systems can be found in Chapter 16.

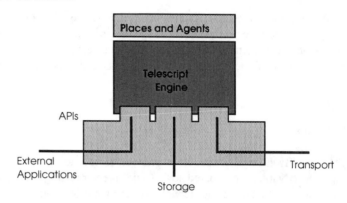

Figure 3.3 Telescript Engine (White 1997) (©1997 AAAI).

Although the mechanisms which have been developed solve many of the problems of implementing roaming software, they do not solve any of the important problems related directly to agents, which leads Nwana and Ndumu (1999) to state: "[...] useful mobile agent applications may only emerge when we have fully developed their static counterpart."

3.5.3 Reactive Agents

We mentioned reactive agents in Section 3.3.4 as systems which only react to external stimuli and do not communicate directly or at all (except for control signals).

In the mid eighties, Minsky introduced the concept of the Society of Mind, a "scheme in which each mind is made of smaller processes. These we'll call *agents*. Each mental agent by itself can only do some simple things that need no mind or thought at all. Yet, when we join these agents in societies—in certain ways—this leads to true intelligence" (Minsky, 1985). Along the same line, Brooks (1990) argued against founding artificial intelligence upon complex symbol systems which are too difficult to construct and to manipulate. He offers, on the contrary, a "new methodology [which] bases its decomposition of intelligence into individual behavior generating modules, whose coexistence and co-operation let more complex behavior emerge." Brooks demonstrated the approach by building mobile robots as interconnected simple systems involving sensors reacting to external conditions and driving actuators. The corresponding architecture comprises several layers of automata (finite state machines) augmented with timing elements. This became known as the *subsumption architecture* (see Chapter 6 for a brief description). Each automaton implements a simple behavior, and their interaction lead to the often complex global behavior of the robot.

Ferber (1996) developed a reactive agent system in which each agent has to choose among several tasks that it can execute at a given time. Tasks are weighted and the weights can change during the course of action. This leads to specialization among the agents and allows simulation of the behavior of agent societies (insects for example). The main goal of this approach was to study the emergence of social structures from the overall interaction of the agents. As in the case of Brooks's robots, the agents interact via their environment (here a simulated version of a physical environment).

Reactive agents have also been studied by Ferguson (1992), Giroux et al. (1996), and Davis (1997). See also Müller et al. (1995) for a discussion of layered architectures. There is much current work in multilayered architectures for hybrid agents, i.e., for systems of reactive agents with reasoning capabilities.

3.5.4 Cognitive Agents

In contrast to the reactive agents described above, cognitive agents (also called deliberative agents) possess an explicit knowledge of their environment. The heritage of cognitive agents is clearly artificial intelligence. They use sophisticated knowledge representations, have expertise, goals, and plans, and so on. In addition to their traditional mechanisms they must interact and engage in cooperation with other agents. Many types of cognitive agents have been proposed. Some are very simple and emphasize autonomy and cooperation, others are quite complex and are used to test cooperative mechanisms or learning.

Cognitive agents can be used for: (1) solving problems too large to be solved by a single agent; (2) addressing naturally distributed problems; (3) integrating legacy systems; (4) building modular systems. Examples range from early systems like DVMT (Durfee et al., 1987), and MACE (Gasser et al., 1987), to more recent ones like ADEPT for business planning (Jennings et al., 1996). Other examples will be found in this book, and some examples in engineering design and manufacturing are described by (Parunak, 1998).

Cognitive agents have also been studied from a theoretical perspective. One of the best known agent architectures for reasoning agents is the belief-desire-intention (BDI) model (Rao and Georgeff, 1992). Quoting from Jennings et al. (1998): "Beliefs correspond to information that the agent has about its environment. Desires represent options available to the agent – different possible states of affairs that the agent may choose to commit to. Intentions represent state of affairs that the agent has chosen and has committed resources to. An agent's practical reasoning involves repeatedly updating beliefs from information in the environment, deciding what options are available, filtering these options to determine new intentions, and acting on the basis of these intentions." Several systems have been developed using the BDI model.

According to Nwana (1996) several issues still remain to be addressed:

- methods for designing systems of cognitive agents;
- models of coordination;

- scalability;
- methods for integrating legacy systems;
- learning;
- evaluation of the resulting systems.

3.5.5 Hybrid Agents

In some cases (as with robots), a pure reactive architecture is not enough to ensure a consistent permanent behavior in agents or the necessary machinery to build symbolic internal representations to use for symbolic planning is too complex to be of any practical use. For such situations, researchers have proposed new mechanisms for introducing goals and goal directed behaviour into reactive agents. Maes (1990) introduced the concept of a network propagating energy activation, for selecting among competing goals while taking into account external stimuli. Noteworthy is the anytime algorithm used for the level of the threshold controlling activation of actions. Other researchers have introduced a multi-layered architecture with a reactive lower level and more deliberative upper levels, as in the InteRRaP approach, of Müller et al. (1995). Such architectures are quite promising but may be difficult to implement.

3.5.6 Interface Agents

Interface agents facilitate the interaction between a user and a computer system. They are intended to improve interaction, e.g., accessing information, assisting with current work, doing learning, or just providing entertainment. According to Laurel (1997), interface agents have four characteristics:

(1) agency, allowing them to appear understanding, helpful and competent;
(2) responsiveness, allowing communication in tune with the user;
(3) competence, from having meta-information about the subjects to be addressed;
(4) accessibility, by providing an adequate physical interface.

Interface agents already developed range from easy-to-program simple services for users (Malone et al., 1997) to quite complex reasoning assistants (Ball et al., 1997). A major problem with interface agents is their acceptance by users. They should ideally behave like English butlers (Negroponte, 1997) knowing their master's or mistress's preferences and offering services in a tactful fashion. To be able to do so, however, they require a sophisticated model of their owner, and need to be able to acquire and maintain such a model. Maes (1997) proposed using parametric learning for certain types of situations, as well as automatic learning (help) by transferring expertise found in other people's interface agents.

Currently two approaches are competing: (1) specialized agents attached to a particular type of service, e.g., the gathering and filtering of information; and (2) assistants attached to a particular user and interfacing the user with a complex system. In the future, agent systems may use a basic 2-layer architecture: an agent-enabled architecture (Bradshaw, 1997), with one layer composed of specific services, and the other layer of interface agents.

The reader will find a good discussion of interface agents in Bradshaw (1997). Also, as reported by Huhns and Singh (1998), FIPA (Foundation for Intelligent and Physical Agents) is trying to standardize some generic aspects of agents, and has been devoting some efforts to human computer interaction.

3.5.7 Middle Agents

Middle agents (also called intermediate agents) in agent systems are internal agents which play a particular role, providing intermediate services. They are found under various names, e.g., brokers, controllers, facilitators, mediators, or matchmakers. Their role may range from that of providing a yellow pages service (a simple type of broker) to dealing with the intricacies of high level protocols (facilitator) for matching a request to a particular service. Ontology servers are appearing as a new type of such middle agents. More information on these middle agents is given in subsequent chapters.

3.6 AGENTS IN CONCURRENT DESIGN AND MANUFACTURING

The case for using agents in industry is well defended by Parunak (1998). Noting that the main trends are increased product complexity and diversity, changes in supply networks, and increased product variety over time, Parunak has analyzed where agent technology can best be used in design and operation activities. His answer: "agents are best suited for applications that are modular, decentralized, changeable, ill-structured, and complex." Most of the reasons often given for adopting an agent approach are linked to their being pro-active object systems and to the simplification of the architecture of the software systems, knowing the cost of software is a major component of the total cost of constructing or operating a factory. The real gain obtained from an agent approach, however, often results from a better description of the real world by focusing on objects rather than functions. Used appropriately, this leads to the desired modularity, allowing flexible simulation, and to better response and improved software reusability. In addition, the fact that agents can cope with a dynamically changing world by performing dynamic linking, allows handling of ill-structured or rapidly changing situations in a more economical way.

Although agent technology has recently been recognized as a promising approach for collaborative design and manufacturing systems, those agents that have so far been implemented in various prototype and industrial applications are not actually very "intelligent." However, these are early days yet in agent technology, and the potential is there for very sophisticated and intelligent systems to be developed.

The use of agents in design has been demonstrated by various research projects (see Chapter 13 for detailed examples) and is considered further through this book. In design, agents have mostly been used for supporting cooperation among designers, providing a semantic glue between traditional tools, or for allowing better simulations. A good overview of multi-agent design systems can be found in (Lander, 1997).

Agents in manufacturing are being used in new ways for manufacturing enterprise integration, supply chain management, manufacturing planning, scheduling and execution control, materials handling and inventory management, and other areas. Many examples are cited in this book, and some examples are detailed in Chapters 14 and 15. A state-of-the-art survey of agent-based manufacturing systems with an extensive annotated bibliography can be found in (Shen and Norrie, 1999). Its extended version in HTML format with direct links to related web sites, is available at: http://imsg.enme.ucalgary.ca/publication/abm.htm.

REFERENCES

Baker, A.D. (1991) *Manufacturing Control with a Market-Driven Contract Net,* PhD thesis, Rensselaer Polytechnic Institute, NY.

Ball, G., Ling, D., Kurlander, D., Miller, J., Pugh, D., Skelly, T., Stankosky, A., Thiel, D., Van Dantzich, M. and Wax, T. (1997) Lifelike Computer Characters: The Personal Project at Microsoft, In Bradshaw, J.M. ed., *Software Agents*, AAI Press / MIT Press, pp. 191-222.

Bobrow, D.G. and Winograd, T. (1977) An Overview of KRL, a Knowledge Representation Language, *Cognitive Science*, 1(1):3-46.

Bradshaw, J.M. (1997) An Introduction to Software Agents, In Bradshaw, J.M. ed., *Software Agents*, AAAI Press / MIT Press, pp. 3-46.

Bradshaw, J.M., Greaves, M. and Holmback, H. (1999) Agents for the Masses? *IEEE Intelligent Systems,* 15(2):53-63.

Brenner, W., Zarnekow, R., Wittig, H. (1998) Intelligent Software Agents. Springer.

Brooks, R.A. (1990) Elephants don't Play Chess, In Maes, P. ed., *Designing Autonomous Agents*, MIT Press / Elsevier, pp. 3-16.

Colbert, E. (1989) The object-oriented software development method: a practical approach to object-oriented development, In *Proceedings of TRI-Ada 89.*

Conry S.E., Kuwabara, K., Lesser, V.R. and Mayer, R.A. (1991) Multistage Negotiation for Distributed Constraint Satisfaction, *IEEE Transaction on Systems, Man and Cybernetics.* 21(6):1462-1477.

Corkill, D.D. and Lesser V. (1983) The Use of Meta-Level Control for Coordination in a Distributed Problem-Solving Network, In *Proceedings of the Eighth International Joint Conference on Artificial Intelligence*, pp. 748-756.

Davis D.N. (1997) Reactive and Motivational Agents: Towards a Collective Minder, In Müller J.P., Wooldridge M.J. and Jennings, N.R. eds., *Intelligent Agents III, Agent Theories, Architectures, and Language.*

Davis, R., Shrobe H. and Szolovitz P. (1993) What is Knowledge Representation? *Artificial Intelligence Magazine*, 14(1):17-33.

Durfee, E.H., Lesser, V.R. and Corkill, D. (1987) Coherent cooperation among Communicating Problem Solvers, *IEEE Transactions on Computers*, 36(11): 1275-1291.

Englelmore, R.S. and Nii, H.P. (1977) A Knowledge-Based System for the Interpretation of Protein X-Ray Crystallographic Data, HPP Memo 77-2, Stanford University.

Erman, L.D., and Lesser, V.R. (1975) A Multi-Level Organization for Problem Solving Using Many Diverse Cooperating Sources of Knowledge, In *Proceedings of 4th International Joint Conference on Artificial Intelligence*, Tbilisi, ASSR, pp. 483-490.

Feigenbaum E. and Feldman. (1963) Computers and Thought. McGraw Hill.

Ferber, J. (1996) Reactive Distributed Intelligence: Principles and Applications. In O'Hare, G. and Jennings, N. eds., *Foundations of Distributed Artificial Intelligence*, Wiley Interscience, pp. 287-314.

Ferguson I. (1992) Toward an Architecture for Adaptive, Rational, Mobile Agents, In Wermer, E. and Demazeau, Y. eds., *Decentralized Artificial Intelligence 3*, North-Holland, pp. 249-261.

Fikes, R., and Kehler, T. (1985) The role of frame based representation in reasoning, *Communications of the ACM*, 28(9):904-920.

Fischer, K., Muller, J.P., Pischel, M. and Schier, D. (1995) A Model for Cooperative Transportation Scheduling, In *Proceedings of ICMAS'95*, AAAI Press/MIT Press, San Francisco, CA, pp. 109-116

Fox, M., Wright, M. and Adam, D. (1986) Experiences with SRL: An Analysis of a Frame-based Knowledge Representation, In Kerschberg, L. ed., *Expert Database Systems*, The Benjamin/Cummings Publishing Company Inc., pp. 161-172.

Franklin, S. and Graesser, A. (1997) Is It an Agent, or Just a Program? A Taxonomy for Autonomous Agents, In Müller J.P., Wooldridge M.J. and Jennings, N.R. eds., *Intelligent Agents III, Agent Theories, Architectures, and Languages*, pp. 21-35.

Gasser, L., Braganza, C. and Herman, N. (1987) MACE: A Flexible Testbed for Distributed AI Research, In Huhns, M. ed., *Distributed Artificial Intelligence*, Pitman, London, pp. 119-152.

Giroux S., Marcenac M., Quinqueton J. and Grasso J-P. (1996) Modeling and Simulating Self-Organized Systems In *Proceedings of ESM'96*, Budapest.

Goldstein, I. and Roberts, B. (1977) NUDGE, A Knowledge-Based Scheduling Program. MIT, AI Memo 405.

Greiner, R. and Lenat, D. (1980) A Representation Language Language, In *Proceedings of AAAI-80*, Stanford, pp. 165-169.

Gruber, T. (1992) Ontolingua: A Mechanism to Support Portable Ontologies (version 3.0), Technical Report KSL 91-66, Stanford University.

Hayes-Roth, B., Hewett M., Washington R. and Seiver A. (1989) Distributing Intelligence within an Individual, In Gasser, L. and Huhns, M. eds., *Distributed artificial Intelligence Volume II*, Morgan Kaufmann, pp. 385-412.

Henkel, J. (1999) Making Sense out of Agents. *IEEE Intelligent Systems*, 15(2):32-37.

Hewitt C. (1979) Control Structure as Patterns of Passing Messages, In Winston, P.H. and Brown, R.H. eds., *Artificial Intelligence: An MIT Perspective*, MIT Press, pp. 435-465.

Hewitt C. (1980) Apiary multiprocessor Architecture Knowledge System. In *Proceedings of the Joint SRC/U. of Newcastle upon Tyne Workshop on VLSI Machine Architecture, and Very High Level Languages*, pp. 67-69.

Huhns, M.N. and M.P. Singh (1998) Personal Assistants, *IEEE Internet Computing*, 2(5):90-92.

Huhns, M. Mukhopadhyay, U. and Stephens, L.M. (1987) DAI for Document Retrieval: The MINDS Project, In Huhns, M. ed., *Distributed Artificial Intelligence*, Pitman Publishing, pp. 249-284.

Jamali N., Thati P. and Agha G.A. (1999) An Actor-Based Architecture for Customizing and Controlling Agent Ensembles, *IEEE Intelligent Systems,* 14(2):38-44.

Jennings, N.R., Faratin, P., Johnson, M.J., O'Brien, P. and Wiegand, M.E. (1996) Using Intelligent Agents to Manage Business Processes, In *Proceedings of the First International Conference and Exhibition on the Practical Applications of Intelligent Agents and Multi-Agent Technology*, London, UK, pp. 345-360.

Jennings, N.R., Sycara, K., Wooldridge, M. (1998) A Roadmap of Agent Research and Development, *Autonomous Agents and Multi-Agent Systems*, 1:7-38.

Jennings, N.R., and Wooldridge, M., (Eds.) (1998) *Agent Technology: Foundations, Applications and Markets*, Springer.

Lander, S.E. (1997) Issues in Multiagent Design Systems. *IEEE Expert*, **12**(2), 18-26.

Lange, D.B. and Oshima, M. (1998) *Programming and Deploying Java Mobile Agents with Aglets*, Addison-Wesley.

Laurel, B. (1997) Interface Agents: Metaphors with Character, In Bradshaw, J.M. ed., *Software Agents*, AAAI Press / MIT Press, pp. 67-77.

Lenat, D.B. and Guha, R.V. (1990) *Building Large Knowledge-Based Systems*, Addison-Wesley Publishing Company.

Maes, P. (1990) Situated Agents Can Have Goals, Maes, P. ed., *Designing Autonomous Agent,* MIT Press / Elsevier.

Maes, P. (1997) Agents that Reduce Work and Information Overload, In Bradshaw, J.M. ed., *Software Agents*, AAAI Press / MIT Press, pp. 146-164.

Malone, T.W., Lai, K.Y. and Grant, K.R. (1997) Agents for Information Sharing and Coordination: a History and Some Reflections, In Bradshaw, J.M. ed., *Software Agents*, AAAI Press / MIT Press, pp. 109-143.

Minsky M. (1985) *The Society of Mind*, Touchtone Book, Simon & Schuster.

Monceyron E. and Barthès J-P. (1992) Architecture for ICAD Systems: an Example from Harbor Design, *Revue Sciences et Techniques de la Conception*, 1(1):49-68.

Müller J.P., Pischel M. and Thiel M. (1995) Modeling Reactive Behavior in Vertically Layered Agent Architectures, In Wooldridge M.J. and Jennings N.R. eds., *Intelligent Agents*, Springer, pp. 261-276.

Müller J.P., Wooldridge M.J. and Jennings, N.R. (1997) *Intelligent Agents III, Agent Theories, Architectures, and Languages*, Springer.

Negroponte, N. (1997) Agents: from Direct Manipulation to Delegation, In Bradshaw, J.M. ed., *Software Agents*, AAAI Press / MIT Press, pp. 57-66.

Norrie, D.H. and Kwok, A.D. (1991) Object-Oriented Distributed Artificial Intelligence, In Maurer H. ed., *New Results and New Trends in Computer Science*, Springer-Verlag, LNCS 555, pp. 225-242.

Nwana H.S. (1996) Software Agents: An Overview, *Knowledge Engineering Review*, 11(3):205-244.

Nwana H.S. and Ndumu, D.T. (1999) A Perspective on Software Agents Research, *Knowledge Engineering Review*, 14(2):125-142.

Parunak, H.V.D. (1998) What can Agents do in Industry, and Why? An Overview of Industrially-Oriented R&D at CEC, In Klusch, M. and Weiss, G. eds., *Cooperative information agents II: learning, mobility and electronic commerce for information discovery on the Internet:* Second International Workshop, CIA'98, Paris, France, July 4-7, 1998, Springer, pp. 1-18

Rao, A.S. and Georgeff, M.P. (1992) An Abstract Architecture for Rational Agents, In *Proceedings of Knowledge Representation and Reasoning*, pp. 439-449.

Reddy, R., Erman, L., Fennel, R. and Neely, R. (1976) The HEARSAY Speech Understanding System: An Example of the Recognition Process, *IEEE Transactions on Computers*, C-25:427-431.

Sandholm T. (1993) An Implementation of the Contract Net Protocol Based on Marginal Cost Calculations, In *Proceedings of the Eleventh National Conference on AI*, Washington, DC, pp. 256-262.

Scalabrin E. (1996) *Conception et Réalisation d'environnement de développement de systèmes d'agents cognitifs*, Ph.D. Thesis, Université de Technologie de Compiègne, France.

Shen, W. and Norrie, D.H. (1999) Agent-Based Systems for Intelligent Manufacturing: A State-of-the-Art Survey, *Knowledge and Information Systems, an International Journal*, 1(2):129-156

Smith R.G. (1979) A Framework for Distributed Problem Solving, In *Proceedings of the 6th International Joint Conference on Artificial Intelligence*, pp. 836-841.

Theriault D. (1983) Issues in the Design and Implementation of Act 2, Technical Report 728, MIT Artificial Intelligence Laboratory.

Wayner, P. (1995) *Agents Unleashed*, Academic Press Professional.

White J.E. (1976) A High-Level Framework for Network-Based Resource Sharing, In *Proceedings of AFIPS Conference*, pp. 561-570.

White J.E. (1997) Mobile Agents, In Bradshaw, J.M. ed., *Software Agents*, AAAI Press / MIT Press, pp. 437-472.

Whittington, G. (1993) Using ToolTalk to Implement a Distributed Discrete Simulation Environment, In *Proceedings of the 19th Conference of the ASU*, University of Aberdeen, UK.

Yonesawa, A. and Tokoro, M. (1987) *Object-Oriented Concurrent Programming*, The MIT Press.

Zdonik, S.B. and Maier, K.D. (1990) *Readings in Object-Oriented Databases*, Morgan Kaufmann.

Part II: Important Issues in System Implementation

4

Knowledge Representation in Agent-Based Concurrent Design and Manufacturing Systems

4.1 INTRODUCTION

Production of a complex product involves several processes: the design process produces a prototype; the industrialization process produces the manufacturing tools and the associated organization; and the manufacturing process actually produces the final product. Since specialists work in each process in different parts of the enterprise, an enterprise is naturally distributed. Each group working inside the enterprise or outside (as sub-contractors) is an island of competence and knowledge. Each has its own language and traditions. Each produces its own documents, most of the time quite independently from other groups. Data and knowledge contained in these technical documents (engineering drawings and technical files) *represent* the *product*, the *manufacturing environment*, and the *processes*.

The primary role of the technical documentation is to enable the enterprise to manufacture the designed products. The documents support communication among the people involved in the production chain, from design to delivery and maintenance. These technical documents are also used for information transfer between adjacent groups, each one making the necessary adjustments to understand the other's culture and idiosyncrasies, which usually limits communications to adjacent levels.

A secondary role assigned to documents is serving as a memory for possible future changes, or for designing similar products. Technical documents contain part of the information representing the knowledge used during the design and manufacturing process (other information is retained in human memory and elsewhere). Some of the documents are centered on the product, others address the manufacturing environment (e.g., tools, fixtures, scheduling, testing methods), and others focus on the processes.

Concurrent engineering requires better organization than sequential engineering as well as a global view of the current state of the design and production processes. The current trends are: (1) to improve the overall organization (through Business Process Re-engineering); (2) to dematerialize technical documents through digital support.

Two approaches are competing for better organization of the technical data; a centralized standardized approach exemplified by the CALS initiative, and a distributed approach supported by agents. The standardized approach advocates a uniform representation common to all actors in the design and manufacturing processes, centered on a single product representation; the distributed approach emphasizes inter-group communication of design and manufacturing knowledge. This chapter discusses the necessary representations for a distributed flexible agent-based approach.

4.2 WHAT NEEDS TO BE REPRESENTED

There are basically two types of representation: (1) representation involving humans and (2) representation for computing purposes. The role of the first type is for recording and transferring information among humans, exposing internal mechanisms in the clearest and most intelligible fashion. For this purpose, conventions have been developed to render such representations terse and pithy. Isometric drawing for piping design and layout is one example. The role of the second type is to optimize computations, obeying the constraints associated with a particular computing technique. Meshing conventions for Finite Element Methods provide an example.

Computers do not use drawings, and the internal representation that they use could be referred as a *deep representation*. Humans, on the other hand, must exploit computational results, and control or direct the work of a computer. Human-understandable formats have to be developed for this purpose. These constitute a *surface representation*. Translation processes are necessary to convert surface data into their internal equivalent. The simplest example is provided by early CAD systems, which were simple translators between a geometric surface representation and internal data structures. Unfortunately, some of the representations used by humans are difficult to computerize, as for example, free hand sketching. Conversely, some representations used by computing algorithms are difficult to translate into easily understandable surface representations, as for example, are the results of 3D constraint computations.

The introduction of agents does not change the state of affairs. Some agents will control internal computation processes, for example, an agent providing constraint computations based on kernel methods, and will use only deep representations. Other agents will have the task of producing surface representations to interface with humans, as for example, do personal assistants. Still other agents will have to handle information or knowledge, a task which was usually not done by traditional tools. Thus, in multi-agent systems, various agents will use different types of representation and must also handle related knowledge. This means they face the same representation problems as previously, with the addition of handling translation from one representation format into another automatically, as well as processing knowledge. Clearly, this is a difficult job that can only be done gradually, starting with quite small groups of agents, adding new ones as needed. Nothing guarantees today that a single global multi-agent system can be developed for handling all of the design, manufacturing, and knowledge management functions of the enterprise.

In summary, irrespective of whether agents are used to support the design and manufacturing, different kinds of information must be represented. However, the use of agent technology introduces additional requirements, in particular, in the representation of knowledge. Thus, the current section is organized into two parts: (1) a first part, including sections 4.2.1 to 4.2.5, that discusses the representations that are necessary irrespective of the chosen approach; (2) a second part that considers those representations (essentially of knowledge) that are necessary when agent technology is used.

4.2.1 Model of the Product to be Designed and Manufactured

In order to be made, products need to be represented. Different types of representation are necessary, first to express ideas, then to give more and more specific elements of information. Representations are intended to perform a function. For example:

- Detailed drawings indicate to a human how to manufacture or to assemble a set of parts;
- Schematics tell the engineer what are the functions of a product, for example, the PFS (Process Flow Sheet) or the P&ID (Process and Instrumentation Diagrams) in chemical engineering;
- Isometric drawings give information on how to lay out a pipeline;
- Additional symbols on a geometric drawing indicate surface finish and tolerances, and are necessary for machining a part;
- PERT or Gantt Chart diagrams give information about workshop scheduling;
- 3D drawings, photographs, or videos, are used for training, diagnosis, or maintenance;
- Computer models are used for simulation.

Most schematic or geometric representations are used at the interface between humans, or between humans and machines to allow humans to control what is happening inside the machine. Most of the representations have been computerized, and new types of representations have been introduced following the use of computers. Representations associated with design and manufacturing techniques show a great variation across the range of possible products or technologies. Any given product uses several types of representation during its life and a major problem is to switch smoothly from one representation to another. The rest of this section gives examples of product representation mostly within the domain of mechanical engineering.

The oldest and most basic representation of objects is their geometric representation. Most designed and manufactured artifacts possess precise geometric dimensions. Thus, a geometrical representation has to be available to specify such dimensions. The art of engineering drawing was evolved to prescribe what shape and size a product should be, and to help set up its manufacturing. Additional technical information is typically added as comments on the drawing or as separate technical documents. In the early 70s, computer-based drafting systems were introduced to aid the drawing process, starting with 2D PCBs (Printed Circuit Boards). Soon thereafter, technical information was added in databases, for example, net-lists resulting from component connectivity, and business information such as component libraries and prices. Such representations were mostly used for detailed design and manufacturing. In mechanical engineering applications, 2D drawings were progressively replaced with 3D engineering models, from which more traditional 2D drawings could be derived. Geometric representations are essential even if the corresponding drawings are not strictly useful except for communication with humans. Indeed, a 3D model of a part can be used to generate a machining program automatically, without the need to display the results except for human supervision.

Later, CSG (Constructive Solid Geometry), FEM (Finite Element Methods), and feature-based design (see Rosen (1993) for a discussion) introduced new types of representation. Thus, although geometry remains a fundamental aspect of product modeling, additional symbolic data are today also handled routinely. The introduction of symbolic information allowed modeling to be addressed earlier in the life of products, (e.g., at the level of conceptual design) using techniques like: bond graphs (Ulrich and Seering, 1989); Petri nets for parametric design (McMahon and Meng, 1996); and behavior graphs (Welsh and Dixon, 1994). Formalizing relationships between form and function can be done with chain models (Palmer and Shapiro, 1993) or through the sketching of abstractions (Mukherjee and Liu, 1995). Design languages were developed for combining shapes using grammars (Brown et al., 1996). It was also shown how parts could be produced by combining constraints (Fu and de Pennington, 1993).

The existence of various proprietary formats in early CAD systems led to the development of the exchange standards IGES (MIL-D-28000, 1987), and PDES/STEP (Langenbach, 1993). Description languages were also developed like IDEF1x (Integrated computer-aided manufacturing DEFinition), NIAM (Nijssen's Information Analysis Method), and EXPRESS for certain types of geometry (Williams A.L., 1989). Standards for text documentation such as SGML (ISO 8879, 1988) were also created. More specialized approaches have been developed for the non-standard geometry of molded parts, stamped parts, and sculptured objects (Mitchell et al., 1995). Complex representations have been proposed for modeling assembled products (Mäntylä, 1989; Shah and Rogers, 1993), and product families (McKay et al., 1996), or complex products (Baldwin and Chung, 1995). New types of representation have also been proposed for combining geometrical data with technical information, modeling constraints, for including variant or version information, and for including comments and process information useful for future knowledge management. Such representations are usually derived from the concept of *frame* proposed by Minsky (1975) in artificial intelligence. Complex programs can act on such representations to organize or access their information (for example, as *cases*), to propagate changes automatically (El Dahshan and Barthès, 1989), to instantiate members of a family of products, or to do kinematic analysis (Joskowicz and Neville, 1996) (Kiriyama et al., 1989). Complex objects are often thought of as being an assembly of predefined components. This is not the case in certain domains like architecture, where objects can be created as the result of geometrical constraints (construction components) as was shown by Zamanian and Fenves (1994). And lastly, meta-models represent a sort of ultimate generic model from which specific part representations can be derived (Bijl, 1989; Bridge, 1989; Henderson and Taylor, 1993).

The reader has now probably been convinced that product representation is a complex issue. A sufficiently complete representation has to be selected which will be adequate for the particular task at hand. For most common needs, there are suitable representations that can be selected. For special purposes, this may not be the case. Ad hoc representations are therefore being developed quite regularly to address increasingly difficult modeling problems. Most such representations are for human communication, although some only support computer algorithms. Whether all of the existing representations are really useful and how one translates from one representation to another are still open research issues.

4.2.2 Model of the Process (or the State of Achievement of the Tasks)

Although modeling the product is a necessary first step, it only provides a static perspective. Dynamic aspects of the production process also require modeling. Modeling the design process allows us to improve it by having the machine take care of tedious details and will allow us to control the sequence of design steps. Modeling the manufacturing process allows us to better control it and to improve its efficiency.

Different types of models are available for design and manufacturing processes:

- general theoretical models;
- models for analysis, simulation, or optimization;
- models that handle knowledge explicitly.

General theoretical models have been introduced mostly in the context of design. Such models described the design process as a set of successive steps and tried to give a mathematical axiomatization to ground the design activity. In the General Design Theory (GDT), Yoshikawa (1982) considered the design process as a mapping from a function space to an attribute space. The GDT approach was later considered further by Reich (1995), who proposed extending its scope. Other general models were developed, such as an evolutionary model for design in an attempt to tackle complexity and uncertainty, incorporating chunking and feedback (Williams D., 1989). More recently, Suh (1998) has proposed a new axiomatic theory of design. General theoretical models are intended as frameworks to describe and to guide the design activity. However, more specific operational models have also been developed.

The earlier operational models of processes were developed for designing manufacturing facilities. They were mostly linked to Operations Research and were for dimensioning facilities or optimizing their operation. Such models are still being used, and are essentially numeric, being discrete, continuous, or mixed (Ravindran, 1987). There are well-known models for production planning, job shop scheduling, and similar problems. New types of model, emphasizing performance evaluation rather than optimization have also been proposed, for example, using signal flow graphs (Eppinger et al., 1997). Artificial intelligence techniques have been used to develop more flexible models in this area, for example, the OPGEN system for planning (Freedman and Frail, 1986), and intelligent databases for engineering applications (Dong and Goh, 1998). There are thus new symbolic methods to compete with Operations Research models. However, such techniques can be combined to produce generalized representations of engineering design that involve qualitative and quantitative aspects (Cagan et al., 1997).

In some of the modeling approaches, knowledge is handled explicitly. At the design level, this tends to produce rich and complex environments allowing the designers to be more productive. At the manufacturing level, this produces complex production models that at the same time handle both flows of matter and information.

Earlier design systems aimed at combining knowledge-based environments with technical tools like solid modelers. Various systems were developed ranging from those that integrated basic AI techniques into traditional CAD systems (which basically amounted to adding rules to the database) to those in which the CAD facilities were redeveloped within an AI environment (for example NuDraw at MIT/LMI). The ARCHIX system is an example of combining a knowledge-based environment and traditional technical tools. In the ARCHIX system (Thoraval, 1991), the various steps for designing the front wheel drive train of the car are known and fixed; the design process is modeled and used as a backbone for helping the engineer to design new configurations; and traditional CAD systems are used as geometric modelers to check various dimensional constraints. The NuDraw system developed at MIT and LMI to design computers, and the MOLE environment proposed by Tweed and Bijl (1989), implement design environments within knowledge systems. This latter approach aims at providing digital assistants for the designer.

For the overall production process, specific models and methods have been developed. Methods based upon the General System Theory of Le Moigne are used to model enterprises or complex processes. The GRAI method was first introduced for modeling SME (Small and Medium Enterprises) and for detecting malfunctioning in their management, in particular in the information paths (Doumeingts et al., 1987). The GRAI method was used in several European projects and extended to cover full enterprise models. Other similar methods have been proposed such as (CIMOSA, 1991), a result from the AMICE European ESPRIT project (ESPRIT-AMICE, 1991). Such methods are currently popular in the context of Business Process Reengineering. Other models were developed as a support for knowledge management, in order to record best practices or past experience. Knowledge associated with the design of the artifact or of the manufacturing resources, as well as experience gained during design and operation are formalized, recorded, and indexed for later re-use. The approaches advocated by Grundstein (1996) for representing activities, or by Fox and coworkers (1995) for business models, offer frameworks for the organization of knowledge. Formalizing information for knowledge management purposes has taxing technical consequences for the underlying computer models. Design and manufacturing contexts have to be modeled and trial and error attempts have to be recorded, which leads to a heavy usage of version mechanisms. See, for example, Taura and Kubota (1999) for a discussion about an engineering history base, or Duffy and Kerr (1993) for the expressing of different perspectives or past designs.

In some cases, the modeling of design or manufacturing activities has led researchers to propose specific languages, for example, for manufacturing (Brown et al., 1995). And in an effort to define more general representations, some researchers propose organizing product models from the process point of view, focusing on concurrent engineering (Molina et al., 1995) or on assembly (Mantipragada and Whitney, 1998).

4.2.3 Domain Knowledge

No design or manufacturing activity can be organized without domain knowledge. Indeed, product models or process models are rarely completely generic, but rather are indirectly dependent on the domain to which they apply. This results from the extreme diversity of the types of objects to be produced and from the variety of approaches that can be used to do so. Thus, a large part of the useful knowledge is domain dependent. Among the domain dependent types of knowledge, one may find:

- domain ontologies;
- knowledge related to product representation;
- knowledge related to processes;
- knowledge related to knowledge management.

Domain ontologies are necessary to define the essential concepts and the associated vocabulary of a particular domain. Useful ontologies (see Chapter 11) partially overlap more generic ontologies (in particular those addressing common sense concepts) and complement them with specialized concepts. They are necessary for establishing the basic vocabulary and the domain semantics. As will be seen, ontologies are used for communication among humans, as a basis for developing knowledge-based systems, for indexing multimedia documents or case bases, and sometimes as a bootstrap for full text indexing. Concurrent engineering cannot be developed without strong ontological support.

Most domains have a particular approach to representing products. In addition, as we already saw, representation may change radically between conceptual design and manufacturing or assembly. Many efforts have been undertaken to develop generic models for whole domains from which more specific models could be derived, as well as to construct libraries of components that can be assembled to produce product models. Thus, the concept of the meta-model is used by many researchers (Veerkamp, 1992; Tomiyama et al., 1991), while others have developed libraries, as Regli and co-workers (1995) did for machining features. An excellent example of the use of the various types of knowledge can be found in (Otto and Wood, 1998) who discuss the redesign of a wok.

Domain knowledge can also be related to a particular tool, or more precisely to how to use the tool for a class of problems. For example, in the preliminary design phase of a harbor, Monceyron and Barthès (1992) had to represent a way to model the use of the residual wave motion program, which required modeling the harbor in iso-depth basins before the program could be used. More generally, a similar situation will apply to legacy tools that are being encapsulated as independent processes.

Domain knowledge can also be related to the design or to the manufacturing process. Many examples can be found where knowledge-based systems are used. In such a case, specific knowledge is usually modeled as a set of rules.

Domain knowledge can also be recorded for future needs. In this case, various techniques may be used for knowledge management. For example, Hua and co-workers (1996) use case databases and Chakrabarti and Bligh (1994) use graphs. Effective practices in design transfer are discussed by Busby (1998). To be useful in later work, recorded knowledge must, in particular, include design rationale.

4.2.4 Design Rationale

Design rationale, whether for products or for the manufacturing environment, is comprised of particularly important types of knowledge. In fact, it basically consists of all the choices that have been made during the design process together with the justification for these choices. Most often, design rationale is not even formalized and stays as tacit knowledge in the heads of people. When formalized and recorded, it is usually written in technical documents separately from engineering drawings and models. It is very important to know design rationale for subsequent reuse in manufacturing, maintenance, and for redesign. Therefore, the design rationale should desirably be included in the various representations of the product and its processes. Early work by Henderson and Taylor (1993), developed a meta-model (called the meta-physical model) for recording conceptual information: context, product definition unit, alternatives, relations, constraints, design intent, decisions. At the same time, Reich and co-workers (1993) were considering machine learning for supporting design practice and developed a prototype environment, BRIDGER, to test the machine learning approach. This led to an information system, *n*-dim (The n-dim Group, 1995), being developed "to address issues observed in actual design practice." However, problems like the accumulation of knowledge or the creation of a large shared memory still remain to be solved. Taking a different approach, Kawakami and co-workers (1996) utilize explanation-based learning to acquire conceptual design knowledge. They use a hierarchical model for the designed object and a multi-layered structure for the design knowledge. Design is axiomatized and organized as a design process model, generalizing Suh's approach (Suh, 1998). More recently, Taura and Kubota (1999) proposed using an engineering history base for retrieving both the teleological and the causal explanation for the process, to enable the reuse of product information. Information is attached to elements of the production process, i.e., actions, and can be retrieved by referring to the constraints, alternatives and reasons corresponding to each action.

In practice, one of the major problems in capturing design rationale is the lack of time available for this by the design teams. Another difficulty is maintaining the knowledge in usable form (as new knowledge is added, for example, during maintenance and redesign) once it has been formalized and inserted into an information system.

4.2.5 Knowledge about Manufacturability

Like design rationale, knowledge about the manufacturability of a given product is particularly important. New methods have been developed in recent years to ensure that the design process will lead to easy-to-manufacture or easy-to-assemble products: Design for Manufacturability (DFM), Design for Assembly (DFA). Some attention has also been given to Design for Disassembly (DFD). In this area, Rosen and Peters (1996) have considered a topological approach to realizing the transition between design and manufacturing, and Srinivasan and co-workers (1996) have described a new methodology for taking tolerancing into account. And Gupta and co-workers (1997) have produced an excellent survey of the methods providing automatic manufacturability analysis, as well as related techniques.

The preceding Sections 4.2.1 to 4.2.5 have discussed what needs to be represented in a design or manufacturing system. The following paragraphs go on to discuss knowledge that is more specifically associated with the agent-based approach. Because agents are independent entities, the associated knowledge must be represented locally at the agent level. This means that previously discussed representations will need to belong to some agent in the system and constitute its expertise, even if these refer to a global state of the application. In addition, agents must have specific mechanisms allowing them to communicate, negotiate, and cooperate. Such mechanisms can be "hard-wired", i.e., implicitly contained in the code rather than explicitly represented in a separate knowledge or data base. A minimal representation of the state of the world and of the internal state of the agent must, at the least, also be provided.

4.2.6 Model of the Environment

Since it is usually not very useful to have a system built on a single agent, applications are normally built as multi-agent systems. These agents will be operating in a particular environment which may be physical (machines, robots, agents with sensors), or "intellectual" (robots restricted to exchanging messages). Agents will usually have a representation of their external world, consisting of a representation of the environment in which they live and of the other agents that live in this environment. The corresponding model of *the state of the world* and the *social model* can be thought of as structural models, which will eventually be complemented by models of the global tasks or goals to be achieved, and by information on the degree of achievement.

Structural representations of the world may be quite complex, depending on the degree of interaction of the agents with the external world[1]. In some systems the world is modeled by a specific agent, which simplifies the representation needing to be held by the other agents. In other systems, each agent has its own model of the world and updates it by analyzing the output of its sensors or the messages it can read. When each agent is going to build its own model of the world, the problem is how to do it and how to maintain the world model dynamically. In this situation, each agent representation can be different and global consistency is not guaranteed. Agents involved in problem solving also

---

[1] Note that we consider here the actual external world, not a model of it.

have to prepare plans, implying hypothetical representations of the world at future times. This requires an explicit representation of time, and a way to maintain consistency, both when the state of the world changes and differs from the internal representation, and when backtracking is required during the planning stage. Finally, the world itself rather than a model of it, can be used directly as advocated by Brooks (1990), in situations where there are agents reacting to the state of the world and acting upon it (in which case there is no need for representation of the world).

Social models involve representation of the other agents in the system and of their skills. They range from nothing to complex representations, according to the global organization of the multi-agent system and how many agents there are and what skills they possess. In some agent systems (open systems, in which agents can come and go) agents do not have a representation of how the system is organized; they simply use a broadcast protocol allowing them to send requests for services without knowing which agent will service them. In other systems, agents will have a limited knowledge of their environment and rely on a small set of acquaintances which can include middle agents like matchmakers, brokers, or facilitators (see Chapter 7). The middle agents keep a detailed list of specialized agents in the system and of their skills and can thus connect a requesting agent to a specialized agent capable of servicing the request. In other systems, without middle agents, where the number of agents is predetermined and never changes, each agent will have to maintain a representation of other agents in the system and of their skills. One of the significant problems, especially in open systems, is how to dynamically build and maintain a representation of other agents and of their skills, whether this is done by every agent in the system or by specialized middle (intermediate) agents.

Part of the environment involves the task(s) being processed. Thus, a representation of the global goals or tasks that are active can help an agent to take better decisions. Again, however, such representations are maintained locally, and the problem is how to achieve overall consistency.

4.2.7 Self-Knowledge

As expected, self-knowledge is a representation of the knowledge of the agent about itself. It includes: (1) knowledge about its skills; and (2) control knowledge. Knowledge about skills allows the agent to reason about accepting a request, and control knowledge controls the execution of the program implementing the agent. Here, we may find the main difference between object-oriented systems and agent-based systems. In object-oriented systems, when an object receives a message, two cases can occur: (1) the object does not possess the corresponding method (locally or inherited), in which case an error message is returned to the sender; or (2) the object possesses the requested method (locally or inherited) which is then executed. The responsibility for sending a message with the correct method name lies with the sender. In agent systems, an agent is a continuously active or executing program. When it receives a message requesting a service, it may or may not accept the job, even though it has the capacity to do it. Whether the request will be processed or not depends on the local internal state of the receiving agent, which in turn is governed by its control knowledge. Thus, agents may have goals and perform tasks which may conflict with external requests.

In addition to the representation of complex information pertaining to skills or goals, it is often useful for an agent to keep a record of past exchanges with other agents. This can be done by simply saving the corresponding messages, or it can be more sophisticated, involving a record of past actions. Actions can, in turn, be structured as cases supporting some form of knowledge management.

Representation of self-knowledge can be quite complex and allow the agent to display complex behaviors. For example, an agent could deliberately provide false information. Although such possibilities are interesting in specific cases (e.g., for studying strategies and tactics, or simulating complex social situations), they have no advantages for multi-agent design and manufacturing systems, in which agents are normally assumed to be benevolent, rational and cooperative. In this case, the representation of self-knowledge will remain relatively simple.

4.2.8 Meaning of Terms

As in any large system, the vocabulary necessary to describe the application must be carefully defined. This is a critical issue for agent systems, in particular, when the vocabulary is used for inter-agent communications. Problems can arise if each specialized group of agents is allowed to use its own jargon, part of which may conflict with the vocabulary of other groups. Hence, there is a necessity to chart dialects and to build translation of terms. But note that terms are used to label concepts. Thus, careful organization of the concepts relating to the reference vocabulary must take place and precise definitions of the meaning of terms must be prepared. This is the role of ontologies at an operational level (see Chapter 11). Ontologies may be very large and include common sense knowledge, or on the contrary be quite specialized. They may be implemented within each agent, or delegated to specific agents in charge of maintaining the meaning of terms.

4.2.9 Exchanged Knowledge

The exchanging of knowledge can involve: (1) exchanging information about the state of the world or the state of some ongoing task; or (2) transferring a particular skill from an agent to another. In the first case, exchanges can be factual or involve complex data structures representing the corresponding information. They can be used to inform other agents about the state of the world, or provide information to accelerate a cooperative problem solving process. In the second case, the main reason for the exchange would be to allow an agent to do some task locally, avoiding subcontracting, which amounts to some form of load balancing and minimizing inter agent messages. Exchanges can be one-to-one or one-to-many. Of course, ontologies usually will play an important role in whether the knowledge is understood correctly. Exchanges commonly involve several messages organized as dialogs which have to be modeled precisely.

4.2.10 Model of the User

Users must be interfaced to the agent system, or, better, included in the system (as in a mixed system). A user cannot be expected to be familiar with the complex protocols used by the agents for communication. Thus, interfaces have to be developed allowing a user to obtain information from the system easily and in turn to provide information or to express requests. Since the work is done cooperatively, the relationship is no longer information-command-report but more of an equal-partner type. The easiest solution to this problem is to design *assistant agents* which provide an interfacing service to their owners, translating special protocol idioms into intelligible displays, and conversely translating user input into the normalized expressions of a communication language. Such assistants, used also in other contexts (Huhn and Singh, 1998) are quite different from user assistants or design critics developed previously with knowledge-based systems. Their function is not to give advice on the design or on the manufacturing process, but to play the role of a simple intermediate, equivalent to that of an English "butler" as put by Negroponte (1997). In order to be able to serve their masters or mistresses, assistants must have knowledge profiles of their owners. According to Huhn and Singh (1998), assistants should support the following requirements:

- *heterogeneity*: Assistants should be able to interact with assistants from other vendors[2];
- *autonomy*: Assistants must behave as their users wish, not according to the dictates of other users or assistants;
- *idiosyncrasy*: Profiles should include details relevant to a user or class of user;
- *multiple perspectives*: Each user can maintain an individual perspective about someone else's agent, and disagree with others about substantive details of the profiles;
- *sharing*: When appropriate, assistants should be able to disseminate the knowledge and opinion of their users.

A particular interesting consequence in having a model of a user in the assistant agent is the possibility of the assistant giving simple answers when the user is not available.

4.3 HOW TO REPRESENT KNOWLEDGE IN AGENT-BASED SYSTEMS

If Section 4.2 is about what is to be represented in agent systems, this section is about how to represent it. To do so, one needs formalisms and tools. Today there is no single agent-based system handling the whole design and manufacturing process of a complex product. At best, agent-based systems address some part of the process with a limited approach. This is not, however, a serious limitation since agent systems can grow and expand, to play increasingly important roles.

---

[2] Although this is mainly intended to Internet type of applications, it also applies to open agent systems where parts can be provided by different vendors.

Agents systems are characterized through these perspectives: (1) the role and structure of a particular agent; and (2) the inter-agent organization. The representation of knowledge will be different according to the perspective and will also depend on the type of application.

From the knowledge representation and management point of view, agents can be categorized into four different types (Figure 4.1):

    (a) specialists interacting only through the network;
    (b) information gatherers using sensors;
    (c) agents capable of acting on the world, e.g., robots;
    (d) agents interacting with humans, e.g., assistants.

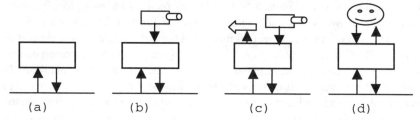

Figure 4.1 Different types of agents.

Regardless of the type of agent we consider, each agent should have the capability to: (1) process messages; (2) process message content; (3) decide what to do; (4) execute expertise; (5) record the transaction. Each of those functions requires knowledge of one or more kinds, as discussed in the previous sections. Processing messages, i.e., receiving, analyzing and answering messages, requires communication knowledge. This knowledge is usually implemented as communication protocols and fixed for a given agent, depending on the software environment. Processing message content requires knowledge about cooperation protocols, models of expertise, task representation, and state representation. Deciding what to do requires mental representation. Executing expertise requires the knowledge about what happens to be represented in an adequate fashion. Recording what is happening involves learning, and requires specific knowledge. These activities can be represented within a Generic Agent structure (from the knowledge representation and management point of view – see Chapter 6 for more details) as in Figure 4.2.

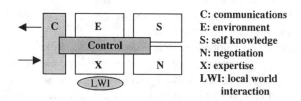

C: communications
E: environment
S: self knowledge
N: negotiation
X: expertise
LWI: local world
      interaction

Figure 4.2 Structure of a generic agent.

In the rest of this section, we review the different forms of required knowledge from the perspective of design and manufacturing applications.

4.3.1 Global Approaches

A multi-agent system for concurrent engineering must be composed of a number of different agents: agents providing specific application services (e.g., computation, visualization, modeling, monitoring, ontologies, databases); control agents (e.g., brokers, matchmakers, facilitators); and agents interfacing with users (assistants). Among the agents providing application services, some are monolithic, such as those encapsulating legacy tools, whereas others work cooperatively. The precise overall organization of each system depends on the particular application and many examples of multi-agent systems are provided in this book. The important point is that *all knowledge is represented locally within some agent*. If we look at each agent in terms of functionality, then a minimal agent will contain a communication module, a control module, and some sort of expertise (C, Control, X in Figure 4.2). The degree to which the corresponding knowledge is explicitly represented depends on the chosen agent platform. In some agent systems, most of the knowledge will be implicit and hidden within procedural software, whereas in other systems it will be explicitly represented using declarative structures. Procedural implementations are efficient but difficult to modify dynamically. Explicit representations of knowledge bring more flexibility, but rapidly become complex.

4.3.2 Product and Process Modeling

If we turn now to concurrent engineering applications, then clearly products and processes have to be somehow represented. However, having a single agent manage the global representation of a complex product, even it were possible, would create a bottleneck, defeating the whole purpose of using agents. Thus, each agent will have a possibly different representation of the product best suited to the specific expertise it will implement (e.g., finite element analysis). The major difficulty will be in coordinating the various types of representations, making sure that they remain relatively consistent. The representations that may be used range from highly symbolic models (usually represented as graphs) to data structures representing, for example, geometry. Translation from one representation into another is a real problem and current standards are not sufficient to support this function. Special agents could be developed for this purpose. The problem of conflicting representations was considered in PACT (Gruber, 1992) where two representations were necessary: one for dynamic computations and the other for kinematics.

Protocols can be developed enabling an agent to build its own internal representation of a complex object by processing messages containing primitives. However, there are some difficulties in doing so: First, building a representation of a complex object from scratch can be quite inefficient in terms of the number of required elementary pieces of information; second, each representation must be maintained, for example, when modifications are brought into the design. Nevertheless, this approach must be taken when encapsulating legacy tools. Another approach involves developing better representation languages to model products and processes, and as well as related knowledge like design rationale. Various proposals have been made, using either object-oriented languages, like ADDL (Veerkamp, 1992), IDDL (Xue et al., 1989), STEP/ EXPRESS, or frame-based approaches (Shen and Barthès, 1996). These do not solve the problem of translating between internal representations and more traditional product representations. Mechanisms to do so have yet to be developed.

4.3.3 Agent Skills

Each agent in a multi-agent system has particular skills. In order for the others to call upon such skills, the agent must have a representation of what it can do, either explicitly, or through a cooperation protocol. Thus, the content of both sent and received messages must be understandable in terms of the skills that the agent is capable of activating. If considered in terms of protocols, this does not look very different from the situation in an object-oriented organization. However, representing what an agent can do explicitly, allows the agent to reason about a requested task, leading it to accept or to refuse it, a much different situation than in the object-oriented context. A minimal solution for representing the skills consists in representing each type of action by a keyword that can have parameters (similarly to methods and arguments in an object-oriented environment). More complex mechanisms can be implemented where the necessary information about how to achieve a particular task is obtained after communication between the receiving agent and the rest of the system. The latter approach is interesting and more flexible, since the specialized agent has the responsibility for searching for and acquiring what it needs to execute a task. One also needs to arrange for the dialog to change automatically when the specialist agent is replaced by a new one having different data requirements. However, the approach implies that agents are capable of conducting dialogs, and hence, that a minimum of understanding exists on the meaning of the exchanged symbols. Ontologies (see Chapter 11) have been developed for this purpose.

Some researchers like Maes (1990) think that decisions can be activated by propagating energy along activation networks, which does not require specific symbolic reasoning. This approach offers some interesting possibilities for activating of agent skills.

And lastly, we note that interface agents (assistants) in addition to specific skills must have a model of their human owner in order to present information in the most suitable fashion. Such agents can also act as technical secretaries (Decker, 1998).

4.3.4 Environment

Knowledge about the external world usually needs to be represented in an agent system. This knowledge is modeled through: (1) a model of the world; and (2) a model of current tasks. The model of the world can either be a model of the external world and/or a model of the social world (a model of other agents and of their skills). We note however, in passing, that some researchers like Brooks (1990) contest the usefulness of having an explicit model of the world for robotics applications. Models of other agents may include qualitative or quantitative judgmental values related to the efficiency of a particular agent for a particular task. In concurrent engineering, a model of the world could, for example, include a partial view of the product, allowing agents to reason about it. Models of tasks in this context are models of tasks shared among several agents (Decker, 1998). A particular task done by a specific agent does not need to be part of a model of the environment, although we found that it can be sometimes useful. Shared tasks are those for which agents are committed and correspond to intentions in the B.D.I. (Belief, Desire, Intention) framework (Rao and Georgeff, 1995). They must be represented using temporal information (dates and durations) to allow recovery (e.g., timeouts).

Strongly interacting agents, i.e., agents cooperating closely to solve a particular problem, can benefit from sharing information about partial results or by sharing plans. Techniques have been proposed for doing so (Mammen and Lesser, 1998).

Representation of the environment, be it the world, other agents, or shared tasks, requires mechanisms for automatically extracting relevant information from the analysis of exchanged messages or from the results of sub-contracted work. And importantly, Abriat (1991) showed in the context of automatic assembly that dating elements of representation with time stamps allows graceful recovery from errors or sudden changes in the external situation.

4.3.5 Self-Representation

The level of detail at which an agent must have self-representation is not clear. An agent can have a representation of its expertise (at minimum a list of keywords corresponding to its skills, associated with a description of the related parameters). An agent may also have a representation of its goals. In design and manufacturing applications, the main goal of an agent is to do its job as best as possible. Researchers like Maes (1990) have, however, claimed that it is not always necessary to represent goals explicitly. An agent can have a list of tasks for which it is not yet committed. In the B.D.I. framework, such tasks correspond to Desires.

In frameworks like B.D.I., the representation of the environment (beliefs), the self-representation (desires), and the task representation (intentions) are closely related and control is implemented by means of a special temporal logic, for example, see CTL* (d'Inverno et al., 1998; Singh, 1998; Mulder et al., 1998).

4.3.6 Negotiation

Interaction among agents can be quite complex and require negotiation. Several formalisms have been developed for specifying protocols (Zeng and Sycara, 1997; Parunak et al., 1998; Nodine and Unruh, 1998), and for representing parametized dialogs (Sierra et al., 1998; Reeds, 1998; Barbuceanu and Fox,

1997; Nishimoto et al., 1998). Special languages like KQML have been designed for implementing negotiation. Usually negotiation is carried out before accepting a task and various mechanisms are used to try to optimize the global outcome. One that is commonly used is the market approach (see Chapter 10 Section 10.3.2).

4.3.7 Communication

At a lower level, communication is facilitated by various protocols. One of the best known protocols for distributing work among agents is the Contract-Net protocol and its many variations (see Chapter 10 Section 10.3.1). See also Barthès and Scalabrin (1997) for the distinction among various levels of protocols.

4.3.8 Local World Interaction

Some agents interact only by means of messages on the network. Others have direct interaction with the external world. Two such agents are of interest; robots and agents monitoring processes, and assistants interfacing with human users. Robot agents roaming the real world or monitoring agents have sensors. Special processing software translates sensor inputs into some numerical or symbolic representation. The corresponding knowledge is either used locally or transmitted to other agents. A characteristic of such knowledge is that it has a birth date and an associated time interval during which it can be considered as valid. Handling such knowledge poses the same problems as in real-time knowledge-based systems controlling external processes.

4.3.9 Knowledge History and Learning

Agents can be programmed to record their history, for example, the sequence of messages they exchanged with the rest of the world. This knowledge can be considered as part of their self-knowledge. Sophisticated techniques derived from artificial intelligence can segment this knowledge and map it onto cases (Burkhardt, 1997). Alternatively it can be processed through learning techniques (see Chapter 5) to improve the expertise of a particular agent. However, this is difficult to implement.

4.4 ADDITIONAL REFERENCES

Knowledge representation problems are quite difficult in design and manufacturing, quite independently of the agent approach. Relevant research in this area can be found in these and other journals:

- Artificial Intelligence for Engineering Design, Analysis and Manufacturing (AI EDAM) (Cambridge University Press)
- Artificial Intelligence in Engineering (Elsevier Science)

- Research in Engineering Design (Springer Verlag)

These problems are also discussed in the IFIP WG 5.2, particularly in its workshop series on Knowledge Intensive CAD (Tomiyama et al., 1996; Mäntylä et al., 1997; Finger et al., 1999).

In the context of agent systems, many types of knowledge representation are needed. The choice of a particular representation depends on the type, scope, and size of system being developed. It is thus useful to examine different agent platforms, and the corresponding representations and languages. Papers on the subject can be found in the book series on Intelligent Agents published by Springer (http://www.springer.de/), as well as in proceedings of conferences like ICMAS (International Conference on Multi-Agent Systems) and Autonomous Agents, and workshops such as CIA (Cooperative Information Agents).

Of research interest is the development of formal languages for modeling agent systems. Papers discussing this issue can be found, in particular, in Intelligent Agents IV: Agent Theories, Architectures, and Languages (Singh et al., 1998).

In agent-based design and manufacturing systems, application knowledge must be acquired and preserved. Learning techniques are being introduced for this purpose (Kawakami et al., 1996). The following chapter will focus on this topic.

REFERENCES

Abriat, P. (1991) Conception et réalisation d'un système multi-agents en robotique permettant de récupérer les erreurs dans les ateliers flexibles, Thèse de Doctorat, Université de Technologie de Compiègne, France.

Baldwin R.A. and Chung M.J. (1995) Managing Engineering Data for Complex Products, *Research in Engineering Design*, 7(4):215-231.

Barbuceanu, M. and Fox, M. (1997) The Design of a Coordination Language for Multi-Agent Systems, In J.P. Müller, M.J. Wooldridge and N.R. Jennings, eds., *Intelligent Agents III: Agent Theories, Architectures, and Languages,* LNAI 1193, Springer, pp. 341-355.

Barthès, J-P. and Scalabrin, E.E. (1997) Cognitive Agents and Exchange Protocols, In *Proceedings of MASTA Conference*, Coimbra, Portugal.

Bijl, A. (1989) Relations, functions and Constraints without Prescriptions, In *Proceedings of the 3rd IFIP WG 5.2 Workshop on Intelligent CAD*, Osaka, pp. 36-54.

Bridge, S.F. (1989) Intelligent Representation of Engineering Design Information for CAD, In *Proceedings of the 3rd IFIP WG 5.2 Workshop on Intelligent CAD*, Osaka, pp. 55-71.

Brooks, R.A. (1990) Elephants don't Play Chess, In P. Maes, ed., *Designing Autonomous Agents*, MIT / Elsevier, pp. 3-16.

Brown, K.N., McMahon, C.A. and Sims Williams, J.H. (1995) Features, a.k.a. The Semantics of a Formal Language of Manufacturing, *Research in Engineering Design*, 7(3):151-172.

Brown, K.N., McMahon, C.A. and Sims Williams, J.H. (1996) Describing process plans as the formal semantics of a language of shape, *Artificial Intelligence in Engineering*, 10:153-169.

Burkhardt, H-D. (1997) Cases, Information and Agents, In P. Kandzia and M. Klusch eds., *Cooperative Information Agents*, Springer, pp. 64-798.

Busby, J.S. (1998) Effective Practices in Design Transfer, *Research in Engineering Design*, 10(3):178-188.

Cagan, J., Grossmann, I.E. and Hooker, J. (1997) A Conceptual Framework for Combining Artificial Intelligence and Optimization in Engineering Design, *Research in Engineering Design*, 9(1):20-34.

Chakrabarti, A. and Bligh, T.P. (1994) An Approach to Functional Synthesis of Solutions in Mechanical Design, Part I: Introduction and Knowledge representation, *Research in Engineering Design*, 6(3):127-141.

CIMOSA (1991) *CIMOSA: Open System Architecture for CIM*, Springer-Verlag.

Decker, K.S. (1998) Coordinating Human and Computer Agents, In W. Conen and G. Neumann, eds., *Coordination Technology for Collaborative Applications*, Springer, pp. 77-97.

d'Inverno, M., Kinny, D., Luck, M. and Wooldridge, M. (1998) A Formal Specification of dMARS, In M.P. Singh, A. Rao, M.J. Wooldridge, eds., *Intelligent Agents IV: Agent Theories, Architectures, and Languages*, LNAI 1365, Springer, pp. 155-176.

Dong, Y. and Goh, A. (1998) An intelligent database for engineering applications, *Artificial Intelligence in Engineering*, 12(1-2):1-14.

Doumeingts, G., Vallespir, B., Darracar, D. and Roboam, M. (1987) Design Methodology for Advanced Manufacturing Systems, *Computer in Industry*, 9(4):271-296.

Duffy, A.H.B. and Kerr, S. (1993) Customized perspectives of past designs from automated group rationalizations, *Artificial Intelligence in Engineering*, 8(3):183-200.

El Dahshan, K., Barthès, J-P.A. (1989) Implementing Constraint Propagation in Mechanical CAD Systems, In V.Ackman, P.J.W. ten Hagen, P.J.Verkamp, eds., *Intelligent CAD Systems II: Implementational Issues*, Springer-Verlag, pp. 217-227.

Eppinger, S.D., Nukala, M.V. and Whitney, D.E. (1997) Generalized Models of Design Iteration Using Signal Flow Graphs, *Research in Engineering Design*, 9(2):112-123.

ESPRIT-AMICE (1991) *CIMOSA: Open System Architecture for CIM*, Springer-Verlag.

Finger, S., Tomiyama, T., and Mantyla, M., eds., (1999) *Knowledge Intensive CAD*, Volume 3, Kluwer Academic Publishers.

Fox, M., Barbuceanu, M. and Gruninger, M. (1995) An Organization Ontology for Enterprise Modeling: Preliminary Concepts for Linking Structure and Behavior, In *Proceedings of the Fourth Workshop on Enabling Technologies, Infrastructure for Collaborative Enterprises*, IEEE Computer Society.

Freedman, R.S. and Frail, R.P. (1986) OPGEN: The Evolution of an Expert System for Process Planning, *AI Magazine*, 7(5):(58-70.

Fu, Z. and de Pennington, A (1993) Constrained-Based Design Using an Operational Approach, *Research in Engineering Design*, 5:202-217.

Gruber, T. (1992) Ontolingua: A Mechanism to Support Portable Ontologies (version 3.0), KSL Memo 91-66, Stanford University.

Grundstein, M. (1996) CORPUS: An Approach to Capitalizing Company Knowledge, In *Proceedings of AIEM 4 - The Fourth International Workshop: Artificial Intelligence in Economics & Management*, Tel Aviv, Israel.

Gupta, S.K., Regli, W.C., Das, D. and Nau, D. (1997) Automated Manufacturability Analysis: A Survey, *Research in Engineering Design*, 9(3):168-190.

Henderson, M.R. and Taylor, L.E. (1993) A Meta-Model for Mechanical Products Based Upon the Mechanical Design, *Research in Engineering Design*, 5(3-4):140-160.

Hua, K., Faltings, B. and Smith, I. (1996) CADRE: case-based geometric design, *Artificial Intelligence in Engineering*, 10(2):171-183.

Huhn, M.N. and Singh, M.P. (1998) Personal Assistants, *IEEE Internet Computing*, 2(5):90-92.

ISO 8879 (1988) *Standard General Mark up Language.*

Joskowicz, L. and Neville, D. (1996) A Representation Language for Mechanical Behavior, *Artificial Intelligence in Engineering*, 10(2):109-116.

Kawakami, H., Katai, O., Sawaragi, T., Konishi, T. and Iwai, S. (1996) Knowledge acquisition method for conceptual design based on value engineering and axiomatic design theory, *Artificial Intelligence in Engineering*, 10(3):187-202.

Kiriyama, T., Ishida, Y., Tomiyama, T. and Yoshikawa, H. (1989) Qualitative Physics for Integrating Models and Representing Physical Features, In *Proceedings of the 3rd IFIP WG 5.2 Workshop on Intelligent CAD*, Osaka, pp. 217-223.

Langenbach, C. (1993) Using STEP: A concept of operations, *CALS Journal*, Summer.

Maes, P. (1990) Situated Agents Can Have Goals, In P. Maes, ed., *Designing Autonomous Agents*, MIT/Elsevier. pp. 49-70.

Mammen, D.L. and Lesser, V. (1998) Problem Structure and Subproblem Sharing in Multi-Agent Systems, In *Proceedings of the Third International Conference on Multi-Agent Systems*, IEEE Computer Society, pp. 174-181.

Mantipragada, R. and Whitney, D.E. (1998) The Datum Flow Chain: A Systematic Approach to Assembly Design and Modeling, *Research in Engineering Design*, 10(3): 150-165.

Mäntylä, M. (1989) WAYT: Towards a Modeling Environment for Assembled Products, In *Proceedings of the 3rd IFIP WG 5.2 Workshop on Intelligent CAD*, Osaka, pp. 249-264.

Mäntylä, M., Finger S., and Tomiyama T., eds. (1997) *Knowledge Intensive CAD*, Volume 2, Chapman & Hall.

McKay, A., Erens, F., Bloor, M.S. (1996) Relating Product Definition and Product Variety, *Research in Engineering Design*, 8(2):63-80.

McMahon, C.A., and Meng, X.Y. (1996) A Network Approach to Parametric Design Integration, *Research in Engineering Design*, 8(2):14-32.

MIL-D-28000 (1987) Military Specification. Digital Representation for Communications Product Data: IGES

Mitchell, S.R., Jones, R. and Hinde, C. (1995) An Initial Data Model, Using the Object-Oriented Paradigm, for Sculptured-Feature-Based Design, *Research in Engineering Design*, 7(1):19-37.

Minsky, M. (1975) A Framework for Representing Knowledge, In P Winston, ed., *The Psychology of Computer Vision*, McGraw Hill.

Molina, A., Al-Ashaab, A.H., Ellis, T.I.A., Young, R.I.M. and Bell, R. (1995) A Review of Computer-Aided Simultaneous Engineering Systems, *Research in Engineering Design*, 7(3):38-63.

Monceyron, E. and Barthès, J.P. (1992) Architecture for ICAD Systems: an Example from Harbor Design, *Revue Sciences et Techniques de la Conception,* 1(1):49-68.

Mukherjee, A. and Liu, C.R. (1995) Representation of Function-Form Relationship for the Conceptual Design of Stamped Metal Parts, *Research in Engineering Design*, 7:253-269.

Mulder, M., Treur, J. and Fisher, M. (1998) Agent Modeling in METAM and DESIRE, In M.P. Singh, A. Rao, M.J. Wooldridge, eds., *Intelligent Agents IV: Agent Theories, Architectures, and Languages*, LNAI 1365, Springer, pp. 193-207.

Negroponte, N. (1997) Agents: From Direct Manipulation to Delegation, In J.M. Bradshaw, ed., *Software Agents*, MIT/AAAI Press, pp. 57-77.

Nishimoto, K., Sumi, Y. and Mase, K. (1998) Enhancement of Creative Aspects of a Daily Conversation with a Topic Development Agent, In W. Conen and G. Neumann eds., *Coordination Technology for Collaborative Applications*, Springer, pp. 63-76.

Nodine, M.H. and Unruh, A. (1998) Facilitating Open Communication in Agent Systems: The InfoSleuth Infrastructure, In M.P. Singh, A. Rao, M.J. Wooldridge, eds., *Intelligent Agents IV: Agent Theories, Architectures, and Languages*, LNAI 1365, Springer, pp. 281-295.

Otto, K.N. and Wood, K.L. (1998) Product Evolution: A Reverse Engineering and Redesign Methodology, *Research in Engineering Design*, 10(4):226-243.

Palmer, R.S. and Shapiro, V. (1993) Chain Models of Physical Behavior for Engineering Analysis and Design, *Research in Engineering Design*, 5:161-184.

Parunak, V., Sauter, J. and Clark, S. (1998) Toward the Specification and Design of Industrial Synthetic Ecosystems, In M.P. Singh, A. Rao, M.J. Wooldridge, eds., *Intelligent Agents IV: Agent Theories, Architectures, and Languages*, LNAI 1365, Springer, pp. 45-59.

Rao, A.S. and Georgeff, M.P. (1995) BDI Agents: From Theory to Practice, In *Proceedings of the first International Conference on Multi-Agent Systems*, AAAI Press / MIT Press, pp. 312-319.

Ravindran, A., Phillips, D.T. and Solberg, J.J. (1987) *Operations Research: Principles and Practice*, Wiley.

Reeds, C. (1998) Dialogue Frames in Agent Communication, In *Proceedings of the 3rd International Conference on Multi-Agent Systems*, IEEE Computer Society, pp. 246-253.

Regli, W.C., Gupta, S.K. and Nau, D.S. (1995) Extracting Alternative Machining Features: An Algorithmic Approach, *Research in Engineering Design*, 7(3):173-192.

Reich, Y. (1995) A Critical Review of General Design Theory, *Research in Engineering Design*, 7(1):1-18.

Reich, Y., Konda, S.L., Levy, S.N., Monarch, I.A. and Subrahmanian, E. (1993) New Roles for Machine Learning in Design, *Artificial Intelligence in Engineering*, 8(3):165-181.

Rosen, D.W. (1993) Feature-Based Design: Four Hypotheses for Future CAD Systems, *Research in Engineering Design*, 5(3-4):125-129.

Rosen, D.W. and Peters, T.J. (1996) The Role of Topology in Engineering Design research, *Research in Engineering Design*, 8(2):81-98.

Shah, J.J. and Rogers, M.T. (1993) Assembly Modeling as an Extension of Feature-Based Design, *Research in Engineering Design*, 5(3-4):218-237.

Shen, W. and Barthès, J-P.A. (1996) An Object-Oriented Approach for Engineering Design Product Modeling, In T. Tomiyama, M. Mantyla, and S. Finger, eds., *Knowledge Intensive CAD*, Volume 1, Chapman & Hall, pp. 19-38.

Sierra, C., Jennings, N.R., Noriega, P. and Parsons, S. (1998) A Framework for Argumentation-Based Negotiation, In M.P. Singh, A. Rao, M.J. Wooldridge, eds., *Intelligent Agents IV: Agent Theories, Architectures, and Languages*, LNAI 1365, Springer, pp. 177-192.

Singh, M.P. (1998) A Customizable Coordination Service for Autonomous Agents, In M.P. Singh, A. Rao, M.J. Wooldridge, eds., *Intelligent Agents IV: Agent Theories, Architectures, and Languages*, LNAI 1365, Springer, pp. 93-106.

Singh, M.P., Rao, A. and Wooldridge, M.J., eds., (1998) *Intelligent Agents IV: Agent Theories, Architectures, and Languages*, LNAI 1365, Springer.

Srinivasan, R.S., Wood, K.L. and McAdams, D.A. (1996) Functional Tolerancing: A Design for Manufacturing Methodology, *Research in Engineering Design*, 8(2):99-115.

Suh, N.P. (1998) Axiomatic Design Theory for Systems, *Research in Engineering Design*, 10(4):189-209.

Taura, T. and Kubota, A. (1999) A Study on Engineering History Base, *Research in Engineering Design*, 11(1):45-54.

The n-dim Group (1995) n-dim - An Environment for Realizing Computer Supported Collaboration in Design Work, Technical Report, EDRC 05-93-95, Carnegie Mellon University.

Thoraval, P. (1991) Systèmes intelligents d'aide à la conception: ARCHIX & ARCHIPEL, Thèse de Doctorat, Université de Technologie de Compiègne, France.

Tomiyama, T., Umeda, Y., Ishii, M., Yoshioka, M. and Kiriyama, T. (1991) Knowledge systematization for a knowledge intensive engineering framework, In T. Tomiyama, M. Mantyla, and S. Finger, eds., *Knowledge Intensive CAD,* Volume 1, Chapman & Hall, pp. 55-80.

Tomiyama, T., Mäntylä, M. and Finger, S., eds., (1996) *Knowledge Intensive CAD,* Volume 1, Chapman & Hall.

Tweed, C. and Bijl, A (1989) MOLE: A Reasonable Logic for Design? In V.Ackman, P.J.W. ten Hagen, P.J.Verkamp, eds., *Intelligent CAD Systems II: Implementational Issues,* Springer-Verlag, pp. 146-167.

Ulrich, K.T. and Seering, W.T. (1989) Synthesis of schematic descriptions in mechanical design, *Research in Engineering Design,* 1(1):5-18.

Veerkamp, P. (1992) On the Development of an Artifact and Design Description Language, CWI, Amsterdam.

Welsh, R.V. and Dixon, J.R. (1994) Guiding Conceptual Design through Behavioral Reasoning, *Research in Engineering Design,* 6(3):169-188.

Williams, A.L. (1989) Comparative Analysis of the Modeling Languages: IDEF1X, EXPRESS and NIAM, McDonnell Douglas Information Technology.

Williams, D. (1989) A Philosophical Basis for Support Systems for Evolutionary Design, In *Proceedings of the 3ʳᵈ IFIP WG 5.2 Workshop on Intelligent CAD,* Osaka, pp. 470-477.

Xue, D., Takeda, H., Kiriyama, T., Tomiyama, T. and Yoshikawa, H. (1989) An Intelligent Integrated Interactive CAD, In *Proceedings of the 3ʳᵈ IFIP WG 5.2 Workshop on Intelligent CAD,* Osaka, pp. 478-497.

Yoshikawa, H. (1982) General Design Theory and a CAD System, Man-Machine Communication in CAD/CAM, In *Proceedings of IFIP WG 5.2/5.3 Working Conference (Tokyo),* North Holland.

Zamanian, M.K. and Fenves, S.J. (1994) A Referential Scheme for Modeling and Identifying Spatial Attributes of Entities in Constructed Facilities, *Research in Engineering Design,* 6(3):142-168.

Zeng, D. and Sycara, K. (1997) How can an Agent Learn to Negotiate? In J.P. Müller, M.J. Wooldridge and N.R. Jennings, eds., *Intelligent Agents III: Agent Theories, Architectures, and Languages,* Lecture Notes in Artificial Intelligence 1193, Springer, pp. 233-244.

5

Learning in Agent-Based Concurrent Design and Manufacturing Systems

5.1 INTRODUCTION

Agents operating in open, dynamic environments must be able to flexibly adapt to changing demands and opportunities. In particular, individual agents are forced to engage with other agents with varying goals, abilities, composition, and life-span. To effectively utilize opportunities presented and avoid pitfalls, agents need to learn about other agents and adapt local behavior based on group composition and dynamics.

According to Russell and Norvig (1995), whenever the designer has incomplete knowledge of the environment the agent will live in, learning is the only way that the agent can acquire what it needs to know. It also provides a good way to build high-performance systems—by giving a learning system experience in the application domain.

For many application tasks, even in environments that appear to be simple, it may be difficult or even impossible to correctly determine the behavioral repertoire and concrete activities of a multi-agent system a priori, that is, at the time of its design and prior to its use. This would require, for instance, that it is known a priori which environmental requirements will emerge in the future, which agents will be available at the time of emergence, and how the available agents will have to interact in response to these requirements. Problems, such as those resulting from the complexity of multi-agent systems, can be avoided or at least reduced by endowing the agents with the ability to learn, that is, with the ability to improve the future performance of the total system or a part of the system.

There has been considerable research on multi-agent learning for particular applications such as the prisoners dilemmas, predator/prey, code warriors, virtual robots, self-playing game-learners, traffic controls, with varying success, but relatively little for learning in agent-based concurrent design and manufacturing systems. Therefore, this chapter will introduce general concepts, methods, and algorithms on multi-agent learning, with explanations and related examples on agent-based concurrent intelligent design and manufacturing systems.

Note that a very exciting research subject on Learning Interface Agents or Personal Assistants is primarily focused on User Modeling and Intelligent User Interface, which can be considered as a classic AI research theme and is out of the scope of this chapter. Integrated readers may refer to (Maes and Kozierok 1993; Akoulchina and Ganascia 1997) for an overview and examples on this topic.

5.2 WHY TO LEARN

Concurrent Design and Manufacturing systems operate in real-time, dynamic environments. In an agent-based engineering design system, new computational tools or human specialists may have to be integrated into the working environment, and existing tools may breakdown or be removed from the running environment. In an agent-based manufacturing system, there are always problems such as failure of machines, tools or transport vehicles, and shortages of required materials; also, new tools, machines or transport vehicles may need to be added into the running manufacturing environment. Thus, an agent-based concurrent design and manufacturing system must be able to adapt to changing environments and handle emergent contexts.

According to Grecu and Brown (1996b), learning adds additional power for agents to deal with real design problems in agent-based design systems. In this context, learning is needed for several important reasons. First, it can adapt a system to the domain and to the design problems it has to solve. And, no less important, it can adapt the agents to each other, to make the problem-solving process smoother. Therefore, learning targets two levels in each agent: the level of its domain knowledge and the level of its interaction knowledge.

A multi-agent design system includes many agents with many potential interactions. Conflicts considerably increase the number of interactions needed during design. The more serious the conflict (i.e., the more difficult it is to resolve), the more messages are exchanged to find an acceptable solution. As the ratio of messages per design decision increases, the efficiency of the system decreases. High conflict situations provide a strong motivation for any technique that leads to reduced overhead. Learning can reduce this kind of overhead, by reducing the number of conflicts and/or by reducing the number of messages during interactions.

On the other hand, learning can improve design quality. One of the most desirable effects of learning is to improve the quality of designs. However, it is often hard to relate design quality change with the design decisions or cooperation decisions that caused them. Learning can also improve the design process. A new multi-agent design system will need a considerable amount of decision revision during its first problem-solving sessions. This is due to the fact that the system is not "tuned" to the domain and the agents have little knowledge about each other. Learning can provide a fair amount of support in this situation, by avoiding conflicts and reducing the interaction needed to find a design solution.

Learning is even more important in agent-based manufacturing systems. A manufacturing system may need to be dynamically re-configured to support reduced batch sizes, increased product mix, or changes in design plans, all of these involving many changes in production setups. Extensive search processes are required to identify the multi-agent interactions for each problem case. These interactions give rise to observable patterns of behavior evolving concurrently within each level in the organization for different types of requirements, constraints, and agents.

According to Goldman and Rosenschein (1996), learning in a multi-agent environment can help agents improve their performance. Agents, when meeting with others, can learn about the partner's knowledge and strategic behaviors. Agents that operate in dynamic environments could react to unexpected events by generalizing what they have learned during a training stage. In cooperative problem solving systems, cooperative behavior can be made more efficient if agents adapt to information about the environment and about their partners. Agents that learn from each other can avoid repeatedly coordinating their actions for similar problems. They will sometimes be able to avoid communication at run-time by using learned coordination concepts, which is especially useful when they do not have enough time to negotiate.

5.3 SINGLE-AGENT LEARNING OR MULTI-AGENT LEARNING

Learning in multi-agent systems is more than a mere magnification of learning in single agent systems. On the one hand, learning in multi-agent systems encompasses learning in single systems, because an agent, although embedded in a multi-agent system, can learn in a solitary way independent of other agents. This is single-agent or isolated learning: learning that does not rely on the presence of multiple agents. On the other hand, learning in multi-agent systems extends learning in single-agent systems, because agents in a multi-agent system can learn in a communal way inasmuch as their learning is influenced (e.g., initiated, redirected, or made possible at all) by exchanged information, shared assumptions, commonly developed viewpoints of their environment, commonly accepted social and cultural conventions and norms which regulate and constrain their behavior and interaction, and so on. This is multi-agent or interactive learning: learning that relies on or even requires the presence of multiple agents and their interaction. Single agent and multi-agent learning constitute the principal categories of learning in multi-agent systems. However, sometimes it is difficult to draw a clear line between these two learning categories, for instance, when one might think of what an agent learns about or models other agents.

When the people talk about learning in multi-agent systems, they usually think of multi-agent instead of single-agent learning. Two usages of the term 'multi-agent learning' can be distinguished:

- In its stronger and more specific meaning, 'multi-agent learning' refers only to situations in which several agents collectively pursue a common learning goal.
- In its weaker and less specific meaning, 'multi-agent learning' additionally refers to situations in which an agent pursues its own learning goal, but is

affected in its learning by other agents, their knowledge, beliefs, intentions, and so on.

Independent of its underlying meaning, multi-agent learning is a many-faceted activity, and therefore it is not surprising that many synonyms of this term can be found in the literature. Examples of such synonyms, each stressing one or another facet, are organizational learning, mutual learning, cooperative learning, collaborative learning, shared learning, team learning, and social learning.

5.4 WHEN TO LEARN

In agent-based concurrent design and manufacturing systems, agents should learn in the following situations:

- When the system configuration changes. For example: when new computational tools are introduced to the system; old computational tools are updated with new functionalities or removed from the running system; new manufacturing resources (machines, tools, transport vehicles, workers, etc.) are added to the running manufacturing system; some resources in service breakdown or are removed from the system. In such cases, each agent should learn about the changes in the running system and update its knowledge about its environment and other agents.
- When failures occur in the system. Failures can occur either during the design process or when testing, simulating or manufacturing a designed artifact. Both situations represent learning opportunities.
- When there exist differences with respect to the same design or manufacturing object. Differences, especially between different points of view, are an important source of conflicts. Learning triggered by noticing differences helps anticipate and alleviate conflicts.
- When a project or a task is terminated with success. In design systems, each successful design or design decision is a learning opportunity, insofar as the result represents an improvement compared to previous cases. In manufacturing systems, successful process planning (an efficient combination of manufacturing resources for a special manufacturing task) is a good learning case.
- When there is a need to improve abilities. The need to improve designs, design processes or manufacturing process plans can be translated into a requirement to improve agent abilities. This means that an agent can use learning about design situations, manufacturing resource availability or other design/resource agents, to help it make more informed decisions in the future. Observing patterns, dependencies or other relations can be useful for improving abilities even though it is not motivated by events falling in any of the previous four categories (situations).

5.5 WHERE TO LEARN

In agent-based concurrent design and manufacturing systems, the way that agents make information available to other agents determines where learning opportunities arise. Varying the representation and the context in which knowledge is communicated favors some types of learning in comparison to others.

Direct communication implies transfer of knowledge between agents through direct channels. Knowledge is encapsulated in a message with a special format and sent to other agents. The transferred knowledge is not altered in this process. When an agent receives this type of message, it extracts the useful information from the message and updates its related knowledge bases. This type of communication can be found in the DIDE project (see Chapter 13, Section 13.5).

Indirect communication assumes the presence of an intermediary agent, conveying information from one agent to another. Knowledge may be pruned or abstracted in this process and may also be submitted to ontological transformations. Examples of this type of intermediary agent are the facilitator in PACT (see Chapter 13, Section 13.2), the central node in First-Link (see Chapter 13, Section 13.4), the mediator in MetaMorph (see Chapter 14, Section 14.2), and the design board in SiFAs (see Section 5.8.1 of this chapter).

Some multi-agent learning systems are implemented with training agents (tutor agents). In this case, the training agents or tutor agents are the main knowledge source of learning for the system.

5.6 WHAT IS TO BE LEARNED

What kind of knowledge can, and shall, be learned in multi-agent systems? The most important are: successes and failures in the past; the usefulness of different pieces of knowledge with respect to different tasks and situations; the capabilities and accountability of other agents in the system; the relationship between the multi-agent system and its environment, etc.

While the range of the learnable is wide, there are specific areas where learning could have a major impact on design and manufacturing, as well as for agent interactions. Some of the following areas of learning target the level of design reasoning or manufacturing resource allocation, while others target the meta-level of agent interaction.

- Constraints related to parameters or, in general, to elements of the design: By recording constraint information from other agents, it becomes possible to avoid conflicts through anticipatory evaluations.
- Dependencies between design parameters: Dependencies act as soft constraints. They are critical for deciding on which agents an impact should be made.

-    Support in favor of or against a decision: Such rationale can primarily help in weighing proposals of different agents and in deciding which agent should revise its decision.
-    Design rules: As agents have different functionality and points of view, sharing of design rules is desirable among agents having the same target or at least the same domain.
-    Preferences: If agent preferences are known, they can help set up better initial proposals made by one agent to another.
-    Preconditions and post-conditions for rules, actions and tasks: This area of learning is important for a better adaptation of decisions to specific contexts. Agents decide by themselves when to act, and therefore should be able to recognize favorable situations for getting involved.
-    Prediction of decisions of other agents: This amounts to building a behavioral model of other agents, by relating those factors which influence an agent with the agent's decisions.
-    Types of conflicts: Conflicts can be seen, indexed and classified differently by the different types of agents. Enabling agents to classify and recognize conflicts enhances the power of a conflict anticipation and resolution mechanism.
-    Heuristics to solve conflicts and to negotiate: Such heuristics, being less dependent on the specific functionality of an agent, can be learned or transferred from one agent to another and refined according to local necessities.
-    Combinations of manufacturing resources for specific tasks: These combinations of resources at the group level or the system level can be learned as cases for manufacturing process planning. Such cases may be used in subsequent process planning and scheduling to reduce communication and negotiation among resource agents.
-    Manufacturing system's behavior: By learning the system's behavior and propagating it into the near future through simulation forecasting (see Chapter 14, Section 14.2 MetaMorph), the system can learn from future emergent behavior and prevent otherwise unforeseen perturbations and changes in production priorities on the shop floor.

5.7 HOW TO LEARN

5.7.1 Organizational Learning Theory

Organizational learning has been mainly studied in the areas of management and economics. In these areas, organizational learning is recognized as a fundamental requirement for an organization's competitiveness, productivity, and innovativeness in uncertain and changing technological and market circumstances. Thus, organizational learning is considered to be essential to the flexibility and sustained existence of an organization.

Organizational learning refers to an organization's capability to identify and to store knowledge derived from both individual and organizational experiences, and to modify its own behavior according to feedback received from its environment. Organizational learning supposes that an organization is able to control its behavior with respect to its own goals and objectives, to perform self-monitoring activities, to filter out the relevant information from environmental scanning processes, and to adapt itself to changes in its social, economic, and political environment.

Organizational learning is performed at three interacting levels: individual level, micro-organizational level (group level) and macro-organizational level (system level). Individual learning may contribute to organizational learning if it is not obstructed by organizational constraints, such as responsibilities or well-established information processing procedures. On the micro-organizational level (group level) the members of an organization negotiate and integrate their individual experiences in order to build up group-level knowledge. The results of this (permanently evolving) process (i.e., whether the group performs better than the best of its members) largely depends upon both the intra-group and the inter-group relationships within an organization. It is important to note that, to a large extent, micro-organizational learning evolves informally, i.e., is not primarily determined by fixed organizational rules and procedures. Macro-organizational level (system level) learning evolves in relation to an organization's macrostructure, i.e., its performance, successes and failures are largely determined by the structure of its inter-group relationships.

The ability of an organization to learn from individual knowledge and both individual and organizational experiences depends on whether the following are true:

- The knowledge to be learned is described in terms of a common ontology.
- Conflict resolving mechanisms are available to help decide which knowledge shall be included into the body of organizational knowledge, and which knowledge shall be excluded from it.
- Knowledge management tools are available, which guarantee that organizational knowledge is accessible to (and will be accessed by) exactly those members of an organization who need it for doing their work.
- Pieces of organizational knowledge are related to each other by a network of relationships in order to provide for organizational reasoning. There are the basic mechanisms within an organization (analogous to forward or backward chaining in rule-based systems) from which more complex organizational problem-solving procedures are created.

Organizational learning has been recognized as one of important approaches for multi-agent learning (Weiss 1993; Kirn 1996; Nagendra Prasad et al., 1997), and has been used in some agent-based intelligent design and manufacturing systems (Grecu and Brown 1996a, 1996b; Maturana 1997). Learning in agent-based concurrent design and manufacturing systems needs to be performed at three levels: single agent level (individual level), group level and system level.

5.7.2 Classification of Multi-Agent Learning

There are various possible forms of learning in multi-agent systems. According to Weiss (1993), two criteria can be applied to structure this variety: learning method and learning feedback.

Classification by Learning Method

This classification is based on the learning method or strategy used by a learning entity (a single agent or a group of agents). The following methods are usually distinguished:

- *Rote learning:* direct implementation of knowledge and skills without requiring further inference or transformation from learner;
- *Learning from instruction and by advice taking:* operationalization - that is, transformation into an internal representation and integration with prior knowledge and skills – of new information like instructions or advice that is not directly executable by the learner;
- *Learning from examples and by practice:* extraction and refinement of knowledge and skills like a general concept or a standardized pattern of procedure from positive and negative examples or from practical experience;
- *Learning by analogy:* solution-preserving transformation of knowledge and skills from a solved problem to a similar but unsolved problem;
- *Learning by discovery:* gathering new knowledge and skills by making observations, conducting experiments, and generating and testing hypotheses or theories on the basis of the observational and experimental results.

A major difference among these methods lies on the amount of learning effort required by them (increasing from top to down).

Classification by Type of Learning Feedback

The second classification is based on the learning feedback that is available to a learning entity and that indicates the performance level achieved so far. This criteria leads to the following distinctions:

- *Supervised learning[1]:* the feedback specifies the desired activity of the

---

[1] According to Russell and Norvig (1995): Any situation in which both the inputs and outputs of a component can be perceived is called *supervised learning*. In contrast, suppose that when learning the condition-action component, the agent receives some evaluation of its action (such as a hefty bill for rear-ending the car in front) but is not told the correct action (to brake more gently and much earlier). This is called *reinforcement learning*, and the hefty bill is called *reinforcement*. In fact, drawing the line between supervised and reinforcement learning is somewhat arbitrary; reinforcement learning can also be thought of as supervised learning with a less informative feedback signal.

learner and the objective of learning is to match this desired action as closely as possible;

- *Reinforcement learning*: the feedback only specifies the utility of the actual activity of the learner and the objective is to maximize this utility;
- *Unsupervised learning*: no explicit feedback is provided and the objective is to find out useful and desired activities on the basis of trial-and-error and self-organization processes.

In all three cases, the learning feedback is assumed to be provided by the system environment or the agents themselves. This means that the environment or an agent providing feedback acts as a "teacher" (tutor) in the case of supervised learning and as "critic" in the case of reinforcement learning; in the case of unsupervised learning, the environment and the agents just act as passive "observers".

It is important to see that different agents do not necessarily have to learn through the same learning method or the same type of learning feedback. Moreover, in the course of learning an agent may employ different learning methods and different types of learning feedback. Both of the above classifications directly or indirectly lead to a distinction between learning and teaching agents, and they show the close relationship between multi-agent learning on the one hand and teaching and tutoring on the other. Learning can also be classified according to the following criteria:

The purpose and goal of learning. This criterion allows distinguishing between the following extremes (and many graduations in between them):

- Learning that aims at an improvement with respect to one single agent, its skills and abilities.
- Learning that aims at an improvement with respect to the multiple agents as a unit, their coherence and coordination.

This criterion could be refined with respect to the number and compatibility of the learning goals pursued by the agents. Generally, an agent may pursue several learning goals at the same time, and some of the learning goals pursued by the agents may be incompatible while others are complementary.

The decentralization of a learning process (where a learning process consists of all activities carried out by one or more agents in order to achieve a particular learning goal). This criterion concerns the degree of distribution and parallelism of learning, and there are two obvious extremes:

- only one of the available agents is involved in the learning process, and the learning steps are neither distributed nor paralleled;
- all available agents are involved, and the learning steps are maximally distributed and paralleled.

An agent's involvement in a learning process. With respect to the importance of involvement, one can identify the following two extremes:

- the involvement of the agent under consideration is not a necessary condition for achieving the pursued learning goal (it may be replaced by another equivalent agent);

- the learning goal cannot be achieved without the involvement of exactly this agent.

Other aspects of involvement that could be applied in order to refine this criterion are its duration and intensity. It also has to be taken into consideration that an agent may be involved in several learning processes, because it may pursue several learning goals.

The agent-agent and agent-environment interaction required for realizing a learning process. Two obvious extremes are the following:

- learning requires only a minimal degree of interaction;
- learning would not be possible without extensive interaction.

This criterion could be further refined with respect to the frequency, persistence, level, pattern and type of interaction.

Many combinations of different values for these criteria are possible. For instance, one might think of a small group of agents that intensively interact (by discussing, negotiating, etc.) in order to understand why the overall system performance has decreased in the past, or of a large group of agents that loosely interact (by sometimes giving advice, sharing insights, etc.) in order to enhance the knowledge base of one of the group members.

5.8 EXAMPLES

Here we give two examples of learning in agent-based concurrent design and manufacturing systems. The first is related to learning in an agent-based engineering design system, and the second concerns learning in an agent-based manufacturing system.

5.8.1 Learning by Single Function Agents (SiFAs) During Spring Design

In the SiFAs system (see Chapter 13, Section 13.6 for the description of the SiFAs system), single function agents generate a significant number of interactions while deciding the value of a parameter. This overhead arises because each single function agent is responsible for only a portion of a decision, and the final decision has to gain approval from all points of views. The number of single function agents involved in a parameter decision is not trivial. The SiFAs experiment involved three types of agents (selectors, critics and praisers) for material selection during spring design. The selectors are the only agents allowed proposing parameter values. The critics express objections to the choices made by the selectors. The praisers are complementary agents to critics. Their role is to highlight every selection which has a very good rating from their point of view.

As interactions consume a large portion of the computational effort expended on value selection, the primary goal of learning is to improve the knowledge which the agents have about each other. This learning results from agent interactions, and is aimed at predicting the further behavior of the interaction partners. Since agents act based on the their functionality, learning can take advantage of this knowledge. Therefore, the learning strategies in SiFAs were made dependent on the type of the agent whose behavior is to be predicted.

Learning Environment in SiFAs Spring Design System

The spring design system is composed of several single function agents and a design board (Figure 5.1). The design board of SiFAs is a module which is visible to all the agents participating in design. It is subdivided into three parts:

(1) *design specifications*, including all the requirements provided by the user for the current design;
(2) *design state*, which describes the design and records the parameter values decided so far;
(3) *exchange board*, where agents make their proposals, engage in negotiations and reach agreements.

Every agent can post design proposals, objections or modifications to the *exchange board*, and send the design results to the design board to update the *design state*. Every agent can also see the information describing the *design state* and the *design specifications*. Because there are many differences between agents (e.g., functionality, point of view, specific domain target), it is assumed that an unfiltered observation of an agent by another agent is not possible.

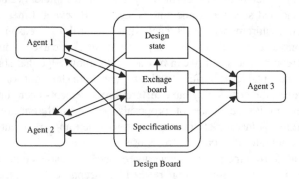

Figure 5.1 Learning environment in SiFAs spring design system.

What is Learned about Each Agent

What is learned about an agent depends on its type:

- *Learning about praisers*. Praisers point out parameter values which are particularly suitable from their point of view. Whenever a praiser praises a parameter value, the information is recorded by selectors. For example, a material may be considered excellent from the point of view of hardness regardless of any other conditions or design values. Thus learning is merely storing praised values associated with praisers. If context-sensitive praisers are used, the inductive technique used for critics applies to praisers too. Praisers provide only one learnable item at a time, and can only act when a selector proposes a value. This can occur during a negotiation. Consequently, knowledge about the praisers is acquired only to the content to which proposals are praised during negotiations.

- *Learning about critics.* Critics can be context-insensitive, or context-sensitive. The values they do not accept can depend on the design requirements, e.g., the resistance of a spring material to impact loading can be good or bad, depending on the conditions under which the spring is required to work. The information recorded about a critic is the set of values it considered unacceptable and the design conditions that the critic used to make those decisions. The classes correspond, in this case, to the individual material a critic objects to. For any particular requirement, the information learned about the critic is complete (i.e., the critic makes all its objections known).

- *Learning about selectors.* Selector behavior is the most difficult to predict, since, for a given design context/state, a selector can use different methods or sets of prestored values. Even if the set of values remains constant, the preferences among those values can be variable. The classes that partition a selector's behavior are defined by sets of parameter value preferences. Assume that selector A encounters selector B several times, with the same design requirements for B each time. A still has reason to record B's responses every time, as they might offer sparse sequences of B's preferences. The sparsity can be due to B's knowledge that some values will be unacceptable to critics and are therefore left out of its proposals. However, the critics may have different requirements each time and thus the 'holes' in B's sequence of proposals will be different. In addition, the recording is incomplete because it stops when an agreement is reached. A will integrate the current response sequence of B with the sequence compiled from the previous encounters under the same conditions. The main issue is that, while being refined, concepts can 'migrate'. Different sequences of responses can become identical after several interaction sessions, merging into a single concept. Alternatively, concepts can be split if under some particular subconditions the refined responses become different.

How Agents Use the Learned Knowledge

The knowledge learned by an agent about the other agents is used differently in SiFAs system, depending on the type of the agent to which it refers:

- *Knowledge about selectors.* During a negotiation, a selector will decide on its next proposal. Before posting its proposal the selector will see whether it has knowledge about the behavior of the other selectors under these design conditions. If so, the selector will anticipate their proposals and prepare a new proposal. A new evaluation of the next-best proposals of the other selectors will be used to determinate their responses. The predictive reasoning continues until either one of the following conditions is fulfilled: (1) a value is found which is likely to be agreed on by all the selectors; (2) for one of the other selectors no further preferences are known.
- *Knowledge about critics.* Knowledge about the critics is used to eliminate proposals known to be unacceptable. This holds true for the proposals which the current selector intends to submit, as well as for the proposals it anticipates from the other agents. Even if there is no knowledge about how a critic will respond, the selector continues its anticipatory reasoning. The critic will eventually 'protest' about an unacceptable agreement of the selectors.
- *Knowledge about praisers.* The knowledge about praisers is used in a very similar manner to that of the critics. During the anticipation of the other selectors' proposals, praiser opinions will be taken into account to find out which agent has to revise its counterproposal, assuming that no agreement is anticipated.

These methods attempt to ensure that what has been learned is used effectively, and that information is exchanged in a cooperative fashion.

5.8.2 Learning in the MetaMorph Multi-Agent Manufacturing System

To enable a mediator to learn about its environment, other mediators and agents, a simple registration mechanism can be used to capture, manager and update its knowledge which is then reused in its future interactions and coordinations with other mediators and agents. This section focuses on the two learning mechanisms implemented in the MetaMorph multi-agent manufacturing system (see Chapter 14, Section 14.2 for a detailed description of this system) to enhance system performance and responsiveness. First, a mechanism that allows mediators to learn from history is placed "on top" of every multi-agent resource to capture significant multi-agent interactions and behaviors. Second, a mechanism for propagating the system's behaviors into the future is implemented to help mediators "learn from the future." The combined action of these two learning mechanisms provides a promising approach for multi-agent manufacturing learning.

5.8.2.1 Learning from History

The context of manufacturing requests is established under static and dynamic variations. Static variations relate to the physical configuration of products. Dynamic variations depend on time, system loads, system metrics, costs, customer desires, etc. These two main sources of information relate to a wide spectrum of emergent behaviors, which can be separated into specific behavioral patterns. A 'learning from the history' mechanism based on distributed case-based learning approach was developed during the MetaMorph project for capturing such behavioral patterns at the resource mediator level and storing in its knowledge base. Such knowledge is then reused for subsequent manufacturing requests through an extended case-based reasoning mechanism.

A manufacturability or manufacturing request sent to a resource community is first filtered by the respective resource mediator to decide whether the request can be recognized as associated with a previously considered product or is unknown. If it is recognized, the resource mediator retrieves the learned patterns to send to a selected group of agents identified in the patterns. For an unknown request, the resource community's mediator uses its standard matchmaking actions to specify the primary set of resource agents to be contacted regarding their capability to satisfy this request. The solution to this unknown request then proceeds through the propagation of coordination clusters and decision strategies previously defined. During this process, the resource mediator involved learns from partial emergent interactions at the coordination cluster level. This learning is distributed among several coordination clusters. The plan aggregation process then enables the classification of various feature-machine-tool patterns, which are encoded and provided to the community's mediator for storage and future reuse.

The approach described above was implemented to show proof-of-concept of distributed deliberative learning for advanced manufacturing systems. However, the mechanisms are centralized through the dynamic mediators and will become increasingly slow for larger amounts of information, i.e. with increased product types and larger factory floors. The mechanisms in this learning system would, in such cases, need to be replaced by other mechanisms which allow for storing, pattern matching, and rapid information retrieval in a single action.

5.8.2.2 Learning from the Future through Forecasting

In MetaMorph, the system's behavior is extended into the near future for learning from future emergent behaviors through the implementation of simulation forecasting mechanisms.

In conventional systems, the effect of unknown future perturbations can only be estimated through probabilistic mechanisms that add in some perturbations to the scheduling calculations. MetaMorph used a 'learning from the future' mechanism to uncover dynamic interactions likely to perturb plans and schedules during execution stages. The system maintains a virtual model of the shop floor, which can be used as the base for propagating the system's behavior into the near future. The system also maintains performance evaluation parameters used to decide whether to accept or reject future interactions.

Shop floor mediators monitor system status by conducting global performance evaluations. Various forecast-triggering and adjusting parameters are used based on system load, extension of the scheduling horizon, and adjustment periodicity. The thresholds for triggering are arbitrarily set by the users according to high-level policies.

A composite function called the System Heuristic Function is used to partially measure the system's performance. The function measures average flow time, maximum completion time, average due time, number of late jobs and total number of jobs, as shown in the following equation.

$$H = w_f * \overline{T}_{flow} / \overline{T}_{due} + w_c * T_{maxcpl} / \overline{T}_{due} + w_j * J_1 / J_t$$

where :

w_f – average flow time weight

w_c – maximum completion time weight

w_j – late jobs weight

$\overline{T}_{flow}$ – average flow time

$\overline{T}_{due}$ – average due time

T_{maxcpl} – maximum completion time

J_1 – number of late jobs

J_t – total number of jobs

The function value H is called the system heuristic. A lower H value indicates that the system has a higher performance with a lower cost. Its primary value is for assessing deviation towards an undesirable situation of increasing production time and lateness. Each component in this function is weighted according to the system's management policies. Other components can be added to this function so as to measure system's performance more completely. The heuristic ratio (R_h) is calculated as follows:

$$R_h = \frac{H'}{H}$$

H'–forecast simulation performance

H – system performance before adjustment

Lower and upper thresholds $(-\sigma, +\sigma)$ have been used to define acceptance and rejection heuristic-ratio regions. These thresholds define a square-shaped region with an amplitude equal to 1 (dimensionless). For ratios falling within the square region, the system's current state of commitments is accepted and remains unaltered. If the ratio falls outside the square region, the forecast behavior (scheduling and activities) is accepted to replace the current scheduling state.

Forecasting the system's behavior consists of capturing the current state of the multi-agent system (plans) and simulating it for a period of time into the future. The forecasting process is thus a simulation of future interactions based on the intelligent agents' commitments.

The forecasting simulation period is established by the system's manager according to high-level policies. However, the determination of this parameter may be customized to automatically adapt to the system's requirements. There are two approaches that can be used for the forecasting period:

1) The forecasting simulation may be run for a period of time sufficiently large to permit the completion of every committed job. But for large simulation periods, this criterion incurs accumulative deviations, since jobs may arrive in the system during the forecast period that affect intermediate allocation slots. Because these intermediate allocation slots are not preempted during simulation, this approach only produces rough estimations.

2) The second approach attempts to avoid the accumulation of deviate behaviors, limiting the forecasting simulations to short periods of time only. This approach is more accurate, since the duration of the simulation is much less than the jobs' arrival frequency. Here, the multi-agent system can easily be adjusted while including the intermediate requirements.

Whichever the approach, the forecast produces an estimated heuristic value, which is used to decide whether to introduce the forecast schedule tuning or to continue with the existing system's scheduled commitments.

A detailed description of above mentioned learning mechanisms in MetaMorph can be found in (Maturana 1997).

5.9 RESEARCH LITERATURE AND FURTHER REFERENCES

5.9.1 Research Literature

Despite learning in multi-agent systems being recognized as a very important topic by researchers in the related areas, it has only been given real attention in recent years. Some researchers have applied Machine Learning techniques to multi-agent learning systems, and others proposed new approaches especially for multi-agent learning, but only a few examples could be found for agent-based engineering design and manufacturing systems. Thus, we will review not just multi-agent learning in engineering design and manufacturing, but also some other interesting multi-agent learning systems in other application domains.

5.9.1.1 Multi-Agent Reinforcement Learning

Much of the literature of multi-agent learning relies on reinforcement learning. Some of the earliest attempts at multi-agent learning used these methods because they are mature techniques developed by the Machine Learning community and they are simple to implement and have elegant theoretical formalisms. More importantly, these algorithms rely on weak reinforcement learning signals provided as feedback by the environment. These signals impose few, if any, demands in terms of their content. In most cases they are just numerical measures of the "goodness" of the performance of the system as a whole.

Tan (1993) conducted three case studies of multi-agent reinforcement learning involving cooperation by sharing sensations, episodic experience, (i.e., sequences of sensation, action, and reward triples), and learned decision policies. According to Tan (1993), agents that learn according to the Q-learning algorithms[2] (Watkins and Dayan 1992) and also cooperate with other agents by exchanging information (partial solution, plans of action) can learn more quickly than agents that do not cooperate.

Gu and Maddox (1996) developed a learning model called DRLM (Distributed Reinforcement Learning Model) that allows distributed agents to learn multiple interrelated tasks in a real-time environment. DRLM consists of a hidden task model (HTM) used for dealing with incomplete perception, a composite state model (CSM) for interdependency between tasks, and Q-learning subsystem (QLS) for updating action merit. DRLM was implemented in a Flexible Manufacturing System (FMS) where sensors (modeled as agents) have to learn to communicate with humans about the material handling activities using graphical actions such as displays and animation.

Littman (1994) proposed to use Markov games as a framework for multi-agent reinforcement learning.

In the experiment of Sen et al. (1994), two agents have to push a block to a goal position. Agents learned complementary policies without sharing any information with the other agent. The agents are actually not even aware of the existence of the other; they learn how to push the block by receiving reinforcement from the environment regarding the distance of the block from the optimal path.

Sen, Skaran and Hale (1994) realized a reinforcement learning system where two agents learn to push a box cooperatively to a pre-designed location. The agents learn complementary policies to move a box from a start location to a goal location without any knowledge about each other. They studied interesting learning issues like transfer of knowledge and complementary learning. Agents in this system are simple and without domain knowledge or planning capabilities. Their knowledge is encoded in the learned policy matrix.

Sandholm and Crites (1996) investigated the use of the Q-learning algorithm in the Iterated Prisoner's Dilemma game. The learning agent was able to develop optimal strategies against opponents with static strategies, but when both the players were learning concurrently, the learners were less effective.

Mataric (1994) introduced the concept of progress estimators. In addition to the reinforcement upon reaching a goal, partial internal critics called progress estimators, when active, provide a metric of improvement in the context of specific goals. For example, a progress estimator associated with homing behavior gives a positive reinforcement if the distance to home is decreased upon picking up a puck. Movement away from home generates a negative reinforcement.

---

[2] Q-learning is a well known reinforcement learning algorithm in the literature of Machine Learning (cf. Watkins and Dayan 1992).

5.9.1.2 Classifier Systems as Learning Algorithms

Shaw and Whinston (1989) proposed a classifier system based multi-agent learning system. Each agent contains a set of condition-action rules and a message list. The rules that can be triggered in the present cycle in an agent can bid, based on their strengths. The rule with the highest bid is chosen and executed to output a message as an action, perhaps to other agents. Its strength is reduced by the amount it bids and is redistributed to the rules that were responsible for the messages that triggered it. These rules could be distributed among many agents. An agent choosing to perform an action on the environment receives an external reinforcement. A classifier system has the effect of reinforcing chains of effective action sequences. This classifier system was tested in a Flexible Manufacturing domain. It functioned in a contract-net framework where inter-agent bidding was used as a feedback mechanism that drives inter-agent learning. An agent announces a task with a series of operations to be performed on it. Other agents in the system bid for performing this task based on their past usefulness in performing similar jobs, the capability for the announced job and their readiness for the job. The winning agent increases its strength by an amount proportional to the bid and the agent announcing the task decreases its strength by the same amount. This represents a kind of market mechanism where the announcing agent pays the winning agents for its services by the amount bid. In addition, the agents themselves can use genetic operators to improve and reorganize their own capabilities for performing various jobs. This type of learning is purely local and does not involve any cooperative control component. In the Flexible Manufacturing domain, the genetic learning capability can be mapped on to the need for reconfiguration to avoid unbalanced utilization of high-performance cells.

Another related work using classifier system for learning suitable multi-agent hierarchical organizational structuring relationships was presented by Weiss (1993).

5.9.1.3 Case-Based Learning and Knowledge Refinement Systems

Grecu and Brown (1996a) studied learning among Single Function Agents (SiFAs) that are highly specialized and very competent in making design decisions about a particular aspect of parametric design. Each agent has a particular function like `select`, `critic`, `praise`, `estimate` or `evaluate` and deals with a single parameter from a given point of view like "reliability" or "durability". Agents use an interaction board (see Section 5.8.1 in this chapter), where they post their proposals, decisions and partially developed designs. The nature of these agents requires a large number of interactions in order to arrive at a mutually acceptable design. Grecu and Brown investigated the use of learning to reduce interactions among agents. Their domain deals with the parametric design of helical springs.

Ohko, Hiraki and Anzai (1996) proposed a LEMMING learning system to use Case-Based Reasoning (CBR) to reduce communication cost for task negotiation on Contract Net protocol. By using CBR, LEMMING can derive useful knowledge from messages in Contract Net protocol and find a suitable agent that should receive task announcement.

MetaMorph (Maturana 1997) used Case-Based Learning as one of its learning mechanisms at the mediator level of the multi-agent manufacturing system to reduce distributed search space and therefore to improve production planning.

Nagendra Prasad, Lesser and Lander (1995) presented one of the attempts at multi-agent case-based reasoning. Each of the agents contributed a subcase to an overall composite case. Assembling the subcases is treated as a distributed constraint optimization problem where the conflicts between subcases are resolved through search rather than avoided through elaborate indexing using goals and subgoals representing the context of the subcases. Each agent retrieves local subcases using local knowledge and then the agents cooperatively search through the space of partial cases to assemble a composite case that is mutually acceptable to all of them. The experiments were based on a mechanical design problem – a steam condenser design. The results showed that case-based reasoning improved both solution quality and the processing time compared to blind search.

Sugawara (1995) proposed a hierarchical planner that can use past plans to minimize messages at planning stage.

Haynes and Sen (1997) proposed using case-based learning to resolve conflicts in a multi-agent system. Their domain is a predator-prey domain similar to that used by Tan (1993). The work of Haynes and Sen represents one of the few attempts at empirically demonstrating the power of case-based learning in multi-agent cooperative control.

Byrne and Edwards (1996) proposed refining the knowledge bases of individual group members to improve the effectiveness of the entire group. They utilized a refinement facilitator agent that uses KQML messages to coordinate refinements that benefit the group.

5.9.1.4 Mutually Supervised Learning

Goldman and Rosenschein (1996) proposed mutually supervised learning for multi-agent systems. In their system, each agent acts as the teacher of its partner. The agents are trained by receiving examples from a sample space; they then go through a generalization step during which they have to apply the concept they have learned from their instructor. The agents in this model first go through a training step, and are then able to choose their actions by generalizing what they have learned. The agents do not need to re-coordinate their actions for every new situation or problem. The main issue in this research is how to train the agents in a way that minimizes the number of mistakes in the generalization step. This distributed learning approach to coordination is useful whenever agents do not have enough time to negotiate, when they exist in a dynamic environment and will benefit by adapting to unpredictable situations, and when the agents face similar problems repeatedly. But it may not be usable in complex applications because each agent has to model the entire sensing environment of its partner.

5.9.1.5 Other learning systems

In Sian's multi-agent learning system (Sian 1991), every agent has a learning module. The agents evaluate various hypotheses while having only partial knowledge about them. The agents can know more about a specific hypothesis by receiving messages from other agents expressing their level of confidence in that hypothesis. The general objective of the system was for agents to settle on the best hypothesis they can, given the knowledge they have.

Sen and Sekaran (1996) proposed a new reciprocity approach for agents to adapt to others. The authors argued that the traditional choice of deterministic reciprocity is insufficient, and proposed a probabilistic reciprocity scheme to decide on whether or not to help another agent. By this type of mechanism, agent A may decide to help agent B even if the later had refused help in the previous time-step. The decision is based only on the balance of their exchanges (the balance between the savings obtained from and extra cost incurred by agent B from agent A over all their previous exchanges), not on when requests for help were accepted or denied.

Joshi (1996) used a combination of neuro-fuzzy learning and static adaptation to coordinate the activity of multiple agents in the PYTHIA project whose objective was to develop a system that will accept a description of a problem from the user, and then automatically select the appropriate numerical solver and computing platform along with values for the various associated parameters.

Nagendra Prasad, Lesser, and Lander (1997) proposed to deal with agents learning about their roles[3] in an organization and about the local and joint search spaces in group decision making. They used different supervised learning schemes, including a form of instance-based learning, in building a group of agents that learn to effectively design artifacts. They concluded that even though learning by itself does not allow the agents to produce the same solution quality as can be obtained by direct negotiation, it does provide for significant savings in communication cost.

5.9.2 Further References

Learning in multi-agent systems, in particular for concurrent engineering design and intelligent manufacturing, is a relatively new research topic. Here we list some interesting references or pointers:

More details concerning theories and methods of machine learning:

Russell, S. and Norvig, P. (1995). Artificial Intelligence: A Modern Approach. Prentice Hall. (Part VI: Chapters 18, 19, 20 and 21)

---

[3] An organizational role is a task or a set of tasks to be performed in the context of a single solution.

More recent information on multi-agent learning:

Learning in Multi-Agent Systems Webliography [http://dis.cs.umass.edu/research/agents-learn.html]

Recent publications on multi-agent learning:

Weiss, G. and Sen, S. (Ed.) (1995). Adaption and Learning in Multi-Agent Systems, Lecture Notes in Artificial Intelligence, Volume 1042, Springer-Verlag.

Weiss G., (Ed.) (1997) Distributed Artificial Intelligence Meets Machine Learning: Learning in multi-agent environments, Lecture Notes in Artificial Intelligence, Volume 1221, Springer-Verlag.

Special issue on Multi-agent Learning of the Machine Learning Journal, 1997.

Special issue on Learning in DAI Systems of the Journal of Experimental and Theoretical Artificial Intelligence, 1997.

Proceedings of the Workshop on Agents that Learn from Other Agents, (held as part of the 1995 International Machine Learning Conference) [http://www.cs.wisc.edu/~shavlik/ml95w1/procs.html]

Proceedings of AAAI Spring Symposium Series 1996: Adaptation, Co-evolution and Learning in Multiagent Systems, Stanford University, CA, March 25-27, 1996.

Proceedings of AAAI-97 Workshop on Multiagent Learning, Providence, Rhode Island, July 28, 1997.

REFERENCES

Akoulchina, I. and Ganascia, J.-G. (1997) SATELIT-Agent: An Adaptive Interface Based on Learning Interface Agents Technology, In *Proceedings of the Sixth International Conference*, UM97, Springer-Wien-New York. Available online at http://www.um.org/

Byrne, C. and Edwards, P. (1996) Refinement in Agent Groups, In Weiss G. and Sen, S. eds., *Adaption and Learning in Multi-Agent Systems*, Lecture Notes in Artificial Intelligence 1042, pp. 22-39, Springer-Verlag.

Goldman, C.V. and Rosenschein, J.S. (1996) Mutually Supervised Learning in Multiagent Systems, In Weiss G. and Sen, S. eds., *Adaption and Learning in Multi-Agent Systems*, Lecture Notes in Artificial Intelligence 1042, pp. 85-96, Springer-Verlag.

Grecu, D. and Brown, D. (1996a) Learning to design together, In *Proceedings of the 1996 AAAI Spring Symposium on Adaptation, Co-evolution and Learning in Multiagent Systems*, Stanford, CA.

Grecu, D. and Brown, D. (1996b) Learning by Single Function Agents during Spring Design, *Artificial Intelligence in Design '96*, Gero, J.S. and Sudweeks, F. (eds), Kluwer Academic Publishers, Netherlands.

Gu, P. and Maddox, B. (1996) A Framework for Distributed Refinement Learning, In Weiss G. and Sen, S. eds., *Adaption and Learning in Multi-Agent Systems*, Lecture Notes in Artificial Intelligence 1042, pp. 97-112, Springer-Verlag.

Haynes, T. and Sen, S. (1997) Learning Cases to Compliment Rules for Conflict Resolution in Multiagent Systems, *Special Issue on Evolution and Learning in Multiagent Systems of International Journal of Human-Computer Studies*.

Joshi, A. (1996) To Learn or Not to Learn, In Weiss G. and Sen, S. eds., *Adaption and Learning in Multi-Agent Systems*, Lecture Notes in Artificial Intelligence 1042, pp. 127-139, Springer-Verlag.

Kirn, S. (1996) Organizational Intelligence and Distributed Artificial Intelligence, In O'Hare, G.M.P. and Jennings, N.R. eds., Foundations of Distributed Artificial Intelligence, pp. 505-526, Wiley-Interscience.

Littman M.L. (1994) Markov games as a framework for multi-agent reinforcement learning, In *Marching Learning 1994*, pp. 157-163.

Maes, P. and Kozierok, R. (1993) Learning Interface Agents, In Proceedings of the Eleventh National Conference on Artificial Intelligence, Washington, D.C., MIT Press, pp. 459-465.

Mataric, M. J. (1994) Reward functions for accelerated learning. *In Proceedings of the Eleventh International Conference on Machine Learning,* San Francisco, CA.

Maturana F.P. (1997) *MetaMorph: An Adaptive Multi-Agent Architecture for Advanced Manufacturing Systems*, Ph D thesis, Dept. of Mechanical and Manufacturing Engineering, The University of Calgary.

Nagendra Prasad, M.V., Lesser, V. R., and Lander, S. E. (1995) On retrieval and reasoning in distributed case bases, In *Proceedings of 1995 IEEE International Conference on Systems Man and Cybernetics*, Vancouver, Canada.

Nagendra Prasad, M.V., Lesser, V.R., and Lander, S.E. (1997) Learning Organizational Roles for Negotiated Search in a Multi-agent System, *Special issue on Evolution and Learning in Multiagent Systems of International Journal of Human-Computer Studies.*

Ohko, T., Hiraki, K., and Anzai, Y. (1996) Learning to Reduce Communication Cost on Task Negotiation, In Weiss G. and Sen, S. eds., *Adaption and Learning in Multi-Agent Systems*, Lecture Notes in Artificial Intelligence 1042, pp. 177-190, Springer-Verlag.

Russell, S. and Norvig, P. (1995) *Artificial Intelligence: A Modern Approach*, Prentice-Hall.

Sandholm T.W. and Crites R.H. (1996) On Multi-Agent Q-Learning in Semi-Competitive Domain, In Weiss G. and Sen, S. eds., *Adaption and Learning in Multi-Agent Systems*, Lecture Notes in Artificial Intelligence 1042, pp. 191-205, Springer-Verlag.

Sen, S., Sekaran, M. and Hale, J. (1994) Learning to coordinate without sharing information, In *Proceedings of AAAI'94*, pp. 426-431.

Sen, S. and Sekaran, M. (1996) Using reciprocity to adapt to others, In Weiss G. and Sen, S. eds., *Adaption and Learning in Multi-Agent Systems*, Lecture Notes in Artificial Intelligence 1042, pp. 206-217, Springer-Verlag.

Sen, S. (1996) IJCAI-95 Workshop on Adaptation and Learning in Multiagent Systems, *AI Magazine*, **17**(1), 87-89.

Shaw, M. J. and Whinston, A. B. (1989) Learning and adaptation in DAI systems, In Gasser, L. and Huhns, M., editors, *Distributed Artificial Intelligence*, volume 2, pp. 413-429. Pittman Publishing/Morgan Kauffmann Publishers.

Sian, S. (1991) Adaptation based cooperative learning multi-agent systems, In Demazeau, Y. and Muller, J.-P., eds., *Decentralized AI 2*, pp. 231-243, Elsevier Science Publishers, Amsterdam, Netherlands.

Sugawara, T. (1995) Reusing past plans in distributed planning, In *Proceedings of the First International Conference on Multi-Agent Systems (ICMAS'95)*, pp. 360-367, San Francisco, CA, AAAI Press.

Tan, M. (1993) Multi-Agent Reinforcement Learning: Independent vs. Cooperative Agents, In *Proceedings of the Tenth International Conference on Machine Learning*, pp. 330-337.

Watkins, C. and Dayan, P. (1992) Technical note Q-learning, *Machine Learning*, 8:279-292.

Weiss, G. (1993) Learning to coordinate actions in multi-agent systems, In *Proceeding of the International Joint Conference on Artificial Intelligence (IJCAI'93),* San Matco, CA.

6

Agent Architectures

6.1 INTRODUCTION

The internal architecture of an agent is essentially the description of its modules and how they work together. It is necessary to define the internal architecture of an agent, to allow the designer of an agent-based system to integrate the special requirements for the application, and to define the trade-offs between them.

Ferber (1995) compared the internal architecture of an agent with the structure of a computer. An agent's units of perception and action and its deliberative system are analogous to the computer's input/output, memory, and calculation units. However, an agent is a less generalized 'computer' because one can simulate an agent and even a multi-agent system in a computer; and an agent needs to be more evaluative than a computer if it is to be able to reach its goals autonomously. The analogy is interesting, and may be useful to the readers of this book in understanding the internal architecture of an agent, and even in constructing an agent according to a given architecture.

Agent architectures used in various agent-based systems (including agent-based concurrent design and manufacturing systems) range from the very simple (a single-function control unit with an input and an output) to the very complex (even human-like, see (Minsky, 1987)). In this chapter, we first depict the desirable characteristics of an intelligent *complex* agent and then describe the necessary modules (components) needed to implement such desirable characteristics. Note, however, that in many agent-based systems only a subset of these characteristics are needed for the agents, hence only a subset of their modules are needed in constructing these agents. The internal architecture of an agent specifies how the selected modules are organized in relation to each other. We thus consider approaches described in the literature for organizing these modules in different ways.

6.2 DESIRABLE CHARACTERISTICS OF AN AGENT

Although agents in various agent-based systems range from very simple to very complex, here is a brief list of what we see as desirable characteristics of intelligent *complex* agents in agent-based concurrent design and manufacturing systems:

- *Network-centric*: distributed and self-organizing;
- *Communicative*: able to interact with other agents through communication;

- *Semiautonomous*: neither always under the control of humans or other agents, nor completely autonomous, i.e., humans should be able to control the amount of agent autonomy;
- *Reactive*: able to react timely and appropriately to unforeseen events and to changes in its environment;
- *Deliberative*: capable of sufficient deliberation to perform tasks in a goal-directed manner;
- *Collaborative*: able to interact with humans and other machine agents (Collaborative interactions allow agents to resolve conflicts and inconsistency in information, current tasks, and world models, thus improving their decision-support capabilities);
- *Pro-active*: Capable of taking the initiative; not driven only by events, but capable of generating goals and acting rationally to achieve them;
- *Predictive*: able to make predictions about the future, and in particular, able to predict the effect of actions;
- *Adaptive*: able to accommodate changing user needs and task environments;
- *Flexible*: able to deal with the heterogeneity of other agents and information resources;
- *Persistent*: capable of long periods of unattended operation;
- *Mobile*: able to move around a network if necessary.

However, as earlier noted, agents in a specific agent-based design and manufacturing system may only need to have a subset of these characteristics.

6.3 BASIC MODULES (COMPONENTS) FOR AGENTS

To implement the agent characteristics described in Section 6.2, appropriate modules (components) are needed. Simple agents may only need a small number of modules (such as those for perception, reasoning and action) while more complex agents will need more. This section describes commonly needed modules for agents in agent-based concurrent design and manufacturing systems:

- Communication Interface
- Perception
- Execution
- Social Knowledge
- Self Knowledge (Self Representation)
- Domain Knowledge (Project Representation)
- Knowledge Management
- Learning
- Reasoning
- Problem Solving Models
- Coordination
- Planning and Scheduling

- Control
- Conflict Management
- Application Interfaces

Note that an agent in a specific agent-based system may only be composed of a subset of these modules. In real implementations, several modules may conveniently be combined into one (e.g., the Communication Interface, Perception and Execution modules can be integrated into one World Interface module, as in INTERRAP (Müller, 1996)). Also, one module can be split into a number of more low-level detailed modules (e.g., the Social Knowledge module can be split into several modules containing separate knowledge about communication protocols, interaction conventions, conversation policies, and other agents' capabilities).

6.3.1 Communication Interface

The Communication Interface of an agent is its only interface to other agents. It is usually implemented using methods or functions for receiving messages from and sending messages to other agents.

A message box is commonly set up for temporarily storing received messages. Processing of incoming messages requires two steps: (i) receiving, storing, and sorting messages; (ii) decoding message content for further processing by the agent within the context of a particular task. Processing of out-going messages requires encoding of the information to be transmitted, and finally mailing it according to the exchange protocol.

6.3.2 Perception

The Perception module is one of an agent's interfaces to its environment (the outside world). Commonly, the Perception module obtains signals from the agent's sensors and may also receive messages from other agents. Here we consider only the function of obtaining signals from the sensors, leaving messages from other agents to be handled by the Communication Interface module (Section 6.3.1).

In a typical Perception module, a sensor is represented as being an object having methods for sensor calibration, for enabling and disabling sensor activity, and for reading its current value. These current values are then made available to the agent's control module through a perception buffer.

The real problem for the Perception module is not to be overwhelmed by the constant stream of information. Sophisticated mechanisms are therefore needed to filter useful information out from the large amount of raw sensory input. 'Sensor-fusion' techniques may be used to assist in this task.

6.3.3 Execution

The Execution module is another interface for an agent to its environment. Once an agent has perceptively recognized that a significant event has occurred, the next step is to take some action. This action could be to realize that there is no further action to take, or it could be to send a message to another agent to take an action on the sender agent's behalf. However, as for the Perception module, we exclude the function of sending messages to other agents from the Execution module. Thus, we consider the Execution module only controls the physical or external actions the agent may perform.

6.3.4 Social Knowledge

An agent's social knowledge allows it to interact with other agents. This social knowledge may include the following components:

- *Communication*: Knowledge about the means of inter-agent communication (e.g. information about communication channels and the locations of agents) and about an inter-agent communication language.
- *Interaction*: How to interact with other agents in order to procure services, perform tasks for others, etc. An agent may, for example, know how to use a cooperation strategy such as the Contract Net protocol.
- *Agent Models*: An agent can use these models to identify other agents with whom it is useful to interact, and to make this interaction more effective. For example, an agent may wish to determine which agents have the skills necessary to perform a particular task. A model may contain information such as the skills of other agents and their level of authority in the group.

6.3.5 Self Knowledge (Self Representation)

The agent's self knowledge contains agent's knowledge about itself, including its physical state, location and skills, etc. This self knowledge is used by the agent to participate in tasks, and to reply to other agents' request about its competence. The reasoning module described below needs this knowledge to match the agent's skills with received requests and activate necessary tasks.

6.3.6 Domain Knowledge (Project Representation)

The agent's domain knowledge concerns its problem-solving domain and environment. Usually, this module contains the description of the working projects (problems to be solved), partial states of the current project (problem), hypotheses developed and intermediate results, etc. Such domain knowledge is imperative for proactive agents if they are to take the initiative in a working project.

6.3.7 Knowledge Management

As described above, an intelligent agent should have social knowledge, self knowledge, domain knowledge and may even need other types of knowledge (e.g., in some complex systems, agents may require common-sense knowledge).

Having knowledge is not sufficient: an agent needs to store and process such knowledge locally. Services such as content-based information distribution[1] impose requirements on how local knowledge is to be processed. For example, deductive inference is needed for content-based distribution, truth-maintenance capabilities are needed for belief revision, and so on. Other requirements may come from the enterprise modeling domain: the ability to represent complex enterprise models or to reason about common-sense notions like time. The following mechanisms may be needed in the knowledge management module:

- *Terminology construction and verification*: defining the terms of a domain by logical composition mechanisms, automatically organizing terms in taxonomies, ensuring consistency of definitions (see Chapter 11 for details).
- *Model construction*: building actual models of domains using the defined terminologies; ensuring consistency of these models.
- *Multi-agent belief revision*: supporting belief revision among multiple agents in a manner preserving consistency among the participating agents.
- *Common sense representation and reasoning about time.*
- *Model change management*: monitoring the dynamic modification of models by notifying interested agents when changes in their areas of interest occur.

Note that agents in most existing agent-based systems, even in very complex BDI agent-based systems, do not yet have all of these mechanisms.

6.3.8 Learning

An agent working in a dynamic environment needs to adapt to its changing environment. It needs to learn, for updating its knowledge about its environment, about other agents and about the problem(s) to be solved. Chapter 5 gives a detailed introduction to learning in agent-based concurrent design and manufacturing systems.

---

[1] Content-based information distribution: routing information to interested agents on the basis of the information content and the interests of the agents.

6.3.9 Reasoning (Decision Making)

A reasoning (or decision making) module is needed for almost all agents from the very simple to the very complex. An agent needs such a module to make decisions according to its knowledge (including self knowledge, social knowledge and domain knowledge) and its received information. Various reasoning mechanisms may be employed in any one agent, including:

- *Deductive knowledge processing*: using deduction to organize knowledge and answer queries;
- *Constraint based reasoning*: inferring features of models from the constraints they satisfy;
- *Case-based reasoning*: using historic task patterns existing in the agent's case base(s) to fulfil new tasks;
- *Schema-based reasoning*: using schemas[2] representing contextual problem-solving knowledge existing in the agent's knowledge base(s) to resolve new problems;
- *Other reasoning mechanisms*: such as neural networks and fuzzy logic.

6.3.10 Problem Solving Models

Many application domains, such as design, diagnosis, and production planning, have developed specific problem-solving models using hierarchical refinement, heuristic classification, case-base reasoning, or other solution techniques. Such models may be integrated in modules described above, such as the Domain Knowledge module and the Knowledge Management module. They may also be implemented through an independent knowledge base (model base) which can be used by a case-based reasoning or model-based reasoning mechanism.

6.3.11 Coordination

The coordination module contains coordination models which are shared conventions about exchanged messages during cooperative action. Coordination models are often generically defined in terms of rules about cooperation and situation assessment that

---

[2] Schemas are packets of actions that should be performed together in some situation to achieve an agent's goal or to take some action. Schema-based reasoning is a generalization of case-based reasoning; it extends the usual idea of case-based reasoning to encompass all aspects of the reasoner's behavior, and it extends it to make use of generalized "cases" (i.e., schemas) rather than particular cases, thus saving efforts needed to transfer knowledge from an old case to a new situation (Turner, 1994).

are applicable to most (if not all) real applications. Some mechanisms for using these models may be general, while others may be more application specific.

6.3.12 Planning and Scheduling

The planning and scheduling module maps the agent's goals into operational actions. This module can also be implemented by using two separate modules for planning and scheduling respectively. The planning module decides what to do in order to achieve the agent's goals. It maps the agent's current goals into local plans or joint plans. The scheduling module decides when to do what. It will thus be involved in merging partial or complete plans for various goals into an execution schedule, which takes compatibility and constraints among different plans into account.

6.3.13 Control

In some agent architectures such as Blackboard architectures (Section 6.4.2.3) and Layered architectures (Section 6.4.2.5), the control module is a very important component for controlling the information flow, mapping of the agent's goals into operational actions, and even monitoring the execution of actions. In these cases, the Planning and Scheduling module is part of the control module.

6.3.14 Conflict Management

The conflict management module enables the agent to make decisions when it is confronted with contradictory information derived or received from other agents. In real industrial applications, contradictory information occurs quite often and being able to cope with it increases the robustness of agents.

In most agent-based systems, conflict detection and resolution is done at the system level (or multi-agent level), and in some cases, with human interventions. Chapter 10 will focus on this topic.

6.3.15 Application Interfaces

An agent may control a number of non-agent applications which can be legacies or even purpose built. It must therefore provide an interface allowing data, parameters and control specifications to be transmitted to and from the applications. Since an agent may have to integrate more than one different application, a generic application interface and several specific application interfaces may be needed. A generic application interface is desirable to allow integration of diverse applications during the agent's life time.

6.4 AGENT ARCHITECTURES

Many different agent architectures have been described in the literature for agent-based concurrent design and manufacturing systems. In this section, we discuss some of the different approaches, using two types of classification: by behavior and by internal organization.

6.4.1 Classification by Behavior

Agent architectures described in the literature may be classified into the following four categories according to the agent's behavior: deliberative, reactive, collaborative and hybrid architectures[3].

6.4.1.1 Deliberative Architecture

Deliberative agents are also called *cognitive* agents, *intentional* agents, or *goal-directed* agents.

Deliberative agents have domain knowledge and the planning capability necessary to undertake a sequence of actions with the intent of moving towards or achieving a specific goal. Deliberative agents may proactively cooperate with other agents to achieve a given task. They may use any or all of the symbolic artificial intelligence reasoning techniques which have been developed over the past forty years.

Most deliberative agent models are based on Newell and Simon's physical symbol system hypothesis (Newell and Simon, 1976). This assumes that agents which maintain an internal symbolic representation of themselves and their world, will thereby have an explicit mental state which can be modified by some form of symbolic reasoning. For deliberative agents, real-world information must be translated into accurate symbolic knowledge representing real-world entities and processes. Deliberative agents use their symbolic knowledge to infer actions in the real-world environment and to interact with peer agents.

One well-known deliberative agent architecture is the BDI (beliefs, desires, intentions) architecture. The basic idea here is to describe the internal state of an agent by means of a set of mental categories (beliefs, desires and intentions), and to define a control architecture by which the agent rationally selects its course of action based on these representations. The BDI architecture thus contains explicit representations of beliefs, desires and intentions:

-    The beliefs of an agent express its expectations about the current state of the world and about the likelihood of a course of action achieving certain effects. Beliefs may be modeled using possible-worlds semantics, where a

---

[3] This classification is similar to that proposed by Müller (1996). Collaborative architectures, however, are called 'architectures for interacting agents' in this publication.

set of possible worlds is associated with each situation, denoting the worlds that the agent believes to be possible. Other representations for beliefs can also be used. Beliefs may be false.

- Desire is an abstract notion that specifies preferences for future world states or courses of action. An important feature of desire is that an agent is allowed to have inconsistent desires, and that it does not have to believe that its desires are achievable. We do not expect an agent to act on all its desires.
- Intentions are those things the agent is either committed to doing (intending to) or committed to bringing about (intending that). Usually, an agent cannot pursue all its goals or options at once. Even if the set of goals is consistent, it is often necessary to select a certain goal (or a set of goals) to commit to. This process is called the formation of intentions. Thus, the current intentions of an agent can be described by a set of selected goals together with their state of processing.

The Intelligent Resource-bounded Machine Architecture (IRMA) proposed by Bratman et al. (1988) is a classic BDI architecture. It prescribes how an agent selects its course of action based on explicit representations of its perceptions, beliefs, desires, and intentions. The architecture incorporates a number of modules including an intention structure, which is basically a time-ordered set of partial, tree-structured plans, a means-end reasoner, an opportunity analyzer, a filtering process, and a deliberation procedure. As soon as the agent's beliefs are updated by its perceptions, the opportunity analyzer is able to suggest options for action based on the agent's beliefs. Further options are suggested by the means-end reasoner from the current intention structure of the agent. All available options are run through a filtering process where they are tested for consistency with the agent's current intention structure. Finally, options that pass the filtering process successfully are passed to a deliberation process that modifies the intention structure by adopting a new intention, i.e., by committing to a specific partial plan.

6.4.1.2 Reactive Architecture

Because it has been found that many real world problems cannot be tackled satisfactorily by using the symbolic deliberative approach, some researchers have proposed a totally different approach. In this reactive approach, the resulting architectures are called *reactive* architectures (and also known as *behavior-based* or *situated* agent architectures).

Reactive agents respond in an event-condition-action mode. They do not have internal models of the world. They respond solely to external stimuli and the information available from their sensing of the environment (Brooks, 1986). Like neural networks, reactive agents exhibit emergent behavior, which is the result of the interactions of these simple individual agents. When reactive agents interact, they share low-level data, not high-level symbolic knowledge. One of the fundamental tenets of reactive agents is that they are grounded in physical sensor data. They do not use symbolic representations. Application of this architecture has been remarkably successful in fairly simple robots which use sensors to perceive the world.

In reactive agent architectures, deliberative reasoning is replaced by emergent behavior, which adapts to changes in the real-world environment in a timely way. Here, intelligent behavior can be obtained without explicit abstract reasoning.

Three important precepts concerning reactive architectures were put forward by Maes (1990b; 1991). First, the concept of emergent functionality: that the dynamics of behavior interaction lead to complexity of interaction among dissimilar behaviors. This emergent complexity occurs since there is no *a priori* plan to regulate the behavior of reactive agents. Second, a reactive agent can be viewed as a collection of autonomous modules responsible for specific tasks, with a tendency to minimize communication among modules. Third, reactive agents tend to operate on representations near the level of sensor data.

Various architectures have been described for reactive agents. The simplest reactive architecture is provided by a finite number of rules which relate perceptions to actions and provide a means to describe reactions to situations. This approach usually supposes that an agent does not have any memory of past states, and that action comes only from perception. It is also generally assumed that the rules are mutually exclusive and that the system does not contain any rule conflicts. This architecture is very useful for simple behaviors, but it appears to be inadequate for more complex forms of action control. The Production System Agent architecture can be considered as being an extension of this simplest reactive architecture. It was proposed by Ishida (1994) and will be briefly described in Section 6.4.2.4.

The subsumption architecture proposed by Brooks (1986) is one of the earliest of the reactive architectures. In this architecture, multiple behaviors associated with a specific level of activities (e.g., moving an arm or turning a camera) compete to win control over the activities. The subsumption architecture involves a hierarchy of activation, inhibitions, and responses to the external world. This architecture will be detailed later in Section 6.4.2.2.

Other types of reactive agent architectures include those based on task competition, neural networks, and other approaches (Ferber, 1995).

6.4.1.3 Collaborative Architecture

Collaborative agents are also called *interacting* agents or *social* agents.

Collaborative agents work together to solve problems. Communication between agents is an important element, and although each individual agent is autonomous, it is the synergy resulting from their cooperation that makes collaborative agents interesting and useful. Collaborative agents can solve large problems which are beyond the capability of any single agent and allow a modular approach based on specialization of agent functions or domain knowledge. For example, collaborative agents may work as design assistants on large, complex engineering projects.

Individual agents may be called upon to verify different aspects of the design, but their joint expertise is applied to ensure that the overall design is consistent.

Collaborative attitudes are incorporated in such agent systems to facilitate interaction, communication, task decomposition, distribution, cooperation, and negotiation. At the macro-level, collaborative agent systems emphasize the concept of "agency." Agency relates to how agents interact to form subsystems of agents (emergent organizations). At the micro-level agency relates to the characteristics and behaviors of individual agents operating in a multi-agent system. This chapter primarily deals with the micro-level. Issues related to the macro-level will be taken up in the following chapters.

Agents using the Contract Net (Smith, 1980) as the basis for their inter-agent negotiation protocol have a simpler collaborative architecture, whereas agents using extended or modified versions of the Contract Net Protocol (detailed in Chapter 9) have more complex internal architectures.

Agents using the same version of the Contract Net as their negotiation protocol may have different internal architectures composed of different numbers of the modules (components) described in Section 6.3. However they will have similar decision making mechanisms, considered from the agent collaboration point of view. Figure 6.1 shows the decision making mechanisms in a Manager[4] agent and a Bidder agent.

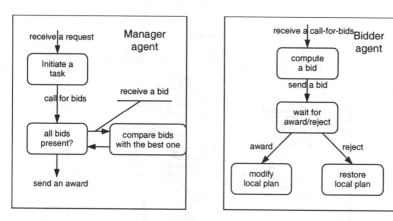

Figure 6.1 A Manager agent and a Bidder agent using the Contract Net Protocol.

Purely collaborative architectures are not very interesting though they may sometimes be useful. However, when collaborative agent architectures incorporate features from other architectures, we move towards more powerful and promising capabilities. Examples of collaborative architectures which incorporate deliberative features are the GRATE* architecture of Jennings (1992), the MECCA agent

---

[4] Using the Contract Net protocol, a Manager agent sends call-for-bids to Bidder agents.

architecture of Steiner (1996), and the COSY agent architecture of Burmeister and Sundermeyer (1992)). Collaborative architectures which incorporate both deliberative and reactive features are usually called hybrid architectures, and are described in the next subsection.

6.4.1.4 Hybrid Architecture

Recent research suggests that neither a completely deliberative agent system nor a totally reactive system can cope with every requirement of a dynamic environment, because purely reactive architectures cannot implement general goal-directed behavior, while purely deliberative architectures have difficulties in large complex systems due to their potentially huge symbolic knowledge representations. As noted in the previous subsection, a collaborative architecture should incorporate other architectural features in order to be useful and effective in industrial application. The foregoing suggests the combination of two or more of the above approaches, and the resulting architectures are called *hybrid* architectures.

Franklin's general agent architecture (Figure 6.2) (Franklin, 1997) can be considered to be a *hybrid* architecture according to this definition. This architecture is constrained by the tenets of the action selection paradigm and by agent design principles proposed by the author for Attention, Internal Models, Coordination and Knowledge.

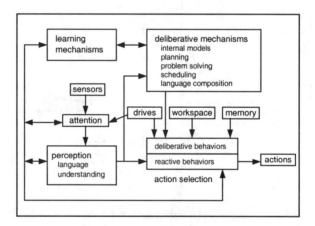

Figure 6.2 A general agent architecture proposed by Franklin (1997).

6.4.2 Classification by Internal Organization

From the internal organization perspective, the following agent architectures can be distinguished: modular, layered, blackboard, subsumption and production system agent architectures.

6.4.2.1 Modular Architecture

A modular agent architecture is organized as an assembly of agent modules (see Section 6.3 for the basic modules needed for agents in agent-based concurrent design and manufacturing systems) with predefined connections. Each module realizes a particular agent function.

Various modular agent architectures described in the literature range from very simple ones that only consists of a few modules to very complex ones involving a large number of modules. Several of the modules described in Section 6.3 can be combined as one 'big' module. One module can also be split into several 'small' modules.

In this type of architecture, all of the connections among the modules are typically fixed, i.e., the information flow in an agent is predefined by the designer. Figure 6.3 shows an example of a simple modular architecture. The signals obtained from outside world through sensors or receptors or from other agents through the agent's message box are received and filtered by the Perception module. The Interpretation module recognizes, translates and decomposes the messages. The Decision Making module makes decisions according to the information received and the agent's own goals. The Planning module carries out the action needed to reach the goals selected. The planning result, i.e., agent's plan, is then transmitted to the Execution module for execution.

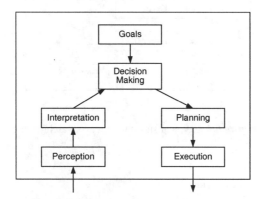

Figure 6.3 A Modular Agent Architecture.

A very complex modular agent architecture can be found in (Moulin and Chaib-draa, 1996). Other examples are the MAGES agent architecture of Bouron et al. (1991), the DIDE agent architecture of Shen and Barthès (1996), and the DESIRE compositional agent architecture of Brazier et al. (1998).

Modular architectures are widely used in multi-agent systems. Most of the deliberative agent architectures (see Section 6.4.1.1) described in the literature are modular.

6.4.2.2 Subsumption Architecture

The subsumption architecture was initially proposed by Brooks (1986) for constructing reactive agents (see Section 6.4.1.2). In fact, it is a specific type of modular architecture having vertically linked modules, each of which is only responsible for a very limited function (behavior).

In a subsumption architecture, a set of vertically arranged modules, represented by augmented finite-state machines, are linked through a master/slave relationship of inhibition. A higher-level module can inhibit the flow of input values into a lower module or the output information sent to actuators. As shown in Figure 6.4, modules are arranged in layers, with each layer being responsible for some level of competence of the agent.

The subsumption architecture has been used to implement a variety of robots with different abilities. It could also be used to implement AGV (automated guided vehicles) and other relatively simple 'robotic' agents in agent-based manufacturing systems.

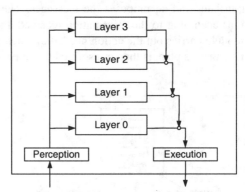

Figure 6.4 A typical subsumption agent architecture.

6.4.2.3 Blackboard Architecture

The blackboard architecture is a classic architecture developed initially within the AI community and used for developing single agent systems. It has also been widely used as the architecture for various multi-agent systems, particularly those involving symbolic deliberative agents as described in Section 6.4.1.1. The well-known DVMT system described in Chapter 2 uses this type of agent architecture, with two blackboards (one for data and the other for the agent's goals) (Lesser and Corkill, 1983). In the MINDS project, Huhns (1987) also used two specialized blackboards.

The traditional blackboard system is distinguished by these three features:

- a global database called the blackboard, which incorporates the partial states of the current problem, hypotheses, and intermediate results, and usually all of the information exchanged among knowledge sources (KSs);
- independent knowledge sources (KSs) which generate solution elements for the blackboard;
- a scheduler to control KS activity.

A key feature of the blackboard architecture is the global database (blackboard) that functions as the memory of the system, where all computational and solution-state data for the KSs are held.

The BB1 architecture is a well-known blackboard architecture developed at the Stanford KSL by Hayes-Roth (1985). It has been used by a number of researchers for developing agent architectures. There are two blackboards in the BB1 architecture: one for control and the other for domain problem solving. These two blackboards provide the only place where the KSs are permitted to communicate and interact while solving a problem.

Traditionally, the blackboard architecture has been used for independent knowledge based systems. More recently, it has also been used for agents. In such cases, the KSs are knowledge based modules as in the modular architecture (Section 6.4.2.1), and one or more blackboards are used for data and knowledge sharing by knowledge base modules, as well as for reasoning based on this knowledge, and for controlling information flow in the agent and its resultant actions.

6.4.2.4 Production System Agent Architecture

The production system agent architecture was proposed by Ishida (1994). A production system agent is essentially a production system capable of interacting with other agents. As shown in Figure 6.5, a production system agent comprises the following three components:

- A *production system interpreter*, which continuously repeats the production cycle. In a collection of distributed production system agents, rules are asynchronously fired by the distributed agents. Since no global control exists, interference among the rules is prevented by local synchronization between individual agents.
- *Domain knowledge*, which consists of domain rules and domain data. Since agents asynchronously fire rules, domain data in different agents can be temporarily inconsistent.
- *Organizational knowledge*, which represents relationships among agents (also called agent-agent relationships). An agent that has such a relationship with a particular agent is called that agent's neighbor. Agent-agent relationships are initially obtained by analyzing domain knowledge at compile time, and are dynamically maintained during reorganization. Since agents asynchronously perform reorganization, organizational knowledge can be temporarily inconsistent across agents.

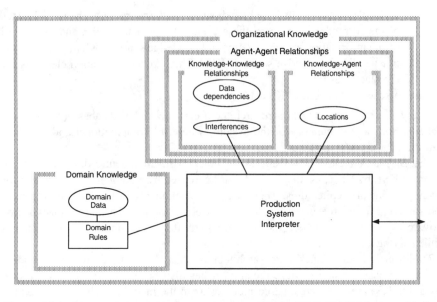

Figure 6.5 Architecture of a production system agent (Ishida, 1994) (©1994 Springer-Verlag).

6.4.2.5 Layered Architecture

Recently, layered agent architectures have been described by a number of researchers. A layered agent architecture is usually organized with components for perception and action at the bottom and reasoning at the top. The perception feeds the reasoning subsystem, which governs the actions, including deciding what to perceive next (Huhns and Singh, 1998). Most layered agent architectures are hybrid architectures as discussed in Section 6.4.1.4.

> Layering is a powerful means for structuring functionalities and control, and thus is a valuable tool for system design supporting several desired properties such as reactivity, deliberation, cooperation, and adaptability. The main idea is to structure the functionalities of an agent into two or more hierarchically organized layers that interact with each other in order to achieve coherent behavior of the agent as a whole.
>
> Müller 1996

An agent with a layered architecture is composed of several modules, with each being structured into a number of layers. Each higher layer typically represents an increased level of abstraction in terms of knowledge, so as to facilitate reasoning and control in this agent.

The INTERRAP agent architecture described by Müller (1996) is a typical layered architecture (see Figure 6.6).

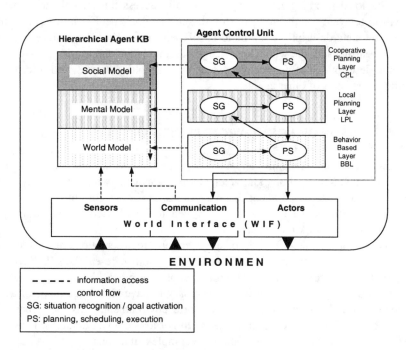

Figure 6.6 The INTERRAP agent architecture (Müller, 1996) (©1996 Springer-Verlag).

An INTERRAP agent consists of three modules: a World Interface (WIF), a Knowledge Base (KB), and a Control Unit (CU). The world interface provides the agent's perception, communication, and action links to its environment. The KB and the CU are each structured in three layers. The Control Unit layers are the Behavior-Based Layer (BBL), the Local Planning Layer (LPL), and the Cooperative Planning Layer (CPL). Each of these consists of two processes: SG (situation recognition and goal activation) and PS (planning, scheduling, and execution). The knowledge base is partitioned into a World Model (WM), a Mental Model (MM), and a Social Model (SM).

This is a vertically layered architecture: control is triggered by the behavior-based layer, which has immediate access to the agent's sensory system, and is moved upward incrementally until a suitable layer competent for execution has been found. From there, control activity spreads back downward to the behavior-based layer, which is the only layer with direct access to the actor functions defined in the agent's world interface.

Each Control Unit layer is allowed to access a specific portion of the agent KB. This access is organized incrementally: each control layer may use information stored in its corresponding KB layer, and in lower KB layers, but is not allowed to access information stored in higher layers. Thus, the behavior-based layer has access only to the world model part of the KB containing the agent's object-level beliefs about the world. The local planning layer may additionally access the mental model where information relevant for planning is stored, such as goal-specific information, plan libraries, or meta-descriptions of patterns of behaviors. The cooperative planning layer may reason using the whole knowledge base, including the agent's social model containing descriptions of negotiation protocols, joint plan libraries, and information referring to the goals of other agents.

6.5 COMPARISON OF AGENT ARCHITECTURES

Deliberative symbolic reasoning agents with large real-world environments require correspondingly large symbolic knowledge bases. Reasoning with this large symbolic knowledge causes increased agent response times. However, for small, well-known domains, the deliberative approach has proven to be efficient at adapting agent behavior to specific working domains.

Reactive agent architectures do not use symbolic representations and have rapid response times. For reactive agents, Brooks and other researchers believe that intelligent behavior can be generated without explicit symbolic-level representations. In other words, they argue that it is not necessary to have elaborated plans of action for agents to efficiently adapt to the real-world environment. This thesis has been shown to be true for relatively simple agents in not too complex domains. It remains to be seen how well it holds up in much more complex situations.

Collaborative agent architectures facilitate interaction, communication, task decomposition, distribution, cooperation, and negotiation among intelligent agents. However, it has been found that collaborative architectures only become really useful for real world application when they incorporate other features such as deliberation and reaction.

According to Müller (1996), layered hybrid architectures offer the following advantages:

- They allow modularization of an agent; different functionalities are clearly separated and linked by well-defined interfaces.
- They make the design of agents more compact, increase robustness, and facilitate debugging.
- Since different layers may run in parallel, the agent's computational capability can be in principle increased by a linear factor.
- The agent's reactivity can be maintained or increased in parallel operation: while planning, a reactive layer can still monitor the world for contingency situations.
- Since different types and partitions of knowledge are required for the implementation of different functionalities, it is often possible to restrict the amount of knowledge an individual layer needs to consider in order to make its decisions. This restriction allows the agent to operate more efficiently and, in practice, can help increase it reactivity.

6.6 RESEARCH LITERATURE AND FURTHER REFERENCES

6.6.1 Research Literature

The following section is provided for those requiring more information on this topic, from the research literature.

6.6.1.1 Earlier Research on Planning Agents

Artificial agents originated in the early 1970s in the AI community. One aim in artificial agent design was to provide an automatic programming methodology for artificial intelligence planning. In these early planning systems, the real world and the goals of the system were represented through symbolic descriptions, as in the STRIPS planning system (Fikes and Nilsson, 1971). This system matched the pre- and post-conditions of actions to achieve its goals. STRIPS was a simple system and ineffective in dealing with highly complex problems. More powerful techniques were subsequently developed for other areas of planning, such as hierarchical and nonlinear planning (Sacerdoti, 1975). By the mid-1980s, however, Chapman showed that some of these theoretically based techniques are unusable in any time-constrained system (Chapman, 1987).

Various attempts were also made to construct planner-based agents which could cope with non-linear planning. For example, the Integrated Planning, Execution and Monitoring (IPEM) system was based on a nonlinear planner (Ambrosio-Ingerson and Steel, 1988).

6.6.1.2 Deliberative Agent Architectures

As noted in Section 6.4.1.1, most deliberative agent models are based on Newell and Simon's physical symbol system hypothesis (Newell and Simon, 1976). Recently, research on deliberative agents has explored the modeling of agents based on beliefs, desires, and intentions. Architectures following this paradigm have become known as BDI (Belief, Desire, and Intention) architectures. Basic concepts of the BDI architectures date back to Bratman (1987) and to the Procedural Reasoning System (PRS) by Georgeff and Lansky (1986). Subsequently, this has become a strong research area for agent design, see e.g., (Rao and Georgeff, 1991; 1992; 1995).

6.6.1.3 Reactive Agent Architectures

In the mid-1980s, guided by researchers such as Brooks (Brooks, 1986), Chapman (1987), Kaelbling (1990), and Maes (1990a), architectures were developed for agents that were called behavior-based, situated, or reactive. These agents make their decisions at run-time, usually based on a very limited amount of information, and using simple situation-action rules or the equivalent. Some researchers, most notably Brooks, denied the need for any symbolic representation of the world; instead, reactive agents make decisions directly based on sensory input. The design of reactive architectures is partly guided by Simon's hypothesis (Simon, 1981) that the complexity of the behavior of an agent can be a reflection of the complexity of the environment in which the agent is operating rather than a reflection of the agent's complex internal design. The focus of this class of systems is directed towards achieving robust behavior instead of correct or optimal behavior.

6.6.1.4 Collaborative Agent Architectures

Research in this area has been primarily focused on coordination and cooperation among distributed intelligent agents. Such research is found in: contract network coordination (Contract-Net) (Smith, 1980); the Mid-deck Active Experiment MACE (Gasser et al., 1987); the Distributed Monitoring Vehicle Testbed (DMVT) (Lesser, 1990); the Michigan Intelligent Coordination Experiment (MICE) (Durfee and Montgomery, 1989); MCS (Doran et al., 1991); and MAS/DAI planning and game theories (Rosenchein and Zlotkin, 1994). Recent research on incorporating cooperative abilities into an agent framework includes: the MAGSY agent architecture by Fischer (1993); the GRATE* architecture by Jennings (1992); the MECCA agent architecture by Steiner (1996); and the COSY agent architecture by Burmeister and Sundermeyer (1992).

6.6.1.5 Hybrid Agent Architectures

Hybrid agent architectures are relatively new architectures proposed for reconciling deliberation and reaction. In addition to the INTERRAP agent architecture described in Section 6.4.2.5, other examples of layered hybrid agent architectures include: the integrated agent architecture of Hayes-Roth (1991); the planner-reactor architecture of Lyons and Hendriks (1992); the Touring Machines architecture of Ferguson (1992); the 3T architecture of Bonasso et al. (1996); and the SIM_AGENT agent architecture of Sloman and Poli (1996).

6.6.2 Further References

An extensive and detailed review of significant agent architectures can be found in (Müller, 1996). A bibliography on agent theories, architectures and languages can be found in (Wooldridge et al., 1996b).

The ATAL (Agent Theories, Architectures and Languages) workshops have included presentations on many agent architectures proposed or developed. These workshop notes have been published by Springer-Verlag in its LNAI (Lecture Notes in Artificial Intelligence) series (e.g., Wooldridge and Jennings, 1995; Wooldridge et al., 1996a; Müller et al., 1997; Singh et al., 1998; Müller et al., 1999)[5]. Other

---

[5] The ATAL99 (and following workshops) proceedings are not yet available from Springer-Verlag in its LNAI series at the time of writing this book.

references to agent architectures can be found in the conference proceedings of ICMAS (published by The AAAI Press/The MIT Press), Autonomous Agents (published by the ACM Press), and the MAAMAW workshops (the first three volumes were published by North-Holland and the following volumes published by Springer-Verlag).

REFERENCES

Ambros-Ingerson, J. and Steel, S. (1988) Integrating planning, execution and monitoring, In *Proceedings of the Seventh National Conference on Artificial Intelligence*, St. Paul, MN, pp. 83-88.

Bonasso, R.P., Kortenkamp, D., Miller, D.P. and Slack, M. (1996) Experiences with an architecture for intelligent, reactive agents, In Wooldridge, M.J., Müller, J.P. and Tambe, M. eds., *Intelligent agents II: Agent Theories, Architectures, and Languages*, Lecture Notes in Artificial Intelligence 1037, Springer, pp. 187-202.

Bouron, T., Ferber, J. and Samuel, F. (1991) MAGES: A multi-agent testbed for heterogeneous agents, In Demazeau, Y. and Müller, J.-P. eds., *Decentralized Artificial Intelligence* 2, Elsevier/North-Holland, Amsterdam, pp. 195-216.

Bratman, M.E. (1987) *Intentions, Plans, and Practical Reason*, Harvard University Press.

Bratman, M.E., Israel, D.J. and Pollack, M.E. (1988) Plans and resource-bounded practical reasoning, *Computational Intelligence*, 4(4):349-355.

Brazier, F., Jonker, C., Treur, J. and Wijngaards, N. (1998) Compositional Design of Generic Design Agent, In *Proceedings of AI & Manufacturing Research Planning Workshop*, Albuquerque, NM, The AAAI Press, pp. 30-39.

Brooks, R.A. (1986) A robust layered control system for a mobile robot, *IEEE Journal of Robotics and Automation*, 2(1):14-23.

Burmeister, B. and Sundermeyer, K. (1992) Cooperative problem-solving guided by intentions and perception, In Demazeau, Y. and Werner, E. eds., *Decentralized AI 3*, North-Holland.

Chapman, D. (1987) Planning for conjunctive goals, *Artificial Intelligence*, 32:333-378.

Doran, J., Carvajal, H., Choo, Y. and Li, Y. (1991) The MCS multi-agent testbed: Developments and experiments, In Deen, S. ed., *Cooperative Knowledge Based Systems*, Springer-Verlag, Heidelberg, 240-251.

Durfee, E.H. and Mongomery, T.A. (1989) MICE: A flexible testbed for intelligent coordination experiments, In *Proceedings of the Distributed Artificial Intelligence Workshop*, pp. 25-40.

Ferber, J. (1995) *Les systèmes multi-agents*, InterEditions, Paris.

Ferguson, I.A. (1992) Toward an architecture for adaptive, rational, mobile agents, In Werner, E. and Demazeau, Y. eds., *Distributed Artificial Intelligence 3*, Elsevier/North-Holland, Amsterdam, pp. 249-261.

Fikes, R.E. and Nilsson, N.J. (1971) STRIPS: A new approach to the application of theorem proving to problem solving, *Artificial Intelligence,* 2:189-208.

Fischer, K. (1993) The Rule-Based Multi-Agent System MAGSY, In *Proceedings of the CKBS'92 Workshop*, Keele University, UK.

Franklin, S. (1997) Autonomous Agents as Embodied AI, *Cybernetics and Systems*, 28(6):499-520.

Gasser, L., Braganza, C. and Herman, N. (1987) MACE: A flexible testbed for distributed AI research, In Hunhs, M. ed., *Distributed Artificial Intelligence, Research Notes in Artificial Intelligence*, Pitman, London, 119-152.

Georgeff, M.P. and Lansky, A.L. eds. (1986) Reasoning about actions and plans, In *Proceedings of the 1986 Workshop*, Morgan Kaufmann Publishers, San Mateo, CA.

Hayes-Roth, B. (1985) A Blackboard Architecture for Control, *Artificial Intelligence*, 26(3):251-321.

Hayes-Roth, B. (1991) An integrated architecture for adaptive intelligent systems, *Artificial Intelligence*, 72(1-2):329-365.

Huhns, M.N. ed. (1987) *Distributed Artificial Intelligence*, Morgan Kaufmann, San Mateo, CA/Pitman, London.

Huhns, M.N. and Singh, M.P. (1998) Agents and Multiagent Systems: Themes, Approaches, and Challenges, In Huhns, M.N. and Singh, M.P. eds., *Readings in Agents*, Morgan Kaufmann Publishers, Inc., San Francisco, CA, pp. 1-24.

Ishida, T. (1994) *Parallel, Distributed and Multiagent Production Systems*, Lecture Notes in Artificial Intelligence 878, Springer-Verlag.

Jennings, N.R. (1992) Towards a cooperation knowledge level for collaborative problem solving, In *Proceedings of the 10th European Conference on Artificial Intelligence*, Vienna, pp. 224-228.

Kaelbling, L.P. (1990) An architecture for intelligent reactive systems, In Alien, J., Hendler, J. and Tate, A. eds., *Readings in Planning*, Morgan Kaufmann, pp. 713-728.

Lesser, V.R. and Corkill, D.D. (1983) The distributed vehicle monitoring testbed: A tool for investigating distributed problem solving networks, *AI Magazine*, 4(3):15-33.

Lesser, V.R. (1990) An overview of DAI: Viewing distributed AI as distributed search, In Nakano, R., and Doshita, S. eds., *Journal of Japanese Society for Artificial Intelligence*, Special Issue on Distributed Artificial Intelligence, 5(4):392-400.

Lyons D.M. and Hendriks. A. J. (1992) A practical approach to integrating reaction and deliberation, In *Proceedings of the 1st International Conference on AI Planning Systems (AIPS)*, Morgan Kaufmann, San Mateo, CA, pp. 153-162.

Maes, P. (ed.), (1990a) *Designing Autonomous Agents: Theory and Practice from Biology to Engineering and Back*, MIT/Elsevier.

Maes, P. (1990b) Situated agents can have goals, *Robotics and Autonomous Systems*, North-Holland, 6:49-70.

Maes, P. (1991) The agent network architecture (ANA), *SIGART Bulletin*, 2(4):115-120.

Minsky, M.L. (1987) *The Society of Mind*, William Heinemann Ltd., London.

Moulin, B. and Chaib-draa, B. (1996) An Overview of Distributed Artificial Intelligence, In O'Hare, G.M.P. and Jennings, N.R. eds., *Foundations of DAI*, Wiley & Sons, pp. 1-56.

Müller, J. P. (1996) *The Design of Intelligent Agents: A Layered Approach*, Lecture Notes in Artificial Intelligence 1177, Springer.

Müller, J.P., Wooldridge, M.J. and Jennings, N.J. eds. (1997) *Intelligent Agents III: Agent Theories, Architectures, and Languages: ECAI '96 Workshop (ATAL96)*, Budapest, Hungary, Lecture Notes in Artificial Intelligence 1193, Springer.

Müller, J.P., Singh, M.P. and Rao, A.S. (1999) Intelligent Agents V: Agent Theories, Architectures, and Languages: ATAL'98, Paris, France, Lecture Notes in Artificial Intelligence 1555, Springer.

Newell, A. and Simon, H.A. (1976) Computer science as empirical enquiry: Symbols and search, *Communications of the ACM*, 19(3):113-126.

Rao, A.S. and Georgeff, M.P. (1991) Modeling Agents within a BDI Architecture, In *Proceedings of the 2nd International Conference on Principles of Knowledge Representation and Reasoning (KR'91)*, Cambridge, MA, Morgan Kaufmann, pp. 473-484.

Rao, A.S. and Georgeff, M.P. (1992) An abstract architecture for rational agents, In *Proceedings of the 3rd International Conference on Principles of Knowledge Representation and Reasoning (KR'92)*, Morgan Kaufmann, pp. 439-449.

Rao, A.S. and Georgeff, M.P. (1995) BDI-agents: from theory to practice, In *Proceedings of the First International Conference on Multi-Agent Systems*, San Francisco, CA, The AAAI Press/MIT Press, pp. 312-319.

Rosenschein, J.S. and Zlotkin, G. (1994) *Rules of Encounter: Designing Conventions for Automated Negotiation among Computers*, The MIT Press.

Sacerdoti, E. (1975) The non-linear nature of plans, In *Proceedings of the Fourth International Joint Conference on Artificial Intelligence (IJCAI-75)*, Stanford, CA, pp. 206-214.

Shen, W. and Barthès, J.P. (1996) An Experimental Multi-Agent Environment for Engineering Design, *International Journal of Cooperative Information Systems,* 5(2-3):131-151

Simon, H.A. (1981) *The Science of the Artificial*, MIT Press, Cambridge, MA, 2nd edition.

Singh, M.P., Rao, A. and Wooldridge, M.J. eds. (1998) *Intelligent Agents IV: Agent Theories, Architectures, and Languages: 4th International Workshop, ATAL'97*, Providence, RI, USA, Lecture Notes in Artificial Intelligence 1365, Springer.

Sloman A. and Poli, R. (1996) SIM_AGENT: A tookit for exploring agent designs, In Wooldridge, M.J., Müller, J.P. and Tambe, M. eds., *Intelligent agents II: Agent Theories, Architectures, and Languages*, Lecture Notes in Artificial Intelligence 1037, Springer, pp. 392-407.

Smith, R.G. (1980) The contract net protocol: high-level communication and control in a distributed problem solver, *IEEE Transaction on Computers*, 29(12):1104-1113.

Steiner, D.D. (1996) IMAGINE: An Integrated Environment for Constructing Distributed Artificial Intelligence Systems, In O'Hare, G.M.P. and Jennings, N.R., eds., *Foundations of DAI*, Wiley & Sons, pp. 345-364.

Turner, R.M. (1994) *Adaptive Reasoning for Real-World Problems: A Schema-Based Approach*, Lawrence Erlbaum Associates, Hillsdale, NJ.

Wooldridge, M.J. and Jennings, N.J. eds. (1995) *Intelligent Agents: ECAI-94 Workshop on Agent Theories, Architectures, and Languages*, Amsterdam, the Netherlands, Lecture Notes in Artificial Intelligence 890, Springer.

Wooldridge, M.J., Müller, J.P. and Tambe, M. eds. (1996a) *Intelligent agents II: Agent Theories, Architectures, and Languages,* Lecture Notes in Artificial Intelligence 1037, Springer.

Wooldridge, M.J., Müller, J.P. and Tambe, M. (1996b) Agent Theories, Architectures and Languages: A Bibliography, In Wooldridge, M.J., Müller, J.P. and Tambe, M. eds., *Intelligent agents II: Agent Theories, Architectures, and Languages*, Lecture Notes in Artificial Intelligence 1037, Springer, pp. 408-431.

7

System Architectures

7.1 INTRODUCTION

If we consider the architecture of an agent as being the *pattern of relationships among the agent's modules*, then the multi-agent system architecture can be correspondingly expressed as the *pattern of relationships among the agents*.

Agent architectures look at agents in the small. It is often more important to consider agents in the large. This is where agent system architectures come in.

Huhns and Singh, 1998

In the previous chapter, we introduced agent architectures and discussed the issues related to the architectural design of individual agents. However, individual agents require the support of an overall system within which to interact. This system must provide a means for agents to communicate with one another, and define protocols for the resulting conversations. This chapter focuses on how agents are configured relative to one another. Other issues related to communication, cooperation, coordination, and negotiation will be discussed in the following chapters.

7.2 ORGANIZATION AND SYSTEM ARCHITECTURES

Researchers in DAI usually distinguish an organization from a structure (a system architecture) (Moulin and Chaib-draa, 1996). A multi-agent system structure is the pattern of information and control/collaboration relationships among agents, as well as the distribution of problem-solving capabilities among them. A structure gives a high-level view of how the group solves problems and the role that each agent plays within this structure. In DAI, an organization views the system from a rather different perspective which is more related to organization theory. Malone proposed this definition of organization:

A group of agents is an organization if they are connected in some way (arranged systematically) and their combined activities result in something better (more harmonious) than if they were not connected. An organization consists of: a group of agents; a set of activities performed by agents; a set of connections among agents; and a set of goals or evaluation criteria by which the combined activities of the agents are evaluated.

Malone, 1990

A useful introduction and syntactical discussion on multi-agent system organization (as well as organization within an individual agent) can be found in (Ferber, 1995).

Because this book is concerned with practical applications of agent technology to industrial problems, we primarily focus on structural relationships among agents from a practical point of view.

7.3 CONTROL RELATIONSHIPS VS. COLLABORATION RELATIONSHIPS

Control relates to the degree of autonomy which an agent possesses. An agent whose goals and/or plans and/or actions are prescribed by the imperatives of another agent or agents has little autonomy. The extent of its autonomy may only be in choosing how the goals are to be met or the plans implemented or when the prescribed actions are to be taken. Such an agent is highly controlled. This situation can be observed in hierarchical systems where 'master-slave' relationships may be strongly enforced.

In collaboration, however, the agents involved are free to accept or reject (or modify) those goals, plans, or actions proposed to them. Commonly, they will possess the persistent goal of collaborating with other agents in so far as this does not conflict with other goals they have. Clearly, to collaborate, agents must posses at least some degree of autonomy. Control is concerned with what an agent or other entity is required to comply with. Collaboration focuses on the positive responses of an agent to proposals received or to situations in which it perceives it could assist.

In theory, a truly open multi-agent system need not have any predefined global control. However, a real world industrial application, particularly a real-time manufacturing system, usually needs some degree of global control to ensure its stability. Of course, control mechanisms are also needed inside each agent as described in the previous chapter, as well as for controlling communications among a group of agents.

An example of a fully controlled system is a centralized blackboard system (see Section 7.3.1). Fully controlled systems have found widespread application in industry, but lack flexibility and are difficult to reconfigure. A fully collaborative system, however, is not practical in industry. Most agent-based concurrent design and manufacturing systems are between these two extremes, using some degree of control to coerce the system towards desired goals, but within these constraints, facilitating collaboration because of its benefits in load balancing, fault recovery, flexibility, reconfiguration, and other aspects of performance.

7.3.1 Control in Blackboard Systems

In the literature of concurrent design and manufacturing systems, some researchers have considered their blackboard systems to be multi-agent systems, and others have implemented their agent-based systems using blackboard architectures. Therefore, we give also a brief discussion on control in such systems.

Control can be strongly implemented in blackboard systems. In these systems, control is centralized, but can be implemented in different ways: using a special control expert called *supervisor* as in EXPORT (Monceyron and Barthès, 1992); combining multiple communication modes as in DICE (Sriram et al., 1992); using a shared graphic model as in ICM (Fruchter et al., 1993); or a shared database as in SHARED (Wong and Sriram, 1993); or a shared object model with a system management module as in KAD (Gupta et al., 1996); or through multiple shared workspaces as in MATE (Saad and Maher, 1996).

7.3.2 Local Control vs. Global Control

Local control in an individual agent has been discussed in previous chapter. Here we discuss issues related to the local control of a group of agents and to the global control of the whole multi-agent system.

Agents can be grouped, in which case communication within the group (and between groups) can be controlled by letting a specific agent assume the function of a 'controller'. In the PACT project, this role is played by agents called facilitators (Cutkosky et al., 1993); in CONDOR by collective agents (Iffenecker, 1994); in ABCDE by managers (Balasubramanian and Norrie, 1995); and in MetaMorph by mediators (Maturana and Norrie, 1996). Detailed descriptions of these and other approaches are given in Section 7.4.2. When the controller is used to federate simple agents which provide elementary functionalities, the group of agents can be considered as being a single macro-agent.

As stated earlier, there is often little or no global control in multi-agent systems, particularly in open agent-based concurrent design systems. However, real world agent-based manufacturing systems, particularly manufacturing control systems, usually need some degree of global control. Such systems often use hierarchical architectures (detailed in Section 7.4.1), or partial hierarchical architectures (Brennan and Norrie, 1998).

An exceptional example of global control in multi-agent design system is the ANARCHY project at CMU (Quadrel et al., 1993). The cooperative (building) design environment in ANARCHY is called Asynchronous Team Design Environment. Although some researchers would classify Asynchronous Teams as being simply a particular type of cooperative software environment, we nevertheless consider it to be another example of a multi-agent system, because of its autonomous agents, broadcast communication and concurrent/asynchronous execution. ANARCHY has a global design strategy which uses simulated annealing. However, the classical formulation of simulated annealing had to be modified to make it suitable for the asynchronous design environment. It would be interesting to verify whether this approach can be adapted to other design problems, such as in mechanical design.

7.3.3 Collaboration Relationships vs. Control Relationships

Collaboration takes place through communication and involves coordinating plans and actions among agents. Collaboration may take place only over a portion of the system. Depending on the type of collaboration in each portion (i.e. group of agents), we may find we have one or other of the different types of federated architectures detailed in Section 7.4.2 and elsewhere. Collaboration can, however, be across the whole system in which case we have an autonomous agent system architecture as described in Section 7.4.3 or some other collaborative system architecture. All of these collaborative systems may have static coordination relationships which are relatively unchanging, or have dynamic coordination relationships changing with time.

We distinguish two types of agent relationships, i.e., control relationships and collaboration relationships, depending on whether the focus is:

- Control (in which one agent prescribes or restricts the actions of another agent) or;
- Collaboration (in which agents coordinate their plans and actions to better satisfy their individual goals and/or global system objectives).

Both kinds of relationships can exist in the same system. For example, there could be a hierarchy of subsystems, with each 'parent' sub-system exerting a high degree of control of its children sub-systems, but inside each subsystem the architecture could be based on collaboration. One can also consider architectures from the point of view of how much they change during run-time. Both of the above kinds of relationship can be static or dynamic. For the dynamic case, the question is how much of the system changes and how quickly. A system could be locally dynamic, i.e., only some subsystems of agents have changing relationships, or it could be globally dynamic, in which case all agents in the system may be continually changing their relationships. Adaptive architectures like MetaMorph (Maturana et al., 1999) may be wholly dynamic, in this sense.

7.4 AGENT SYSTEM ARCHITECTURES

Designing an agent community for a particular real-world industrial application often requires a very early commitment to many of the most important parameters, particularly in the choice of infrastructure[1] technologies. Understanding the

---

[1] Agent system architectures provide the organizing frameworks within which agents are designed and constructed, while an infrastructure provides services that are available to the agents as they operate.

downstream ramifications of these early commitments, at the time the decisions are made is often difficult. The following subsections provide a brief overview of several common approaches that designers have used for building agent based systems. Note that we do not intend to describe all existing system architectures for agent-based systems, but only several typical architectures that have been widely used, especially for agent-based concurrent design and manufacturing systems.

Among the key issues related to system architecture is the level of support for abstraction of agent names and physical addresses. Some architectures allow content-based routing of messages, and others have the routing done through centralized routers. All must let information be shared. This includes a commitment to shared knowledge representations and to communication protocols or translation mechanisms. Some architectures store the shared domain-level information centrally, and others distribute or duplicate such information throughout the agents' local databases. Each architecture has particular strengths: for a specific application, choosing the right one involves matching requirements to capabilities. We explore these issues later in Section 7.5.

In this book, we classify agent-based system architectures into three categories: hierarchical architectures, federated architectures and autonomous agent system architectures.

7.4.1 Hierarchical Architectures

A typical modern manufacturing enterprise consists of a number of physically distributed, semi-autonomous units, each with a degree of control over local resources and with different information requirements. In this situation, a number of practical agent-based industrial applications still use the hierarchical architecture, though it may be criticized for its centralized character and the well-known disadvantages of centralized systems.

Following the traditional hierarchical organization of a manufacturing enterprise, a common hierarchical multi-agent architecture for agent based concurrent design and manufacturing systems is shown in Figure 7.1.

The highest level corresponds to the manufacturing enterprise, which may be implemented as a centralized node composed of various databases and knowledge bases, with monitoring, reasoning, decision making and control mechanisms. Subsystem managers (e.g., R&D department managers, service department, and shop floor managers) are subordinate to this central node. Each subsystem manager coordinates a group of resource agents. Such resource agents may be manufacturing devices, human beings (managers, administrators, designers, production engineers, operators, etc.), CAD/CAM tools, or other software and hardware. Depending on the type of resource, the manager may be designated as a 'Department' or 'Shop Floor' Manager. Other names are also used, depending on the particular corporation.

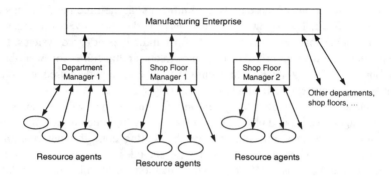

Figure 7.1 A general hierarchical architecture for agent-based design and manufacturing systems.

Figure 7.1 shows only a simple three-layer hierarchical architecture, but the architecture for an agent-based design and manufacturing system may involve many layers.

As an example of a hierarchical architecture for agent-based industrial systems, we give a brief description of the multi-agent system architecture of ADEPT (Advanced Decision Environment for Process Tasks) (Norman et al., 1997). The ADEPT multi-agent architecture consists of a number of autonomous agencies. A single agency consists of, a possibly empty, set of subsidiary agencies, and is represented by a responsible agent. An agency is recursively defined: an agency consists of single responsible (or controlling) agent, a set (possibly empty) of tasks that the responsible agent can perform, and a set (possibly empty) of sub-agencies (see Figure 7.2). Any responsible agent may be approached by other autonomous agents for the provision of a "service". The architecture can model either a hierarchical or a flat organizational structure, or a mixture of the two.

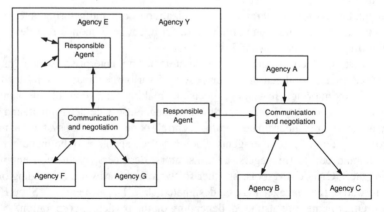

Figure 7.2 The logical hierarchy of agencies in ADEPT (Norman et al., 1997) (©1997 Springer-Verlag).

It should be noted that some aspects of the ADEPT architecture are similar to those of federated architectures described in the following subsection.

7.4.2 Federated Architectures

Because hierarchical architectures suffer from serious problems associated with their centralized character, federated multi-agent architectures are increasingly being considered as a compromise solution for industrial agent-based applications, especially for large-scale engineering applications. A fully federated agent-based system has no explicit shared facility for storing active data; rather, the system stores all data in local databases and handles updates and changes through message passing.

Different federated architectures have been proposed in the literature for agent-based concurrent design and manufacturing systems. Here, we introduce several widely accepted and used approaches.

7.4.2.1 The Facilitator Approach

The facilitator approach was proposed by the SHADE project (McGuire et al., 1993). In this approach, several related agents are combined into a group (Figure 7.3) and communication between the agents always takes place through an interface called the *facilitator*. Facilitators provide a reliable network communication layer, route messages among agents based on the content of the messages, and coordinate the control of multi-agent activities.

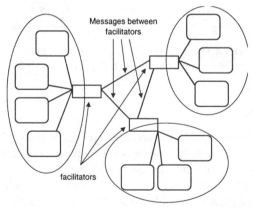

Figure 7.3 Federation multi-agent architecture using facilitators.

Usually, each facilitator is also responsible for providing an interface between a local collection of agents and remote agents, serving four purposes:

(1) providing a layer of reliable message-passing;
(2) routing outgoing messages to the appropriate destinations;

(3) translating incoming messages for consumption by its agents; and

(4) initializing and monitoring the execution of its agents.

In this type of system, communication and coordination occur between agents and facilitators and among facilitators, but not directly between agents. Messages from agents to facilitators are undirected, i.e., they have content but no address. It is the responsibility of the facilitators to route such messages to agents able to handle them. In performing this task, facilitators can go beyond simple pattern matching — they can translate messages, decompose problems into sub-problems, and schedule the work on those sub-problems. The routing can be done interpretively (with messages going through the facilitator) or it can be done by the facilitator setting up specialized links between the individual sending and receiving agents and then stepping out of the picture.

The Desktop Utility agent system architecture proposed by Cranefield and Purvis (1997) improves on the basic facilitator architecture by adding two specialized agents: a planning agent and a console agent (see Figure 7.4). Now, if we integrate the functionality of the planning agent and the console agent into the facilitator, we move close to the mediator approach described in Section 7.4.2.4.

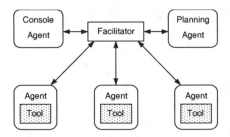

Figure 7.4 The desktop utility agent system architecture (Cranefield and Purvis, 1997) (©1997 Springer-Verlag).

7.4.2.2 The Broker Approach

The brokering mechanism consists of receiving a request message from an intelligent agent, understanding the request, finding suitable receptors for the message, and multicasting the message to the selected group of agents.

Brokers (also called broker agents) have been used in a number of agent based concurrent design and manufacturing systems (Park et al., 1993; Gray et al., 1998; Peng et al., 1998). Broker agents may have somewhat different functions in different agent based systems, Figure 7.5 shows a general system architecture using broker agents.

By comparing Figure 7.3 and Figure 7.5, it is easy to see the functional differences between a facilitator and a broker: a facilitator is responsible for a selected group of agents, while all agents in the system may contact all brokers in the same system for finding service agents to carry out a particular task. Usually, in addition to the functions provided by a facilitator, a broker provides some additional functions such as monitoring and notification.

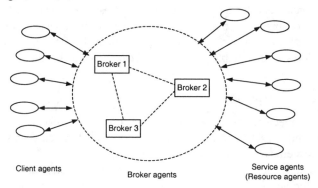

Client agents   Broker agents   Service agents
                                                                  (Resource agents)

Figure 7.5 A general system architecture using broker agents.

7.4.2.3 The Matchmaker Approach

The matchmaking mechanism is a superset of the brokering mechanism, since it uses the brokering mechanism to match agents. However, once appropriate agents have been found, these agents can be directly linked (see Figure 7.6). The matchmaking agent (also called matchmaker) can then step out of the scene to let the agents proceed with their communication themselves.

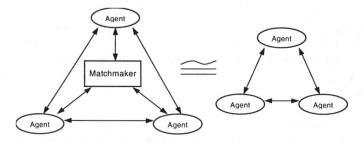

Figure 7.6 Matchmaking mechanism.

The matchmaking mechanism is very useful for information retrieval (searching) within the Internet (Decker et al., 1997). According to Decker et al (1996), "the process of matchmaking allows one agent with some objective to learn the name of another agent that could take on that objective" and "a matchmaker is an agent that knows the names of many agents and their corresponding capabilities." So, the main functionality of a matchmaker is to establish links between client agents (also called requesters, users, etc.) and service agents (also called providers, servers, etc.). After such links are established, the client agents may communicate directly with the service agents without further contacting the matchmakers. Note that the so-called 'yellow-pages agent' is a form of matchmaker. In this case, however, there will be some explicit mechanism which allows agents to register their services on the yellow pages. Agents searching for particular services can then check with the yellow pages agent for the addresses of suitable providers and can then check with them directly.

7.4.2.4 The Mediator Approach

The Mediator architecture draws from approaches used in the blackboard system, the Contract-Net (Smith, 1980), the non-explicitly coordinated systems, as well as from the supervisor system (Kirn and Schneider, 1992).

As described in Section 7.4.2.1, one of the advantages of federation architectures is that many heterogeneous intelligent agents can be statically or dynamically associated within a large distributed multi-agent system (their umbrella organization). Facilitator agents act as message redirectors to interlink dissimilar agents for communication. The Mediator agent (Wiederhold, 1992[2]; Gaines et al., 1995; Luck and d'Inverno, 1995; Ayala and Yano, 1996) has considerable similarity to the facilitator, broker, and matchmaker, but is distinguished by its (additional) coordination role. In this particular type of federation organization, intelligent agents still link to mediator agents to find other agents in the environment. Additionally, however, mediator agents assume the role of system coordinators by promoting cooperation among intelligent agents and possibly also learning from the agents' behavior.

Mediation Behavior

Mediator agents typically provide system associations without interfering with lower-level decisions unless critical situations occur. Mediator agents are able to expand their coordination capabilities to include mediation behaviors, which may focus on high-level policies for situations such as breaking decision deadlocks. Mediation actions are performance-directed behaviors.

Collaborative Clusters

Mediator agents can use brokering and matchmaking mechanisms (see Sections 7.4.2.2 and 7.4.2.3) to find appropriate agents when establishing collaborative subsystems (also called coordination clusters[3]). To efficiently use these mechanisms, mediator agents need to have sufficient organizational knowledge to match agent

---

[2] Note that the notion of Mediator here in federated agent-based concurrent design and manufacturing systems is somewhat different from that proposed by Wiederhold (1992), which is originally designed for distributed database management systems.

[3] In the classical DAI literature, such clusters are called *coalitions*. In more recent literature, the terms cluster or group are now often used instead.

requests with needed resources. Organizational knowledge at the mediator level is often basically just a list of agent-to-agent relationships that is dynamically enlarged.

The brokering and matchmaking mechanisms can be used to generate two relevant types of collaboration subsystems. The first corresponds to an indirect collaboration subgroup, since the requester agent does not need to know about the existence of the other agents that temporarily match the queries. The second type is a direct collaboration subgroup, since the requester agent is informed about the presence and physical location of matching agents so can continue with direct communication.

Common additional activities for mediator agents involved in either type of collaboration are interpreting messages, decomposing tasks, and providing processing times for each new subtask. These capabilities make mediator agents very important elements in achieving the working integration of dissimilar intelligent agents.

7.4.2.5 Other Federated Architectures

Middle agents
Decker et al (1997) put forward the concept of *middle agents* which are used to "support the flow of information in electronic commerce, assisting in locating and connecting the ultimate information provider with the ultimate information requester." According to this definition, the *brokers* and *matchmakers* described above would be two types of middle agents. Other types of middle agents include the *agent name servers* or *directory service agents* in CIIMPLEX (Peng et al., 1998; Steiner, 1996) and the *white page agents* or *yellow page agents* in MIX (Iglesias et al., 1996).

A yellow page agent is, in fact, a passive (may also be active with some special mechanisms) matchmaker. It is a middle agent that stores 'capability advertisements' (services provided by listed agents) that can then be checked by requesters. White page agents can (usually) only be used to find an agent's address from its name. An agent name server is similar to a white page agent.

Central nodes and design boards
Some agent based concurrent design and manufacturing systems do not use any facilitators, mediators, brokers, or other types of middle agents, but they use agents that could also be classified in this category. For example, the *central node* in First-Link (Park et al., 1994) which is a central product model which can be accessed by independent software modules, and the *design board* in SiFAs (Brown et al., 1995), could both be described as middle agents. These two architectures are, in fact, very similar to the blackboard architecture.

Agent-based blackboards
Lander et al. (1996) proposed using agent-based blackboards to manage agent interactions. As shown in Figure 7.7, each local group of agents shares a data

repository that is provided specifically for the efficient storage and retrieval of active shared data. Along with design data, tactical control knowledge can be represented in the shared repository, enabling reasoning about how to proceed with the design process to have the same status and priority as reasoning about the design itself. Within an agent group, a control shell (analogous to a facilitator) notifies appropriate agents of relevant events. The control shell can be programmed to implement a wide range of coordination strategies.

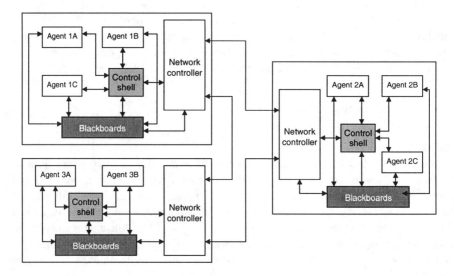

Figure 7.7 A multi-blackboard architecture.

7.4.3 Autonomous Agent System Architectures

We have discussed the concept and definition of autonomous agents in Chapter 3. We argue that an autonomous agent should at least have following characteristics:

(1) it is not controlled or managed by any other software agents or human beings;

(2) it can communicate and interact directly with any other agents in the system as well as with other external systems (Figure 7.8);

(3) it has knowledge about other agents and about its environment;

(4) it has its own goals and an associated set of motivations.

The autonomous agent approach is also called the Agent Network approach (Lander, 1997). In this type of architecture, neither communication nor state knowledge is consolidated – each agent must know where and when messages should be sent, what other agents are available, and what capabilities other agents have. In its pure form, this architecture does not provide for shared knowledge, although in the practical case some common repositories can be advantageous for performance and other reasons.

Such an approach is usually used in large-grained multi-agent applications[4] like ARCHON (Cockburn and Jennings, 1996) and DIDE (Shen and Barthès, 1996).

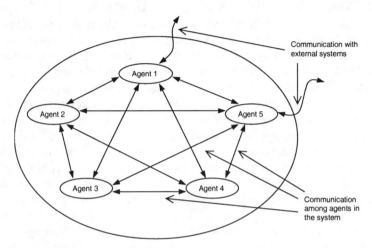

Figure 7.8 Autonomous Agent System Architecture.

7.5 COMPARISON OF DIFFERENT APPROACHES

One of the important features of autonomous agents is their independence. Independence is even more prominent in the concept of Actors, as proposed by Hewitt et al. (1973). In an Actor architecture, modules operate independently but are assumed to cooperate towards a common goal. In contrast, an autonomous agent may have and follow its own goals and associated motivations. However, agents in federated multi-agent systems do not act completely independently. They communicate or interact with other agents through facilitators, mediators, brokers or other types of middle agents, and may be coordinated through agents such as mediators.

In the facilitator architecture, agents interact through facilitators that translate agent-specific knowledge into and out of a standard knowledge interchange language. Each agent can therefore reason (internally) in its own terms, asking other agents for information and providing other agents with information as needed through the facilitators. The facilitator model supports greater flexibility than direct communication used in the autonomous agent systems, because agents do not need to

---

[4] A large-grained multi-agent system is composed of a small number of 'big' agents.

know detailed information about which other agents exist and what their capabilities are.

In the mediator approach, learning (where implemented) is at the group level, while in the autonomous agent approach, learning is at the individual level. Each autonomous agent has to have knowledge about its environment and the other agents, and has to learn in some sense if it is to update its knowledge.

Initially, one may think that the federation approach abandons the decentralization principle of distributed organizations by redefining centralized architectures. This is not so, since the ultimate goal of a federation organization is to release the agents from the burden of establishing their own communication links and assisting them to coordinate their activities with other agents. In such systems, individual agents are then free to concentrate their reasoning capabilities on autonomous planning and responding to the environment's stimuli. When clusters are formed, the organization at any instant may appear to be composed of 'partial hierarchies'. If these structures are dynamically created, however, and destroyed when their tasks are accomplished, it becomes evident that centralization is only being used locally, temporarily, and where it is advantageous. Used appropriately, clustering together with other mechanisms, can be used to obtain an adaptive architecture (Maturana et al., 1999).

In summary, the autonomous agent architecture is well suited for developing distributed intelligent design systems when existing engineering tools are encapsulated as agents and connected to the system for providing special services, and the system consists of a small number of agents. This type of architecture is also very useful for developing autonomous multiple robotic systems. Federated architectures are suitable for developing distributed manufacturing systems which are complex, dynamic, and composed of a large number of resource agents, information agents and service agents. Implemented appropriately, these architectures can provide computational simplicity and manageability. In the short term, hierarchical architectures will still continue to be used for industrial application, but they will be progressively replaced by the federated architectures and the autonomous agent system architectures.

7.6 RESEARCH LITERATURE AND ADDITIONAL READINGS

The following sections are provided for those who wish to know more about the topic of this chapter from previous research, than has been presented so far.

7.6.1 Research Literature

7.6.1.1 Earlier Research

Among the earliest and most significant researches in this area are Hewitt's Modular Actor Formalism (Hewitt et al., 1973) and Smith's Contract Net (Smith, 1980). The autonomous agent approach described in this chapter is, in fact, derived from Hewitt's actor concept. The Contract Net (or its extended or modified versions) has been widely used as a negotiation protocol in different agent based systems. We discuss this issue later in Chapter 10.

7.6.1.2 Hierarchical Architectures

As mentioned in Section 7.4.1, although hierarchical system architectures have some undesirable features, they are still widely used in developing distributed agent based manufacturing systems because of the traditional way that real manufacturing enterprises have been organized. Static or dynamically hierarchical architectures can be found in HMS (Holonic Manufacturing Systems) (Van Leeuwen and Norrie, 1997; Bussmann, 1998), ADEPT (Advanced Decision Environment for Process Tasks) (Norman et al., 1997), ADDYMS (Architecture for Distributed Dynamic Manufacturing Scheduling) (Butler and Ohtsubo, 1992), DAS (Distributed Asynchronous Scheduler) (Burke and Prosser, 1994), and the Production Planning and Control Structure (Fischer, 1994). The Open Agent Architecture proposed by Cohen et al. (1994) is a hierarchical multi-blackboard architecture.

7.6.1.3 Federated Architectures

A number of federated architectures have been proposed in the literature. Three approaches – Facilitators, Brokers, and Mediators – are well known and widely used in concurrent design and manufacturing systems. Although definitions of these architectures, as found in the literature, may differ somewhat from those given in this book, we have tried to adhere to commonly-accepted interpretations.

The Facilitator approach was proposed for the SHADE project (McGuire et al., 1993) and demonstrated in the PACT project (Cutkosky et al., 1993), where several facilitators (facilitation agents) were used to ease the communication and coordination among agents. CIIMPLEX (Peng et al., 1998) and some other agent-based concurrent design and manufacturing systems (especially systems using KQML (Mayfield et al., 1996) as an inter-agent communication language) also use this approach. In a multi-agent refinement system called DRAMA (Distributed Refinement Among Multiple Agents) (Byrne and Edwards, 1996), facilitators were used for the coordination of the refinement process.

Broker agents have been widely used in distributed information management systems and Internet based systems (Decker et al., 1997). They have also been used in agent-based concurrent design and manufacturing systems (Park et al., 1993; Peng et al., 1998; Gray et al., 1998).

The Mediator approach was initially proposed by Wiederhold (1992) for distributed database management systems. Similar approaches have been proposed for developing agent-based mediator-centric manufacturing systems (Gaines et al., 1995; Maturana and Norrie, 1996; Shen et al., 1999). In MetaMorph II (Shen et al., 1999), the matchmaking mechanism was introduced into mediator agents within a hybrid multi-agent architecture for distributed intelligent manufacturing, agent-based manufacturing planning and scheduling, and supply chain management.

In addition to the three approaches to federation described above, there are some other agent-based collaborative engineering design and manufacturing systems that could be classified as federated architectures. The First-Link project (Park et al., 1994) was intended to test a specific software architecture where a central product model (*central node*) was accessed by independent software modules, somewhat like a blackboard architecture. The Next-Link project (Petrie et al., 1994) was the successor to First-Link and was to study how to coordinate the independent agents, with a mechanism called the *Redux'* services developed by Petrie (1993) being added to the system for this purpose. In the Process-Link project (Goldmann, 1996), this type of *central node* was replaced by an *agent manager* which is very similar to the previously described mediator. SiFAs (Brown et al., 1995) used a similar architecture with a *design board* which is visible to all agents participating in design. Lander et al. (1996) proposed using a multi-blackboard architecture for managing agent interactions within a collaborative engineering design system. CAAD (Cooperative Agents and Applications in Design) used a similar approach to allow several designers to cooperate through a multi-blackboard system (Branki and Bridges, 1993).

MIX (Iglesias et al., 1996) is a federated architecture for the design of multi-agent systems that allows an integration of both connectionist and symbolic subsystems. MIX consists of an agent model describing the structure of an individual agent, and a network model defining the communication infrastructure used by agents. A distinction is drawn between agents offering network-related services (so-called network agents) and application agents. The network model offers various facilities including a yellow-page service, coordination facilities (e.g., the contract net protocol), and rudimentary knowledge facilities supporting communication among heterogeneous information agents.

7.6.1.4 Autonomous Agent System Architectures

The Autonomous Agent approach has remained an exciting research topic in the agent research community. It is especially useful for autonomous robotics control systems. Numerous researchers have also used this approach for developing agent-based concurrent design and manufacturing systems and other agent-based industrial applications.

The DIDE project (see Chapter 13) used this approach in implementing agent-based intelligent engineering design environments. DIDE is organized as a population of asynchronous cognitive agents integrating engineering tools and human specialists within an open environment. Each tool (or interface for a human specialist) is encapsulated as an agent. Engineering tools and human specialists are then connected through a local network and communicate via this network. Each can also communicate directly with other agents located in any other local networks, using the Internet. All agents are independent and autonomous. They exchange design data and knowledge via a local network or the Internet. DIDE does not use any facilitators or mediators. There is no static global control structure in the system. DIDE is specially designed to allow dynamic changes in the agent set (i.e., one can add new agents to or remove existing agents from the system without stopping and reinitializing the working environment).

AARIA used the autonomous agent approach in developing agent-based manufacturing systems (Parunak et al., 1997) (also see Chapter 14). In AARIA, a factory is essentially represented as a supply chain. Manufacturing entities (e.g. people, machines, and parts, etc.) are encapsulated as autonomous agents. However, a Manager (also as an agent) oversees all entities and intervenes in their activities as required, and Scheduling manages the flow of Resource engagements and Parts through various Unit Processes so that they are coordinated appropriately with each other. It is therefore not a pure autonomous agent system according to our definition.

The ARCHON project (Cockburn and Jennings, 1996) was originally developed for integrating multiple pre-existing legacy systems and used the autonomous agent approach. In an ARCHON community, there is no centrally located global authority, each agent controls its own IS (Intelligent System) and mediates its own interactions with other agents. The system's overall objectives are expressed through the separate local goals of each community member. Because the agents' goals are often interrelated, social interactions are required to meet global constraints and to provide the necessary services and information. Such interactions are controlled by the agent's AL (ARCHON Layer). Examples of interaction are: asking for information from acquaintances; requesting processing services from acquaintances; and spontaneously volunteering information which is believed to be relevant to others.

7.6.2 Further References

Lander (1997) has reviewed a number of proposed agent system architectures for multi-agent design systems. The ATAL (Agent Theories, Architectures and Languages) workshops have included descriptions of various agent system architectures, in addition to agent architectures (See Chapter 6 for related references to workshop notes). Various architectures have also been presented in the conference proceedings of ICMAS (published by The AAAI Press/The MIT Press), Autonomous Agents (published by the ACM Press), PAAM (published by the Practical Application Company, London), International DAI workshops (e.g., Klein and Sharma, 1994), MAAMAW workshops (e.g., Werner and Demazeau, 1992; Castelfranchi and Werner, 1994; Castelfranchi and Muller, 1995; Boman and Van de Velde, 1997; Van de Velde and Perram, 1997), and some regional DAI Workshops such as the Australian DAI workshops (Zhang and Lukose, 1995, 1996; Pagnucco et al., 1997; Cavedon et al., 1997). A bibliography on agent theories, architectures and languages can be found in (Wooldridge et al., 1996).

REFERENCES

Ayala, G. and Yano, Y. (1996) Interacting with a mediator agent in a collaborative learning environment, In Anzai, Y., Ogawa, K. and Mori, H. eds., *Symbiosis of Human and Artifact: Future Computing and Design for Human-Computer Interaction*, Advances in Human Factors/Ergonomics, Elsevier Science, pp. 890-900.

Balasubramanian, S. and Norrie, D.H. (1995) A Multi-Agent Intelligent Design System Integrating Manufacturing and Shop-Floor Control, In *Proceedings of First International Conference on Multi-Agent Systems*, San Francisco, CA, AAAI press/MIT Press, pp. 3-9.

Boman, M. and Van de Velde, W. (eds.) (1997) *Multi-agent rationality*, 8th European Workshop on Modeling Autonomous Agents in a Multi-Agent World, MAAMAW'97: Lecture Notes in Artificial Intelligence 1237, Springer.

Branki, N.E. and Bridges, A. (1993) An Architecture for Cooperative Agents and Applications in Design, In Beheshi, M.R. and Zreik, K. eds., *Advanced Technologies*, pp. 221-230.

Brennan, R. and Norrie, D. (1998) Evaluating the performance of alternative control architectures for manufacturing, In *Proceedings of IEEE International Symposium on Intelligent Control*, pp. 90-95.

Brown, D., Dunskus, B., Grecu, D. and Berker, I. (1995) SINE: Support for Single Function Agents, In *Proceedings of AIENG'95, Applications of AI in Engineering*, Udine, Italy.

Burke, P. and Prosser, P. (1994) The Distributed Asynchronous Scheduler. In Zweben, M. and Fox, M.S. eds., *Intelligent Scheduling*, Morgan Kaufman Publishers, San Francisco, CA, pp. 309-339.

Bussmann S. (1998) An Agent-Oriented Architecture for Holonic Manufacturing Control, In *Proceedings of First International Workshop on IMS*, Lausanne, Switzerland, pp. 1-12.

Butler, J. and Ohtsubo, H. (1992) ADDYMS: Architecture for Distributed Dynamic Manufacturing Scheduling, In Famili, A., Nau, D.S. and Kim, S.H. eds., *Artificial Intelligence Applications in Manufacturing*, The AAAI Press, pp. 199-214.

Byrne, C. and Edwards P. (1996) Refinement in Agent Groups. In Weiss, G. and Sen, S. eds., *Adaption and Learning in Multi-Agent Systems*, Lecture Notes in Artificial Intelligence 1042, Springer, pp. 22-39.

Castelfranchi, C. and Werner, E. eds. (1994) *Artificial Social Systems* – Selected Papers from MAAMAW-92, Lecture Notes in Artificial Intelligence 830, Springer.

Castelfranchi, C. and Muller, J.-P. eds. (1995) *From Reaction to Cognition* – Selected Papers from MAAMAW-93, Lecture Notes in Artificial Intelligence 957, Springer.

Cavedon, L., Rao, A. and Wobcke, W. eds. (1997) *Intelligent Agent Systems: Theoretical and Practical Issues*, Lecture Notes in Artificial Intelligence 1209, Springer.

Cockburn, D. and Jennings, N.R. (1996) ARCHON: A DAI System for Industrial Applications, In O'Hare, G.M.P. and Jennings, N.R. eds., *Foundations of DAI*, Wiley Interscience, pp. 319-344.

Cohen, P.R., Cheyer, A., Wang, M. and Baeg, S.C. (1994) An Open Agent Architecture, In *Proceedings of the AAAI Spring Symposium on Software Agents*.

Cranefield, S. and Purvis, M. (1997) An Agent-Based Architecture for Software Tool Coordination, In Cavedon, L., Rao, A. and Wobcke, W. eds., *Intelligent Agent Systems: Theoretical and Practical Issues*, Lecture Notes in Artificial Intelligence 1209, Springer, pp. 44-58.

Cutkosky, M.R., Engelmore, R.S., Fikes, R.E., Genesereth, M.R., Gruber, T.R., Mark, W.S., Tenenbaum, J.M. and Weber, J.C. (1993) PACT: An Experiment in Integrating Concurrent Engineering Systems, *IEEE Computer*, 26(1):28-37.

Decker, K., Sycara, K., Williamson, M. (1996) Matchmaking and Brokering, In *Proceedings of the Second International Conference on Multi-Agent Systems*, MIT Press/AAAI Press.

Decker, K., Sycara, K., Williamson, M. (1997) Middle-Agents for the Internet, In *Proceedings of IJCAI-97*, Japan.

Ferber, J. (1995) *Les systèmes multi-agents*, InterEditions, Paris.

Fischer, K. (1994) The Design of an Intelligent Manufacturing System, In *Proceedings of the 2nd International Working Conference on Cooperating Knowledge-based Systems*, University of Keele, pp. 83-99.

Fruchter, R., Clayton, M., Krawinkler, H. and Teicholz, P. (1993) Interdisciplinary Communication Medium for Collaborative Conceptual Building Design, Tech. Report, Dept. of Civil Engineering and Center for Integrated Facility Engineering, Stanford Univ.

Gaines, B.R., Norrie, D.H., and Lapsley, A.Z. (1995) Mediator: an Intelligent Information System Supporting the Virtual Manufacturing Enterprise, In *Proceedings of 1995 IEEE International Conference on Systems, Man and Cybernetics*, New York, IEEE, pp. 964-969.

Goldmann, S. (1996) Procura: A Project Management Model of Concurrent Planning and Design, In *Proceeding of WET ICE'96*, Stanford, CA.

Gray, P.M.D., Embury, S.M., Hui, K. and Prce, A. (1998) An Agent-Based System for Handling Distributed Design Constraints, In *Working Notes of the Agent-Based Manufacturing Workshop*, Minneapolis, MN.

Gupta, L., Chionnglo, J.F. and Fox, M.S. (1996) A Constraint Based Model of Coordination in Concurrent Design Projects, In *Proceeding of WET ICE'96*, Stanfrod, CA.

Hewitt, C., Bisshop, P. and Steiger, R. A (1973) Universal Modular Actor Formalism for Artificial Intelligence, In *Proceedings of the 3rd IJCAI*, Stanford, CA, pp. 235-245.

Huhns, M.N. and Singh, M.P. (1998) Agents and Multiagent Systems: Themes, Approaches, and Challenges. In Huhns, M.N. and Singh, M.P. eds., *Readings in Agents*, Morgan Kaufmann Publishers, Inc., San Francisco, CA, pp. 1-24.

Iffenecker, C. (1994) *Un système multi-agents pour le support des activités de conception de produits*, Ph D thesis, Université Paris VI, France.

Iglesias, C.A., Gonzalez, J.C. and Velasco, J.R. (1996) MIX: A general purpose multiagent architecture, In Wooldridge, M.J., Müller, J.P. and Tambe, M. eds. *Intelligent agents II: Agent Theories, Architectures, and Languages*, Lecture Notes in Artificial Intelligence 1037, Springer, pp. 251-266.

Kirn, S. and Schneider, J. (1992) STRICT: selecting the right architecture, In Belli, F. and Radermacher, F.J. eds. *Industrial and Engineering Applications of Artificial Intelligence and Expert Systems*, Lecture Notes in Artificial Intelligence 604, Springer, pp. 391-400.

Klein, M. and Sharma, K. eds. (1994) Proceedings of the 13th International DAI Workshop. Seatle, WA, USA.

Lander, S.E. (1997) Issues in Multiagent Design Systems, *IEEE Expert* 12(2):18-26.

Lander, S.E., Staley, S.M. and Corkill, D.D. (1996) Designing Integrated Engineering Environment: Blackboard-based Integration of Design and Analysis Tools, *Concurrent Engineering: Research and Applications*, Special Issue on the Application of Multiagent Systems to Concurrent Engineering, 4(1):59-72.

Luck, M. and d'Inverno M. (1995) A Formal Framework for Agency and Autonomy, In *Proceedings of First International Conference on Multi-Agent Systems*, AAAI Press/The MIT Press, pp. 254-260.

Malone, T.W. (1990) Organizing information processing systems: Parallels between human organization and computer systems, In Zachary, W.W. and Robertson, S.P. eds., *Cognition, Computation and Cooperation*, Aalbex, Norwood, NJ, pp. 56-83.

Maturana, F. and Norrie, D.H. (1996) Multi-Agent Mediator Architecture for Distributed manufacturing, *Journal of Intelligent Manufacturing*, 7:257-270.

Maturana, F., Shen, W. and Norrie, D.H. (1999) MetaMorph: An adaptive agent-based architecture for intelligent manufacturing, *International Journal of Production Research*, 37(10):2159-2174.

Mayfield, J., Labrou, Y. and Finin, T. (1996) Evaluating KQML as an agent communication language, In M. Wooldridge, J.P Muller and M. Tambe, (eds.), *Intelligent Agent II*, Lecture Notes in Artificial Intelligence 1037, Springer, pp. 347-360.

McGuire, J., Kuokka, D., Weber, J., Tenenbaum, J., Gruber, T. and Olsen, G. (1993) SHADE: Technology for Knowledge-Based Collaborative Engineering, *Concurrent Engineering: Research and Application*, 1(3).

Monceyron, E. and Barthès, J.P. (1992) Architecture for ICAD Systems: an Example from Harbor Design, *Revue Sciences et Techniques de la Conception*, 1(1):49-68.

Moulin, B. and Chaib-draa, B. (1996) An Overview of Distributed Artificial Intelligence. In O'Hare, G.M.P. and Jennings, N.R. eds., *Foundations of DAI*, Wiley Interscience, pp. 1-56.

Norman, T.J., Jennings, N.R., Faratin, P. and Mamdani, E.H. (1997) Designing and Implementing a Multi-Agent Architecture for Business Process Management, In Müller, J.P., Wooldridge, M.J. and Jennings, N.J. eds., *Intelligent Agents III: Agent Theories, Architectures, and Languages,* Lecture Notes in Artificial Intelligence 1193, Springer, pp. 261-275.

Pagnucco, M., Wobcke, W. and Zhang, C. eds. (1997) *Agents and multi-agent systems: formalisms, methodologies and applications*, Lecture Notes in Artificial Intelligence 1441, Springer.

Park, H., Tenenbaum, J. and Dove, R. (1993) Agile Infrastructure for Manufacturing Systems: A Vision for Transforming the US Manufacturing Base, In *Proceedings of Defense Manufacturing Conference*.

Park, H., Cutkosky, M., Conru, A. and Lee, S.H. (1994) An Agent-Based Approach to Concurrent Cable Harness Design, *AIEDAM*, 8(1).

Parunak, H.V.D., Baker, A.D. and Clark, S.J. (1997) The AARIA Agent Architecture: An Example of Requirements-Driven Agent-Based System Design, In *Proceedings of the First International Conference on Autonomous Agents*, Marina del Rey, CA, ACM Press.

Peng, Y., Finin, T., Labrou, Y., Chu, B., Long, J., Tolone, W.J. and Boughannam, A. (1998) A Multi-Agent System for Enterprise Integration, In *Proceedings of PAAM '98*, London, UK.

Petrie, C. (1993) The *Redux'* Server, In *Proceedings of International Conference on Intelligent and Cooperative Information Systems*, Rotterdam.

Petrie, C., Cutkosky, M., Webster, T., Conru, A. and Park, H. (1994) Next-Link: An Experiment in Coordination of Distributed Agents, Position paper for the AID-94 Workshop on Conflict Resolution, Lausanne.

Quadrel, R., Woodbury, R., Fenves, S. and Talukdar, S. (1993) Controlling Asynchronous Team Design Environment by Simulated Annealing, *Research in Engineering Design*, 5(2):88-104.

Saad, M. and Maher, M.L. (1996) Shared understanding in computer-supported collaborative design, *Computer-Aided Design*, 28(3):183-192.

Shen, W. and Barthès, J.P. (1996) An Experimental Multi-Agent Environment for Engineering Design, *International Journal of Cooperative Information Systems*, 5(2-3):131-151.

Shen, W., Maturana, F. and Norrie, D.H. (1999) MetaMorph II: An Agent-Based Architecture for Distributed Intelligent Design and Manufacturing, To appear in *Journal of Intelligent Manufacturing*.

Smith, R.G. (1980) The contract net protocol: high-level communication and control in a distributed problem solver. *IEEE Transaction on Computers*, 29(12):1104-1113.

Sriram, D., Logcher, R., Groleau, N. and Cherneff, J. (1992) DICE: An Object Oriented Programming Environment for Cooperative Engineering Design, In Tong, C. and Sriram, D. eds., *AI in Engineering Design*, Volume 3, Academic Press.

Steiner, D.D. (1996) IMAGINE: An Integrated Environment for Constructing Distributed Artificial Intelligence Systems. In O'Hare, G.M.P. and Jennings, N.R., eds., *Foundations of DAI*, Wiley Interscience, pp. 345-364.

Van de Velde, W. and Perram, J.W. (eds.) (1997) *Agents breaking away*: 7th European Workshop on Modeling Autonomous Agents in a Multi-Agent World, MAAMAW'96: Lecture Notes in Artificial Intelligence 1038, The Netherlands, Springer.

Van Leeuwen, E.H. and Norrie, D.H. (1997) Intelligent manufacturing: holons and holarchies. *Manufacturing Engineer*, 76(2):86-88.

Werner, E. and Demazeau, Y. (eds.) (1992) Decentralized AI 3, North Holland.

Wiederhold, G. (1992) Mediators in the architecture of future information systems. *IEEE Computer*, 25(3):38-49.

Wong, A. and Sriram, D. (1993) SHARED: An Information Model for Cooperative Product Development, *Research in Engineering Design*, 5(1):21-39.

Wooldridge, M.J., Müller, J.P. and Tambe, M. (1996) Agent Theories, Architectures and Languages: A Bibliography, In Wooldridge, M.J., Müller, J.P. and Tambe, M. eds., *Intelligent agents II: Agent Theories, Architectures, and Languages*, Lecture Notes in Artificial Intelligence 1037, Springer, pp. 408-431.

Zhang, C. and Lukose, D. eds. (1995) *Distributed artificial intelligence: architecture and modelling*, Lecture Notes in Artificial Intelligence 1087, Springer.

Zhang, C. and Lukose, D. eds. (1996) *Multi-agent systems: methodologies and applications*, Lecture Notes in Artificial Intelligence 1286, Springer.

8

Communication, Coordination and Cooperation

8.1 INTRODUCTION

Communication, cooperation and coordination are three key issues in multi-agent systems. Communication enables agents to exchange information and to coordinate their activities. Communication is the most important way in which cooperation and coordination among agents takes place (although it is not the only way as Genesereth et al. (1986) and Fenster et al. (1995) point out).

8.2 COMMUNICATION

8.2.1 Levels of Communication

In agent-based systems, communication can exist in levels of sophistication ranging from primitive communication through to high-level communication (Moulin and Chaib-draa, 1996). In this section, we briefly introduce some of these.

8.2.1.1 No Communication or Primitive Communication

As Genesereth et al. (1986) and Fenster et al. (1995) note, communication is not always necessary in a multi-agent system. In some cases, an agent may rationally refer to other agents' plans or behaviors without communicating with them. Genesereth et al. (1986) proposed a game-theoretic approach using payoff matrices that contain agents' payoffs for each possible outcome of an interaction. This approach also assumes that agents have common knowledge of the payoff matrix associated with the interaction. Rosenschein and Bresse (1989) extended this approach by providing agents with a mechanism for further refining the agent's choice of rational moves: uncertainty about the other players' moves is explicitly represented. However, there are some difficulties with this approach: (1) if all agents are speculating on what the others are going to do, the result could be an infinite nesting of beliefs and the reasoning of each agent would be very difficult; (2) mutual rational deduction could lead to very high computational cost in a situation involving several agents. Even if agents indirectly receive some information about the plans or behaviors of other agents, they still need to rely on sophisticated local reasoning to compensate for the lack of direct communication when deciding on appropriate actions and interactions.

Primitive communication is commonly restricted to some finite set of fixed signals with fixed interpretation. This communication strategy was used by Georgeff (1983) for multi-agent planning, to avoid conflicts between plans involving more than one agent.

8.2.1.2 Message Passing and Plan Passing

In the message-passing approach, agents communicate with each other by sending messages. This approach is widely used in existing multi-agent systems, including agent-based concurrent design and manufacturing systems. The Actor language developed by Hewitt et al. (1973, 1977) is one of the earliest applications to use this approach. Note that Hewitt's intuition is fundamentally correct when he states that the control of multi-agent environments is best looked at in terms of communication structures. In his work, however, actors have a very simple structure, and the model gives no formal syntax, semantics, or pragmatics for communication structures. Consequently, this model is rarely used for communication between complex agents, but is an excellent example for readers of this book to understand the basics of message passing.

In a multi-agent system, the complexity of the communication depends upon the agents' characteristics. Reactive agents use sets of communication rules (or activities) that are triggered when specific states exist or predetermined events occur (these events may be messages received from other agents or changes perceived in the environment). Consideration of the possible types of individual message and the types of expected answer yields protocols for interaction that can extend over more than a single message exchange. Such protocols have been developed for various situations, for example, the contract net protocol (Smith, 1980). Communication and action are usually interleaved. Hence, cognitive (or deliberative) agents must be able to create plans in which both linguistic and nonlinguistic actions are combined.

In the plan-passing approach, an agent A communicates its total plan to agent B and B communicates its total plan to A. Although this method can achieve cooperation, it has some severe problems (Rosenschein, 1986): (1) total plan passing is computationally expensive; (2) there is no guarantee that the resulting plan will be compatible with the recipient's knowledge base; (3) total plans are difficult to derive in real world applications because of uncertainty about the present state of the world, as well as its future. Consequently, for real world situations, total plans cannot be formulated in advance, and general strategies must be communicated to a recipient.

8.2.1.3 Information Exchanges through a Shared Data Repository

Information exchange through a shared data repository is an important strategy for many distributed information systems described in the literature. It has also been used in agent-based concurrent design and manufacturing systems. Commonly, a shared data repository is termed a blackboard but it is also called a Central Node, Design Board, Conversation Server, or Cooperation Domain Server.

In agent-based systems, the shared data repository is used for agents to write messages to or post partial results on, and obtain information from. It is usually partitioned into levels of abstraction appropriate for the problem at hand. Agents working at a particular level of abstraction have access to the corresponding level of the repository along with adjacent levels. Thus, data synthesized at any level can be communicated to higher levels, while higher-level goals filter down to drive the expectations of lower-level agents. The simplicity with which the blackboard paradigm handles the classic problem of data-driven versus goal-driven information flow is perhaps the reason why this approach is widely used in existing distributed artificial intelligence systems.

Message passing and communication through a shared data repository may be combined in complex systems. Each agent may be composed of several subsystems (or "sub-agents") that exchange information using a local blackboard. The agents themselves communicate together by exchanging messages. Several multi-agent systems have adopted this communication architecture. The DVMT uses a collection of identical blackboard-based systems to solve problems of monitoring and interpreting data from sensors at spatially distributed locations within a region (Lesser and Corkill, 1983). In the MINDS system (Huhns, 1987) each user works with a blackboard-based agent to retrieve documents from its local database or to get them from other agents. In DIDE (Shen and Barthès, 1996), a database is used to save all product data. Each design agent obtains product information from this database when necessary and saves this back if modifications are done.

8.2.1.4 High-Level Communication

Issues of high-level communication in multi-agent systems relate to natural language understanding, speech act theory, conversations, and other formal theories. We will introduce speech act theory separately in Section 8.2.4, and agent conversations in Section 8.2.5. Natural language understanding is not described in this book. Interested readers may refer to Allen (1986).

8.2.2 Modes of Communication

Agent communication can be implemented in very different ways, depending on the nature of the agents, the global architecture of the system, the timing of the exchanges, or the number of receivers for a message. Communications can occur between humans and/or machines; they can be direct by means of messages or indirect by posting; they can be synchronous or asynchronous; they can concern a single recipient agent or several agents.

8.2.2.1 Direct and Indirect Communication

Communication among agents can be direct or indirect. Direct communication is realized through message passing as described in Section 8.2.1.2, while indirect communication is realized through a shared data repository such as a blackboard. Such indirect communication is equivalent to a multicast communication strategy, in the case of partitioned blackboard; or to broadcast communication otherwise. In the latter case, the communication mechanisms can be quite simple, as in the SHARED workspace developed by Wong and Sriram (1993).

8.2.2.2 Synchronous and Asynchronous Communication

Communication among agents can be synchronous or asynchronous. Agents (including humans) can work together in these modes via computer networks:

1) *synchronous mode*: they work in the same location and at the same time;
2) *distributed synchronous mode*: they work in different sites but at the same time;
3) *asynchronous mode*: they work in the same location but at different times;
4) *distributed asynchronous mode*: they work in different locations and at different time.

Each mode requires appropriate hardware and software support for effective and efficient operation. It is important to recognize the distinctiveness of each mode, since their protocols, network and storage requirements vary. In complex concurrent design and manufacturing systems, both synchronous and asynchronous communications are required. Synchronous communication can be particularly useful for discussions among product designers and production engineers. Commercial software such as Microsoft's Netmeeting™ could be used for such synchronous discussions. These products typically allow application sharing and audio and video communication, which can improve cooperation among product designers and production engineers. However, asynchronous communication is the primary mode of interaction among agents in agent-based concurrent design and manufacturing systems.

8.2.2.3 Single or Multiple Recipient Agents

In agent-based systems, there are the following communication possibilities:

- point-to-point (one agent sends a message to another specific agent)
- broadcast (one agent sends out a message to all other agents in the system)
- multi-cast (one agent sends out a message to a selected group of agents)

Whether there is a single recipient or several recipients depends on the relationship between the sender and the receiver(s) of the message.

When the receiver is known to the sender, the latter may send messages directly and thus establish direct individual communication. This is point-to-point communication. This approach is widely used for communication among cognitive (deliberative) agents. For example, the following message:

$$A \rightarrow B: \texttt{Request(calculate, p1, p2)}$$

is a point-to-point communication in which agent A asks agent B to carry out a calculation involving two parameters p1 and p2.

If the receiver is unknown to the sender, the message is usually sent to all agents in the system (or subsystems), i.e., is broadcast[1]. This approach is very useful in an open and dynamic environment where agents may appear and disappear. In particular, it is widely used in the task decomposition protocols like the Contract Net (see Section 8.3.4 for more details). In this case, B in the above message is replaced by 'all':

$$A \rightarrow \text{all}: \texttt{Request(calculate, p1, p2)}$$

The first and second types of communication are the extremes. Possibilities in-between include the agent sending messages to a group of agents according to (i) whether the addresses of these agents known to the sender; or (ii) a criterion, for example, all agents of a particular type (e.g. mediator agents), or all agents having a special skill. The above message then becomes:

$$A \rightarrow \{\text{ag1; ag2; ag3}\}: \texttt{Request(calculate, p1, p2)}$$

Or

$$A \rightarrow \{x \mid \text{Criterion(y)}\}: \texttt{Request(calculate, p1, p2)}$$

8.2.2.4 Influence of System Architecture

The global architecture of the agent-based concurrent design and manufacturing system strongly influences the way communications are organized. In blackboard systems, the blackboard architecture has a tendency to impose indirect, synchronous, general (broadcasting) communication. In autonomous agent systems or systems using federation architectures, one finds mostly direct, asynchronous communications with varying number of receivers (point-to-point, multicast, broadcast as discussed in the previous section).

8.2.2.5 Human Participation

Usually, agent-based concurrent design and manufacturing systems involve both human and software agents. There is thus communication (interaction) between human and software agents and communication between human agents via computers. Since issues other than communication are involved, we defer this

---

[1] In an agent-based system using federation architectures as described in Chapter 7, the sender may get the receiver's address from service agents like Yellow Page agents or Agent Name Servers and then send the message directly to the receiver. In other cases, the message may be sent to a middle agent such as a facilitator, mediator, or broker who then forwards the message on to the appropriate receiver(s).

discussion to the section on "Human-Machine Interaction" in Chapter 12 (Section 12.3).

8.2.3 Protocols and Languages Supporting Communication Processes

To support communication processes among agents it is necessary to develop agent communication protocols[2]. There are two levels of these protocols: one for supporting the communication linkages, and another for supporting the exchange of information. Each is supported by distinct communication languages.

Among the communication linkage protocols proposed or developed in the literature, there exist both basic protocols such as the Contract Net (Smith, 1980) and more sophisticated protocols such as the Extended Contract Net Protocol (ECNP) proposed by Fischer et al. (1995), the Market-Driven Contract Net by Baker (1991), the B-Contract-Net by Scalabrin (1996), and the Leveled Commitment Contracting Protocol by Sandholm and Lesser (1996).

Protocols supporting the exchange of information include:

- a protocol with four types of messages, each corresponding to the communication modes among agents (Gaspar, 1991);
- a more complex protocol like SANP (Speech Act Negotiation Protocol) (Chang and Woo, 1991), based on the results of linguistic researches;
- KQML (Knowledge Query and Manipulation Language) (Finin et al., 1993);
- ToolTalk™ (Frankel, 1991).

KQML was defined and developed by the External Interfaces working group of the ARPA Knowledge Sharing Initiative (Finin et al., 1993). KQML is indifferent to the format of the content information. A KQML message is annotated by reserved parameters to describe attributes of the directive and the embedded content. The attributes include who the message is :from, who it is :to, what is the :language of the embedded :content, what :ontology defines the vocabulary used within the :content, and what tag to :reply-with. An example of a KQML subscription is as following:

```
(monitor   :from consumer :to producer :reply-with update-111
           :ontology standard-units-and-dimensions :language KIF
           :content (= (q.magnitude (diameter shaft-a) inches) ?x))
```

A subscription is a request to monitor changes within the set of information defined by the content description. In this case, the embedded content describes the

---

[2] The well-known ISO-OSI (International Standards Organization – Open systems Interconnection) seven-layer reference model with its communications services is the main reference to consult for information on the basic communication infrastructure which can support these protocols. Interested readers will easily find descriptions of this model in reference texts on communication.

magnitude of some shaft diameter. The :reply-with parameter is a tag used to identify the sender.

KQML can enable information-flow architectures through the forwarding, broadcasting, and brokering of messages. Forwarding and broadcasting allow the sending agent to use intermediaries to facilitate transport. Brokering allows the agent to be unconcerned with exactly who can process the message. The broker agent is responsible for finding an appropriate recipient for the request. Brokering agents match KQML requests against KQML advertisements of capabilities.

KQML and its extended or modified versions have been widely used in various agent-based systems in the world. It has become accepted as one of the standard agent communication languages (see Chapter 16). A more detailed description of the 1993 version of KQML can be found in (Finin et al., 1993), and of the proposed 1996 version in (Labrou and Finin, 1997).

ToolTalk™ was developed by SunSoft™ as a part of Solaris™ OpenWindows™ (Frankel, 91). Programs interact with ToolTalk™ by calling functions defined in the application programming interface (API). ToolTalk™ can connect agents located anywhere on the network without having to keep track of where they actually reside. In addition, such connections can be modified dynamically. Communications within ToolTalk™ are done using two types of message: *notices* and *requests*. A *notice* message is informational – a way for a process to announce an event. Processes that receive a *notice* assimilate the message without returning anything to the sender. A *request* message is a call for action, the results of which must be returned to the sender as a reply.

8.2.4 Speech Act Theory

Because many agent communication languages are based on Speech Act Theory, we give a short introduction to this area. Speech Act Theory is primarily concerned with natural language utterances. The notion of speech acts goes back to the philosophical work of Austin (1962), who observed that most things people say are not simply propositions that are true or false, but performatives that succeed or fail. Thus, the sentences uttered by humans during communication do not always simply assert a fact, but actually "perform" some action (utterances "accomplish" actions by having certain "forces"). This notion has been picked up widely in the agent research community as a basis for understanding how agents can communicate with one another, and forms the foundation of KQML (see previous section).

According to Searle (1969), there are three types of action associated with an utterance:
- *Locution*: the physical act of uttering the sentence;
- *Illocution*: the action of conveying the speaker's intent to the hearer;
- *Perlocution*: any actions caused as a result of uttering the sentence, which includes the actions taken by the hearer upon the occurrence of the utterance.

For example, "turn on the machine" is the actual locution, with the illocutionary force that causes the hearer to turn on the machine, and whatever the hearer does as a result is the perlocution.

The categorization of speech acts by message types was initially motivated by Searle's classification (Searle, 1969) of illocutionary forces into five categories:

- *Assertive*: providing information that affirms something (e.g., "The machine is turned on");
- *Directive*: sending directives to receivers (e.g., "Turn on the machine");
- *Commissive*: accomplishing certain actions in the future (e.g., "I will turn on the machine");
- *Permissive*: providing the receivers with an indication of the sender's mental state (e.g., "You may turn on the machine");
- *Prohibitive*: declaring a decision or an announcement (e.g., "I name this machine Asterix").

This classification suggests that speech acts can be grouped into classes such that speech acts in the same class share the same condition for their dissatisfaction. It is very practical and useful, but it is not faultless and has been criticized because its basis of classification is semantically coarse, attempting to put together speech acts of varying strengths, and differing pragmatic effects. Interested readers may refer to (Searle et al., 1980; Haddadi, 1996) for more information.

8.2.5 Conversations

In most agent-based concurrent design and manufacturing systems, communication between agents takes place through the use of messages. A message consists of a packet of information, usually sent asynchronously, whose type is represented by a verb corresponding to some kind of illocutionary act (e.g., *request*, *inform*). Messages are exchanged by agents in the context of conversations. A conversation is a sequence of messages between two agents, taking place over a period of time that may be arbitrarily long, yet is bounded by certain termination conditions for any given occurrence. Conversations may give rise to other conversations as appropriate. Each message is part of an extensible protocol - consisting of both message names and conversation policies (also called patterns or rules) common to the agents participating in the conversation. The content portion of a message encapsulates any semantic or procedural elements independent of the conversation policy itself. The rest of this section introduces the concept of conversation policy and the representation of conversations.

8.2.5.1 Conversation Policies

A major issue for designers of agent-based systems is how to implement policies governing conversational and other social behavior among agents. Walker and

Wooldridge (1995) have termed the two major approaches as: off-line design, in which social laws are hard-wired in advance into agents, and emergence, where conventions develop from within a group of agents. Most researchers use the off-line approach (e.g., Barbuceanu and Fox, 1995; Bradshaw et al., 1997)

Conversation policies prescriptively encode regularities that characterize communication sequences between users of a language. A conversation policy explicitly defines what sequences of which messages are permissible between a given set of participating agents.

8.2.5.2 Representation of Conversations

Researchers have proposed various approaches to representing conversations with the following being particularly significant:

-   Finite State Automata (Finite State Machines – FSM) Winograd and Flores (1986)
-   Petri Nets and Colored Petri Nets (Estraillier and Girault, 1992; Coria, 1993)
-   Enhanced Dooley Graphs (Parunak, 1996)

Finite State Machines
Finite State Machines (FSM) (also called State Transition Diagrams – STD) have a long history of usage for defining network protocols. Winograd and Flores (1986) proposed using Finite State Machines for the modeling, analysis and prototyping of distributed systems. This approach has been used with speech act theory by (Barbuceanu and Fox, 1995; Bradshaw et al., 1997) and others for representing conversations in multi-agent systems. Figure 8.1 shows the representation of the conversation policy for a *Request*.

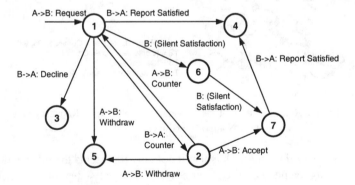

Figure 8.1 Representation of the conversation policy for a Request using the FSM approach (from Bradshaw et al., 1997) (©1997 AAAI).

In Figure 8.1, the circles represent states and the arcs represent speech acts. This approach clarifies the various states through which a discourse may move, but it obscures the identity of the speaker at any point, and loses the evolution of a particular conversation in time. For example, Figure 8.1 indicates that a discourse can include a number of counter-offers between A and B, but does not tell how many such counter-offers actually took place in a given interaction.

Petri Nets and Colored Petri Nets

The FSM approach is a very practical way of describing the structure of conversations when the conversations are isolated, i.e., a conversation can be realized by a single process. However, agents are often engaged in several conversations simultaneously, and they have to manage these multiple conversations. Petri Nets may be a good approach for resolving this problem. Petri Nets have been widely used for protocol modeling in distributed systems (Estraillier and Girault, 1992) and have also been used for modeling conversations and interactions in multi-agent systems (Coria, 1993; Ferber and Magnin, 1994; Lin et al., 1999). Figure 8.2 shows the representation of the conversation policy for a *Request,* using the Colored Petri Nets (CPN) approach (Lin et al., 1999). In Figure 8.2, each *speech act* is represented by a *transition* (T1 ~ T5) in CPN. The *state* is mapped through a combination of CPN *places* and structured *token* holding *messages. Arc expressions* are of two types: (1) *input arc expressions* determine which and how many *tokens* are removed from the *input places*; (2) *output arc expressions* determine which and how many *tokens* are created and input in the *output places*.

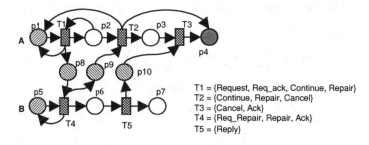

T1 = {Request, Req_ack, Continue, Repair}
T2 = {Continue, Repair, Cancel}
T3 = {Cancel, Ack}
T4 = {Req_Repair, Repair, Ack}
T5 = {Reply}

Figure 8.2 Representation of the conversation policy for a Request using
the CPN approach (Lin et al., 1999)

The advantages of this approach is that it not only represents the information being exchanged among agents, but also describes their internal states, thus describing the conversation process in much more detail.

Colored Petri Nets also facilitate describing an agent's multiple simultaneous communications with different agents. In this case, the diagram represents not only the agent states and transitions, but also the structure of running conversations.

Enhanced Dooley Graphs

The Dooley Graph was proposed by Dooley (1976). Each node of a Dooley Graph represents a *character*, a participant at a distinguished stage of a discourse. Utterances that resolve or complete one another tend to form tightly connected components of the graph, while those that take off in new directions spawn new components. For a more detailed description of both basic Dooley Graphs and Enhanced Dooley Graphs, refer to (Parunak, 1996). Figure 8.3 shows a Dooley graph (Parunak, 1996) based on the sample conversation described in Table 8.1.

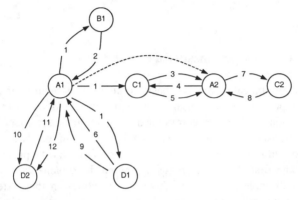

Figure 8.3 Representation of the conversation in Table 8.1 using Enhanced Dooley Graphs (modified from Parunak, 1996) (©1996 AAAI).

Table 8.1 Sample Conversation (modified from Parunak, 1996) (©1996 AAAI).

Seq.	Sender	Receiver	Utterance
1	A	B,C,D	REQUEST: Please send me 50 widgets at your catalog price by next Thursday.
2	B	A	REFUSE
3	C	A	PROPOSE: How about 40 widgets at catalog price by next Friday.
4	A	C	REQUEST: Please send me 40 widgets at your catalog price by next Friday.
5	C	A	COMMIT: I plan to send you 40 widgets at catalog price by next Friday.
6	D	A	COMMIT: I plan to send you 50 widgets at catalog price by next Thursday.
7	A	C	ASSERT: I have found a better supplier, and am not relying on your COMMIT.
8	C	A	REFUSE: I am abandoning my COMMIT.
9	D	A	SHIP: Here are your widgets. Please pay me.
10	A	D	ASSERT+REQUEST: You are five short. Please send the difference.
11	D	A	SHIP: Here are five more widgets. Please pay me.
12	A	D	PAY

According to Parunak (1996), Enhanced Dooley Graphs are useful both for analyzing the performance of existing communities of agents, and in designing new agents. An Enhanced Dooley Graph of the interactions that emerge from a community of agents can provide the basis for a number of quantitative measures relevant to the adaptability of the community. Because the graphs capture information about agent interaction in a way that makes sense to human designers and system operators, they can be applied not only to agents that reason explicitly about speech acts, but also to systems in which humans reason explicitly about speech acts.

8.2.6 Communication and Organizational Knowledge

Intelligent agent communication has two basic components: structural and functional. The structural components comprises the communication facilities provided through the design of the agents, which are communication ports, message encoding, and the interface to the transport medium. Every intelligent agent needs to have these facilities, but two additional functional components must exist to provide effective communication. The first relates to knowledge about the organization. Every agent needs to know or be able to find out about those entities in the system that are interested in receiving information. Second, this organizational knowledge must be complemented with the physical addresses of agents that can act as message receivers.

Defining the organizational knowledge necessary to achieve effective communication and maintain stability in a multi-agent system is a major issue. It would be inefficient to simply provide each intelligent agent with the address of every other agent in the system. Even if this were done, maintaining redundant knowledge at each agent place is not likely to solve all of the communication problems in large multi-agent systems. A lack of organizational knowledge may well result in ineffective behavior of intelligent agents trying to establish cooperation links. A consequence of the additional attempted communications will be an increase in the messaging overheads.

Another issue is how organizational and communication-related knowledge should be represented and managed efficiently. The detailed discussion on knowledge representation and management in agent-based concurrent design and manufacturing systems in Chapter 4 is relevant here.

8.3 COORDINATION

In more detail, the objectives of the coordination process are to ensure that all necessary portions of the overall problem are included in the activities of at least one agent, that agents interact in a manner which permits their activities to be developed and integrated into an overall solution, that team members act in a purposeful and consistent manner, and that all of these objectives are achievable within the available computational and resource limitations.

Lesser and Corkill, 1987

Coordination is the process by which an agent reasons about its local actions and the (anticipated) actions of others to try and ensure the community acts in a coherent manner.

Jennings, 1996

In this section, we first identify the coordination problems existed in agent-based concurrent design and manufacturing systems, then discuss the necessity of coordination in these systems, and finally describe the fundamental coordination mechanisms and techniques proposed in the literature for agent-based systems.

8.3.1 Coordination Problems in Concurrent Design and Manufacturing

Complex artifacts are developed in large manufacturing concerns by complex processes distributed across time, involving many participants and functional perspectives. The design of a commercial jet, for example, requires the integrated contribution of thousands of individuals spread over several continents and spanning possibly decades of time. Effective coordination is critical to the success of this cooperative process since the distributed activities are typically highly interdependent due to shared resources, input-output relationships and so on. The sheer complexity of these interdependencies has begun to overwhelm traditional manual organizational schemes and paper-based coordination techniques, resulting in often huge rework costs, slowed schedules and reduced product quality. As a result, while individual productivity may be high, the failure of existing coordination support practices and technologies has severe impact on the bottom line (Klein, 1995).

8.3.2 Necessity of Coordination in Agent-Based Systems

Coordination is central to the successful operation of agent-based systems, particularly to agent-based concurrent design and manufacturing systems which are very complex and whose stability is essential. Without coordination, a group of agents can quickly degenerate into a chaotic collection of individuals. This situation is absolutely unacceptable in real manufacturing enterprises.

There are important reasons why the actions of multiple agents need to be coordinated (Jennings, 1996; Nwana et al., 1996):

1) There is a need to prevent anarchy or chaos. Coordination is necessary and desirable because, with decentralization in agent-based systems, anarchy can

set in easily. No longer does any agent possess a global view of the entire system to which it belongs. Giving agents a global view is simply not feasible in any community of reasonable size and complexity. Consequently, agents only have their local views, goals and knowledge, which may conflict with those of others. Unless constrained, they can enter into all sorts of arrangements with other agents or agencies. As in any society, such haphazard arrangements are prone to anarchy. To achieve common goals without anarchy, a group of agents need to be coordinated.

2) There are dependencies between agents' actions. Interdependence occurs when goals undertaken by individual agents are related, either because local decisions made by one agent have an impact on the decisions of other agents (e.g., when building a house, designs about the size and location of rooms impact upon the wiring and plumbing) or because of the possibility of harmful interactions among agents (e.g., two AGVs may attempt to pass through a narrow path simultaneously, resulting in collision, damage to the AGVs, blockage of the path, and thus delay in production).

3) There is a need to meet global constraints. Global constraints exist when the solution being developed by a group of agents must satisfy certain overall conditions if it is to be deemed successful. For example, a monitoring system may have to react to critical events within 30 seconds. If individual agents acted in isolation and merely tried to optimize their local performance, then some overarching constraints are unlikely to be satisfied. Acceptable solutions can be developed only through coordinated action.

4) No individual agent has sufficient competence, resources, or information to solve the entire problem. Many problems cannot be solved by individual agents working in isolation, because they do not possess the necessary expertise, resources, or information. For example, it is not possible for an individual design agent (corresponding to a design tool) to complete the design of a complex artifact like a car or a plane; it is usually impossible for a single machine to fabricate a whole product. It may be impractical and undesirable to permanently synthesize the necessary components for solving a problem or satisfying the overall objective into a single entity because of historical, political, physical, or social constraints; therefore, temporary alliance through cooperative problem solving may be the only way to proceed. Different expertises may need to be combined to resolve a very difficult problem that is out of the scope of any one individual agent. Different agents (e.g., manufacturing resource agents) may have different resources which all need to be scheduled to produce a complete product. Finally, different agents may have different information or viewpoints of a problem (e.g., in concurrent design system, the same product may be viewed from a mechanical, electrical, economics, or environmental perspective).

5) Coordination can significantly increase efficiency. Even when individuals can function independently, thereby obviating the need for coordination,

information discovered by one agent can be of sufficient use to another agent that both agents can solve the problem faster.

The easiest way of ensuring coherent behavior would be to provide the group with an agent that has a wider perspective of the system. Such an agent then becomes the central controller of the system. This central controller could gather information from the agents in the group, create plans, and assign tasks to individual agents in order to ensure coherence. In fact, traditional centralized manufacturing systems used this approach, with sub-controllers arranged hierarchically under the central controller. Originally, the controllers were human agents, but more recently many of their functions became carried out by programmable logic controllers (PLC's). The difficulty of reconfiguring such hierarchical control systems, as production demands change, is one of the reasons for industrial interest in multi-agent systems (which have the potential for reconfiguration).

Using a single central controller for a large group of agents has an obvious problem. As the size of the group increases it becomes very difficult for the central controller to be informed of all the agents' beliefs and intentions. Also such a controller can become a severe communication bottleneck and would render the remaining components unusable if it failed. Agent technology provides a natural way of overcoming this problem. We have introduced several related concepts in Chapter 7 when considering system architecture. In the following subsections, we present a number of fundamental mechanisms and techniques that can be used to support multi-agent coordination.

8.3.3 Fundamental Coordination Mechanisms for Agent-Based Systems

Among the various coordination mechanisms described for agent-based systems, five are fundamental:

- *Mutual Adjustment*: Agents share information and resources to achieve some common goal, by adjusting their behavior according to the behavior of the other agents. In order to do so, agents usually need exchange considerable information and make many adjustments. No agent has prior control over the others and the decision-making is a joint process. It is the simplest form of coordination. Coordination in peer groups and in markets is usually by mutual adjustment.
- *Direct supervision:* One agent has some degree of control over others and can control the use of shared resources by the subordinate agents and may also prescribe certain aspects of their behavior. Such a supervision relationship is often established through mutual adjustment (e.g. following acceptance of employment or a contract).
- *Coordination by Standardization:* The supervisor agent coordinates other agents through standardization, by establishing standard procedures for subordinate agents to follow situations they encounter. In mutual adjustments,

such standard procedures are implemented by acceptance, while in direct supervision, they are implemented through (mandatory) requests. Routine procedures in companies or computer programs are examples of coordination by standardization.

- *Mediated Coordination:* A mediator acts as a facilitator (e.g. finds/routes information), and as a broker (a 'go-between' and advisor on resource negotiations or supervisor (exercising some degree of direct supervision). The first role is mandatory, but the others are optional. A mediator facilitates or brokers mutual adjustment between agents and may also use direct supervision.

- *Coordination by Reactive Behavior:* Agents react to particular stimuli ("situations") with specific behaviors ("actions"). With appropriately selected or evolved stimuli-behavior groupings and distributions, system-level patterns of coordinated behavior emerge so as to contribute to the achievement of common or system goals.

There are also other mechanisms for multi-agent coordination described in the literature such as Markov Tracking (Washington, 1998), coordination techniques dealing with commitments and conventions (Jennings, 1996), and knowledge based reasoning techniques (Mazer, 1991).

8.3.4 Techniques Supporting Multi-Agent Coordination

There have been many approaches devised to achieve coordination in agent systems. These can be classified in the following four categories:

- Organizational Structuring
- Subcontracting
- Negotiation
- Multi-Agent Planning

In this subsection, we introduce three categories of coordination techniques: organizational structuring, subcontracting and multi-agent planning. Negotiation will be described separately in Chapter 10.

8.3.4.1 Organizational Structuring

This is the simplest coordination approach, which exploits the properties of the organizational structure. This is because the structure and relationships of the organization implicitly define agent responsibilities, capabilities, connectivity and control flow. It provides a framework for activity and interaction through the definition of roles, communication paths and authority relationships. Durfee et al. (1987) define organizational structuring in terms of the predefined long-term relationships between agents.

Hierarchical organizations are commonly found, using classic master–slave or client-server coordination, for task and resource allocation among slave agents by a master agent. Such coordination can be implemented in the following ways:

- The master agent plans and distributes fragments of the plan to the slaves. The slaves may or may not communicate amongst themselves but must ultimately report their results to the master agent. In this case, while the master has full autonomy with respect to the slaves, the slaves have only partial autonomy with respect to their master.

- Blackboard negotiation which exploits the classic blackboard architecture (Hayes-Roth, 1985) as the mechanism to support coordination. In this scheme, the blackboard's knowledge sources are replaced by agents who post to and read from the general blackboard. The scheduling agent (or master agent) schedules the agents' reads/writes to/from the blackboard. This approach may be used when the problem is distributed, a central scheduling agent is present or when tasks have already been assigned, a priori, to agents.

The second approach above highlights the fact that organizational structures ought not to be considered as being solely hierarchies. For example, in the DVMT system (Lesser and Corkill, 1983) that also exploits a blackboard, coordination occurs amongst peer agents.

In the blackboard coordination scheme, with no direct agent-to-agent communication, a severe bottleneck may result if there are many agents, even in the case of multi-partitioned blackboards as in CAAD (Branki and Bridges, 1993). Further, all agents would need to have a common domain understanding (i.e. semantics). For this latter reason, most blackboard systems tend to have homogeneous and rather small-grained agents, as is the case in the DVMT prototype (Lesser and Corkill, 1983). Even when a master–slave coordination technique is used, the designer should ensure that the slaves are of sufficient granularity to compensate for the overheads that result from goal distribution. Distributing trivial or small tasks can be more expensive than performing them in one location.

Other organizational structures also exist, including the centralized and decentralized market structures. The centralized market structure employs a master–slave coordination approach while a subcontracting technique, described in the next subsection, is more suitable for a decentralized market structure. Market structures can be considered as a special form of group organization based on mutual adjustment. Agents, each of whom controls scarce resources (such as labor, raw material, good, and money), agree to share some of their respective resources to achieve some mutual goal. Valued resources are exchanged, either with or without explicit prices. Once a contract is made, there is an agreement in which the buyer becomes the supervisor (master) of the supplier.

The above strategies are useful where there are master–slave relationships in the multi-agent systems to be modeled. Much control is exerted over the slaves' actions, and hence over the problem-solving process. However, such control, in the extreme, militates against the benefits of DAI: speed (due to parallelism), reliability, concurrency, robustness, graceful degradation, minimal bottlenecks, etc.

Direct supervision is probably the most common form of coordination used when exploiting the properties of an organization structure. Mutual adjustment works very well in small organizations. However, as the size of the group (and the number of tasks) grows, the number of information links and the amount of information to be exchanged become prohibitive very quickly. A large group can, in principle, be coordinated efficiently when divided into subgroups: if most of the necessary information transfer can occur within subgroups, and if the few interactions between subgroups can be handled by supervisors. Subgroups may be coordinated either by mutual adjustment or by hierarchical control (supervision), depending on the application domain and the task characteristics.

8.3.4.2 Subcontracting

A classic technique for coordinating task decomposition and allocation of resources is the Contract Net approach (Smith, 1980) which was briefly mentioned in Chapter 3 (Section 3.3.2). In this approach, an agent can assume two roles:

- A manager who breaks a problem into sub-problems and searches for contractors to do them, as well as monitoring the problem's overall solution;
- A contractor who does a subtask. However, a contractor may become a manager and further decompose the subtask and subcontract its subtasks out to other agents.

These processes are thus recursive. This is a completely distributed scheme where a node can be simultaneously both manager and contractor. This approach has been widely used in many agent-based concurrent design and manufacturing systems.

The contract net is a high-level coordination strategy that also provides a way of distributing tasks, and a means for self-organizing agents in a group. It is best used when the following are true:

- the application task has a well-defined hierarchical nature;
- the problem has a coarse-grained decomposition;
- there is minimal coupling among subtasks.

The advantages of the contract net include the following:

- dynamic task allocation via self-bidding, which leads to better agreements;
- natural load-balancing (as busy agents need not bid);
- allowing agents to be introduced and removed dynamically;
- providing a reliable mechanism for distributed control and failure recovery.

Its limitations include the fact that it does not presume agents have contradictory demands; hence, the approach neither detects nor resolves conflicts (which is one key reason why coordination was needed in the first place). The agents in the contract net are relatively passive, benevolent and non-antagonistic which for many real world problems is unrealistic. In addition, the contract net approach is quite communication-intensive and these costs may outweigh some of its advantages in real world applications. Conry et al.'s generalization of the contract net approach essentially introduces an iterative mechanism for getting agents with conflicting goals to arrive at a consensus (Conry et al., 1986). Other researchers have also proposed extended or modified contract net protocols (see Chapter 10 Section 10.3.1 for details).

8.3.4.3 Multi-Agent Planning

A multi-agent plan is a plan that has been generated for multiple executing agents. Multi-agent planning is the process of creating a multi-agent plan. In order to achieve their goals but avoid inconsistent or conflicting action or interaction, agents build a multi-agent plan that details their required future actions and interactions (up to some time or event horizon) and then may interleave execution with more planning and re-planning.

With this approach, agents know in advance exactly what actions they will take and what interactions will occur. By requiring such a complete specification of behavior, the plan can realistically only have a short-term horizon due to the unpredictability of events in the environment.

Multi-agent planning can be classified into two categories:

- Centralized multi-agent planning;
- Distributed multi-agent planning.

Centralized multi-agent planning

In centralized multi-agent planning, there is typically a coordinating agent who, on receipt of all partial or local plans from individual agents, analyzes them in order to identify potential inconsistencies and conflicting interactions (such as conflicts between agents over limited resources). The coordinating agent then attempts to modify these partial plans and combines them into a multi-agent plan where conflicting interactions are eliminated. In the following, centralized multi-agent planning is illustrated through three examples.

Cammarata et al. (1983) employed centralized multi-agent planning for the air-traffic control problem. Agents (airplanes) in a potential conflict scenario first negotiate among themselves to choose a coordinator (coordinating agent). The coordinator then creates a multi-agent plan to be performed by all involved airplanes.

Georgeff uses the approach where conflicting interactions are identified and grouped into critical regions (Georgeff, 1983, 1984). The plans of individual agents are formed first, and then a central planning agent collects the local plans and analyzes them to identify potential conflicts. This planning agent tries to solve the conflicts by modifying the local plans and introducing communication commands into them so that those agents can synchronize their interactions appropriately.

Jin and Koyama (1990) applied centralized multi-agent planning, in a somewhat different way, in their MATPEN model for coordinating autonomous and distributed agents. Their approach exploits an "expectation based negotiation protocol" which draws from the roles of agents in the organizational structure. When two agents share a conflict, they form a conflict group and initiate a negotiation process. They exchange "expectations" (which represent the expected behavior of other agents in order to resolve the conflict) in order to decide who should play what role in the negotiation process. Eventually, they generate a multi-agent plan to resolve the conflict. Such an approach has been used to construct a collision avoidance system for ships using MATPEN.

Distributed multi-agent planning

In distributed multi-agent planning, the basic idea is to provide each agent with a model of the other agents' plans. Agents communicate in order to build and update their individual plans and their models of others' until all conflicts are removed. The planning activities are therefore divided among the involved agents. This approach is usually used when there is not a single agent with a global view of the activities of all involved agents. In the following paragraphs, we also use three examples to illustrate the distributed multi-agent planning approach.

In Lesser and Corkill's Functionally Accurate and Cooperative (FA/C) approach (Lesser and Corkill, 1981), loosely coupled agents form high level (but possibly incomplete) plans, results and hypotheses which they exchange with each other. Next, they refine these until they converge in some global complete plan. Local inconsistencies are detected and only the part of each agent's results that are consistent with local information are integrated into the local databases. This approach has been used in Lesser and Corkill's Distributed Vehicle Monitoring Testbed (DVMT) which is a system for testing coordination strategies (Lesser and Corkill, 1983).

Another example of distributed multi-agent planning is Durfee's Partial Global Planning (PGP) (Durfee and Lesser, 1987). In this approach, agents execute their local plans while these being modified continuously based on partial global plans (built by exchanging local plans). These agents always try to incorporate potential improvements in their group coordination.

Chu-Carroll and Carberry (1996) propose a recursive *propose-evaluate-modify* framework of actions for multi-agent collaborative planning. In this framework, an agent engaged in collaborative planning proposes modifications to the agents' shared plan under development, while the other agent evaluates and suggests modification to this proposal if unacceptable. This process then recurs. Proposing a modification itself presents two challenges: (i) selecting a focus for modification; and (ii) for the chosen focus of modification, selecting appropriate evidence to justify the modification. The authors present several heuristics to address these issues, and back these up with examples of collaborative negotiation.

In distributed multi-agent planning, two types of activity can be distinguished: task-driven planning and plan coordination. In task-driven planning systems, there is an initial goal or task that is decomposed into subgoals or subtasks and distributed among several planners (top-down problem decomposition). In contrast, plan coordination deals with situations where the agents' plans are pre-existing and the problem is to reconcile the given plans before they are performed in a common environment. As noted, although the extremes can exist, the two activities of task-driven planning and plan coordination are often interleaved.

A brief comparison of two approaches

Generally, multi-agent planning of any form requires that agents share and process substantial amounts of information; hence, it is likely to require more computing and communication resources than other approaches. The centralized multi-agent planning technique shares many of the limitations of the master–slave coordination technique. Naturally, coordination in distributed multi-agent planning is much more complex than in centralized planning as there are typically no agents involved who possess a global view of the distributed system. Since coordination in some multi-agent planning techniques such as PGP is a gradual process, they may be more suitable for some domains than others.

8.4 COOPERATION

Cooperation is often presented as one of the key concepts which differentiates multi-agent systems from other related areas such as distributed computing, object-oriented systems, and expert systems. However it is a concept whose precise usage in agent-based systems is at best unclear and at worst highly inconsistent.

An extensive discussion on definitions of cooperation in multi-agent systems can be found in (Doran et al., 1997). Three significantly different definitions of cooperation are described, ranging from Franklin's broad definition, Doran's weak definition, and Norman's strong definition. (For more detail, interested readers may refer to (Doran et al., 1997) and the references cited therein)).

According to Franklin's broad definition, cooperation exists in almost all multi-agent systems. Agents can even cooperate with no intention of doing so (also called accidental cooperation by some researchers).

According to Doran's weak definition, given a multi-agent system in which the individuals and the various sub-groups may be assigned one or more (possibly implicit) goals, cooperation occurs when the actions of each agent satisfy either or both of the following conditions:

1)   The agents have a (possibly implicit) goal in common, (which no agent could achieve in isolation) and their actions tend to achieve that goal;

2) The agents perform actions that enable or achieve not only their own goals, but also the goals of agents other than themselves.

This definition focuses on the actions of the agents and their goals, irrespective of how they arise. Thus, like Franklin's definition, Doran does not require that the agents deliberate and in some sense "intend" to cooperate for cooperative problem solving to occur. Also the definition does not require that the goals be explicit within the agents.

Norman proposed the strong definition that "to cooperate is to act with another or others for a common purpose and for common benefit", that is, the agents must always have an intention to cooperate with others.

In this book, we adopt Doran's weak definition, which we found more relevant to most agent-based concurrent design and manufacturing systems.

8.4.1 Degrees of Cooperation

The following degrees of cooperation characterize the extent of cooperation among agents:

-   fully cooperative (also called total cooperative);
-   partly cooperative;
-   antagonistic.

Fully cooperative agents that are able to resolve non-independent problems often pay a price in high communication costs. These agents may change their goals to suit the needs of other agents in order to ensure cohesion and coordination. In contrast, antagonistic agents may not cooperate at all and may even block each other's goals. Communication costs required by these systems are generally minimal. In the middle of the spectrum lie traditional systems that often lack a clearly articulated set of goals and whose agents are at least partly cooperative. Note that most real systems have a weak degree of cooperation.

Fully cooperative agents exist in Cooperative Distributed Problem Solving (CDPS) systems where agents work together in a loosely coupled network to solve problems that are beyond their individual capabilities. In this network, each agent is capable of resolving complex problems and can work independently, but the problems faced by the agents cannot be completed without full cooperation. This level of cooperation is necessary because no single agent has sufficient knowledge and resources to solve a given problem, although different agents might have the expertise to separately solve its different parts (i.e., subproblems). CDPS agents cooperatively build a solution to a problem by using their local knowledge and resources to individually solve subproblems, and then by integrating the solutions for these subproblems into an overall solution.

CDPS is particularly useful when distributed control becomes uncertain. The presence of uncertainties, such as incomplete and imprecise information, makes the need for full cooperation even more crucial. CDPS is also useful when each agent has specific knowledge and does not have a clear idea about what it might do or what information it might exchange with others. In this context, Smith's Contract Net (see Chapter 3) provides a cooperative framework to minimize communication, allow load balancing and distributed control, while also maintaining coherent behavior.

A comprehensive discussion of the spectrum from fully cooperative agents to antagonistic agents can be found in (Genesereth et al., 1986). They noted that while DAI researchers commonly accept full cooperation, it is not always true in the real world where agents may have conflicting goals. Such conflicting goals are illustrated by a set of personal meeting-scheduling agents where each agent tries to schedule a meeting at the best time for its particular owner. To study this type of interaction, Genesereth et al. define agents as entities operating under the constraints of various rationality axioms that restrict their choices.

8.4.2 Cooperation Primitives

Cooperation primitives are the basic building blocks of cooperation. They represent a transfer of knowledge from one agent to one or more other agents, with a specific intention. As such, they are represented as shared multi-agent plans, whose preconditions and effects fix the semantics and intention of the primitives and whose procedures are synchronized calls to the communication actions. A cooperation primitive consists of an illocutionary speech act operating upon a goal, a plan, a task, or on arbitrary knowledge (a cooperation object). A sample set of cooperation primitives and their intended use is as follows:

- propose: A proposal (of a cooperation object) starts or continues a discussion among agents. The knowledge transferred by a proposal to other agents is in some sense hypothetical, as the agents sharing this knowledge have not yet committed to it (e.g., "propose: use helical gear as driving gear").
- refine: An agent proposes a further instantiation of the cooperation object (e.g., "refine: use helical gear with module 2.0 as driving gear").
- modify: A counter-proposal of an altered cooperation object (e.g., "modify: use helical gear with module 2.4 as driving gear").
- accept: Indication of commitment to the cooperation object.
- reject: Indication of failure to commit to the cooperation object.
- order: The recipient must accept the cooperation object, usually a task. The recipient may still fail to execute the task, in which case a task-failed result is returned. An order is only applicable if either the object has been previously discussed (and committed to) or an appropriate authority link between sender and recipient is established.
- request: A query for arbitrary knowledge.
- tell: The answer to a previous information request.

The cooperation object plays an important role in determining the semantics of cooperation primitive. For example, if an agent has an active goal it cannot handle by itself, it must get another agent to cooperate by adopting the same goal. Only when this is accomplished, can it initiate cooperation about the plan for achieving the goal.

In order to provide the flexibility required by human participants in a cooperative process, the dynamic definition of new cooperation primitives should be allowed. For example, one could be allowed to introduce message types such as support and oppose in order to more efficiently communicate about intentions.

8.4.3 Cooperation Methods

The semantics of a cooperation primitive cannot rest upon the single message alone, but also upon the history of exchanged messages and the expected replies; that is, the entire dialogue must be taken into consideration. Such a dialogue consisting of a sequence of cooperation primitives is called a cooperation method. Cooperation methods are represented as domain-independent multi-agent plans that can be used to construct and execute domain-specific multi-agent plans. Cooperation methods can be predefined or established dynamically. The latter allows for the flexibility required for human participation.

According to Ferber (1995), we may find six different cooperation methods in the literature:

- Grouping and multiplication
- Communication
- Specialization
- Coordination
- Collaboration by sharing tasks and resources
- Conflict resolution through arbitration and negotiation

8.4.3.1 Grouping and Multiplication

Grouping is one of the simplest methods for agent cooperation, which means that a group of agents get together physically to constitute an assembly or a communication network for solving a specific problem. Such cooperation is widely used in the animal world and in human societies.

Other terms used in the agent community similar to *grouping* are *agent coalition* and *clustering*. Cooperative activity among distributed agents requires the formation of agent coalitions to bond dissimilar agents into harmonious decision-making groups. Agent coalitions are composed of self-interested agents. Since each agent participating in a coalition has its individual representation of the external world, goals, and constraints, the coalitions require the interaction of diverse heterogeneous beliefs through distributed cooperation models.

The effectiveness and performance of cooperative multi-agent systems can be improved through the integration of selected groups of agents previously identified. Such a pre-selection mechanism accelerates agent coalition formation and task resolution. Some researchers have also proposed various mechanisms for dynamic agent coalition formation, for example, selecting dynamically related agents to form agent coalitions (or clusters) for resolving emergent tasks (Lander and Lesser, 1993; Maturana et al., 1999).

Multiplication is a simple augmentation of individual agents in a given multi-agent system to enhance the system's performance. Agent cloning is an important mechanism for implementing multiplication. For example, in an agent-based manufacturing scheduling system (Shen and Norrie, 1998), resource agents are partially cloned as needed in order to reduce scheduling time through parallel computation. These partially cloned agents are included in virtual clusters where agents negotiate with each other to find the best solution for a production task. Decker et al. (1997) used a similar cloning agent approach as one of the possible responses for an information agent to overload conditions.

8.4.3.2 Communication

As mentioned at the beginning of this Chapter, communication is the most important means of facilitating cooperation among agents. Communication can augment the agents' perspectives and capabilities by getting information and know-how from other agents. The different aspects of communication in multi-agent systems have already been discussed in Section 8.2.

8.4.3.3 Specialization

Specialization is a process in which agents become more and more adaptive to their tasks (Ferber, 1995). It is difficult for agents to be specialized in all of the different tasks in a multi-agent system. The successful resolution of a task typically requires certain structural and behavioral characteristics that may not be employed to resolve other tasks effectively. The evidence for this can be easily found in any manufacturing enterprise, where different machines complete different production operations and specialized workers do their specific tasks effectively. The cooperation approach is therefore very useful for developing agent-based manufacturing systems.

8.4.3.4 Coordination

Coordination is one of most important methods of ensuring cooperation in multi-agent systems, though cooperation may not necessarily result in coordination (see Section 8.5 for a detailed discussion). Different aspects of coordination in multi-agent systems have been described in Section 8.3.

8.4.3.5 Collaboration by Sharing Tasks and Resources

Collaboration is an important type of cooperation in multi-agent systems. Collaboration means that a group of agents work together on a common project or a task (Ferber, 1995). It includes all techniques which may be used by these agents for task decomposition and resource allocation. Collaboration can be considered as a response to the question "who does what?" on a given project or a task.

A number of mechanisms and techniques have been proposed in the literature for task decomposition and resource allocation, ranging from centralized to distributed approaches. The best-known technique is the bidding mechanism using the Contract Net protocol (Smith, 1980). Chapter 9 will be entirely focused on different techniques for task decomposition and resource allocation.

8.4.3.6 Conflict Resolution through Arbitration and Negotiation

Arbitration and negotiation are two techniques widely used for conflict resolution in multi-agent systems and for ensuring the stability and performance of these systems. Arbitration is usually a decision-making mechanism based on predefined rules for individual agents in the system and for the whole society of agents. Arbitration is not well suited for deliberative (cognitive) agent systems, whereas negotiation techniques are. Chapter 10 will be entirely focused on technique for conflict resolution through negotiation in multi-agent systems.

8.4.4 Cooperation in Agent-Based Design and Manufacturing Systems

Cooperation exists in almost every agent-based concurrent design and manufacturing systems. In this section, we provide a brief discussion of two cases using agent-based approaches for: (1) concurrent engineering design; (2) shop floor AGV control.

8.4.4.1 Concurrent Engineering Design

Concurrent engineering design projects, particularly concerning large and complex products such as automobiles, locomotives, airplanes, etc., require the cooperation of multidisciplinary design teams using several sophisticated and powerful engineering tools such as commercial CAD tools, engineering database systems, and knowledge-based systems. Individuals or groups of multidisciplinary design teams work in parallel and independently with various engineering tools located on different sites. At any moment, individual members may be working on different versions of a design or viewing the design from various perspectives (such as profitability, manufacturability, resource capability and capacity), at various levels of details.

Agent-based approaches provide a natural way to enable cooperation in these systems and circumstances. Chapter 13 describes in detail a number of systems of this kind.

8.4.4.2 Shop Floor AGV Control

A free-ranging AGV (Automatic Guided Vehicle) system is often used in a flexible manufacturing system for providing material transportation, with both flexibility and efficiency. The operational control of such free-ranging AGV systems can be very complicated since each AGV uses a 'software path' and sensors to locate its position. AGVs can travel in either direction on a path and can go around each other to avoid collision or blocking. There may be different AGVs on the shop floor, some being large, some small. Only the latter could be used on a narrow path. Such systems have many more possibilities for path selection, AGV dispatching and AGV collision avoidance than do the guided-path unidirectional AGV systems. To model such a complex system using the operational research approach becomes extremely difficult. Heuristic approaches may, however, be used to produce approximate results which are suitable for the dynamic environment.

An agent-based approach has been proposed to solve this complex free-ranging AGV problem (Kwok and Norrie, 1993). One of the key issues in implementing such an agent-based AGV system is the cooperation among AGV agents and some other agents like machine agents. In a traditional approach, a centralized program or an algorithm (e.g., search) is usually used to find an 'optimal' solution for a transportation task (or a number of transportation tasks). In an agent-based AGV system, when a transportation task arises, an 'optimal' solution is obtained by cooperation realized through negotiation among AGV agents and other agents in the system.

8.5 COORDINATION, COOPERATION AND COMMUNICATION

In this section, we briefly discuss the relationships between coordination, cooperation and communication according to the definitions used in this book,.

Coordination may require cooperation; but cooperation among a set of agents would not necessarily result in coordination; indeed, it may result in incoherent behavior. This is because for agents to cooperate successfully, they (typically) maintain models of each other as well as develop and maintain models of future interactions. If agents' beliefs about each other are wrong, for example, incoherent behavior may well result. Coordination may occur without cooperation: for example, if a person is running towards you, and you get out of his way, you have coordinated your actions with his. However, you have not entered into cooperation with him. Likewise, non-cooperation among agents does not necessarily lead to incoherent behavior (it may just happen to end up coordinated); however, it is likely to do so. Competition is a form of coordination involving antagonistic agents.

To enable coordination or cooperation, it is usual for agents to communicate with one another. But, agents may achieve coordination or cooperation without communication, provided they possess models of each other's behaviors (Genesereth et al., 1986; Fenster et al., 1995; Franklin, 1997). However, to facilitate coordination, where agents have to cooperate through communication, it is vital that they make known their goals, intentions, results and state to other agents.

8.6 RESEARCH LITERATURE AND FURTHER REFERENCES

Communication, Cooperation and Coordination are three central issues in agent-based systems. They were, are and still will be very exciting and relevant research topics. Numerous papers in these areas can be found in the proceedings of relevant conferences and workshops (ICMAS, Autonomous Agents, ATAL, etc.).

For readers particularly interested in specific topics, we provide here some additional references:

- "Cooperation in Industrial Multi-Agent Systems" (Jennings, 1994)
- "Communication and Cooperation in Agent Systems" (Haddadi, 1996)
- "Coordination in Artificial Agent Societies" (Ossowski, 1999)
- "Coordination in Software Agent Systems" (Nwana et al., 1996)
- Mark Klein's web site [http://ccs.mit.edu/klein/]

REFERENCES

Allen, J.F. (1986) *Natural Language Understanding*, Benjamin/Cummings, Menlo Park, CA.

Austin, J.L. (1962) *How to Do Things with Words*, Oxford University Press.

Baker, A.D. (1991) *Manufacturing Control with a Market-Driven Contract Net*, PhD thesis, Rensselaer Polytechnic Institute, NY.

Barbuceanu, M., and Fox, M.S. (1995) COOL: A Language for Describing Coordination in Multi Agent Systems, In *Proceedings of the First International Conference on Multi-Agent Systems*, The MIT Press/The AAAI Press, pp. 17-24.

Bradshaw, J.M., Dutfield, S., Benoit, P. and Woolley, J.D. (1997) KAoS: Toward an industrial-strength open agent architecture, In J.M. Bradshaw, ed., *Software Agents*, AAAI/MIT Press, pp. 375-418.

Branki, N.E. and Bridges, A. (1993) An Architecture for Cooperative Agents and Applications in Design, In M.R. Beheshi and K. Zreik, eds., *Advanced Technologies*, pp. 221-230.

Cammarata, S. and McArthur, D. and Steeb, R. (1983) Strategies of Cooperation in Distributed Problem Solving, In *Proceedings of the Eighth International Joint Conference on Artificial Intelligence*, pp. 767-770.

Chang, M.K. and Woo, C.C. (1991) SANP: A Communication Level Protocol for Negotiation, In *Proceedings of 3rd MAAMAW*, Kaiserslautern.

Chu-Carroll, J. and Carberry, S. (1996) Conflict Detection and Resolution in Collaborative Planning, In Wooldridge, M.J., Müller, J.P. and Tambe, M. eds., *Intelligent agents II: Agent Theories, Architectures, and Languages*, LNAI 1037, Springer, pp. 111-126.

Conry, S.E., Meyer, R.A. and Lesser, V R. (1986) Multistage Negotiation in Distributed Planning, COINS Technical Report, University of Massachusetts, Amherst, Boston.

Coria, M. (1993) Stepwise Development of Correct Agents: a Behavioural Approach Based of Coloured Petri Nets, In *Proceedings of the Canadian Conference on Electrical and Computer Engineering*, Vancouver, Canada.

Decker, K., Sycara, K. and Williamson, M. (1997) Cloning for Intelligent Adaptive Information Agents, In C. Zhang and D. Lukose, eds., *Multi-Agent Systems: Methodologies and Applications*, LNAI 1286, Springer, pp. 63-75.

Dooley, R.A. (1976) Repartee as a Graph, In R.E. Longacre, ed., *An Anatomy of Speech Notions*, pp. 348-58.

Doran, J.E., Franklin, F., Jennings, N.R and Norman, T.J. (1997) On Cooperation in Multi-Agent Systems, *Knowledge Engineering Review*, 12(3):309-314.

Durfee, E.H. and Lesser, V.R. (1987) Using Partial Global Plans to Coordinate Distributed Problem Solvers, In *Proceedings of the Tenth International Joint Conference on Artificial Intelligence*, pp. 875-83.

Durfee, E.H., Lesser, V.R., and Corkill, D.D. (1987) Coherent Cooperation among Communicating Problem Solvers, *IEEE Trans. Computer*, 11:1275-1291.

Estraillier, P. and Girault, C. (1992) Applying Petri Net Theory to the Modelling, Analysis and Prototyping of Distributed Systems, In *Proceedings of the IEEE International Workshop on Emerging Technologies and Factory Automation*, Cairns, Australia.

Fenster, M., Kraus, S. and Rosenschein, J.S. (1995) Coordination without Communication: Experimental Validation of Focal Point Techniques, In *Proceedings of the First International Conference on Multi-Agent Systems,* San Francisco, CA, AAAI Press/The MIT Press, pp. 102-108.

Ferber, J. and Magnin, L. (1994) Conception de systèmes multi-agents par composants modulaires et réseaux de Petri, In *Actes des jounées du PRC-IA*, Montpellier, France.

Ferber, J. (1995) *Les systèmes multi-agents*, InterEditions, Paris.

Finin, T., Fritzon, R., McKay, D. and McEntire, R. (1993) KQML – A Language and Protocol for Knowledge and Information Exchange, Tech. Report, University of Maryland, Baltimore, Maryland.

Fischer, K., Müller, J.P., Pischel, M. and Schier, D. (1995) A Model for Cooperative Transportation Scheduling, In *Proceedings of the First International Conference on Multi-Agent Systems,* San Francisco, CA, AAAI Press/The MIT Press, pp. 109-116.

Frankel, R. (1991) The ToolTalk Service, A Technical Report, SunSoft.

Franklin, S. (1997) Coordination without Communication. Available through the Internet at: http://www.msci.memphis.edu/~franklin/coord.html

Gaspar, G. (1991) Communication and Belief Changes in a Society of Agents, Towards a Formal Model of Autonomous Agents, In Y. Demazeau and J.-P. Müller, eds., *Decentralized Artificial Intelligence* 2, Elsevier/North-Holland, Amsterdam, pp. 71-88.

Genesereth, M.R., Ginsberg, M.L., and Rosenschein, J.S. (1986) Cooperation without communication, In *Proceedings of 1986 AAAI*, Philadephia, pp. 51-57.

Georgeff, M. (1983) Communication and Interaction in Multi-Agent Planning, In *Proceedings of 1983 National Conference Artificial Intelligence (AAAI'83)*, pp. 125-129.

Georgeff, M. (1984) A Theory of Action for Multi-Agent Planning, In *Proceedings of 1984 National Conference Artificial Intelligence (AAAI'84)*, pp. 121-125.

Haddadi, A. (1996) *Communication and Cooperation in Agent Systems: A Programatic Theory*. Springer.

Hayes-Roth, B. (1985) A Blackboard Architecture for Control, *Artificial Intelligence*, 26:251-321.

Hewitt, C., Bisshop, P. and Steiger, R. (1973) A Universal Modular Actor Formalism for Artificial Intelligence, In *Proceedings of the 3rd IJCAI*, Stanford, CA, 1973, pp 235-245.

Hewitt, C. (1977) Viewing Control Structures on Patterns of Passing Messages, *Artificial Intelligence*, 8:323-364.

Huhns, M.N. (ed.) (1987) *Distributed Artificial Intelligence*, Morgan Kanfmann, San Mateo, CA/Pitman, London.

Jennings, N.R. (1994) *Cooperation in Industrial Multi-Agent Systems*, World Scientific, Singapore.

Jennings, N.R. (1996) Coordination Techniques for Distributed Artificial Intelligence, In O'Hare, G.M.P. and Jennings, N.R., eds., *Foundations of DAI*, Wiley & Sons, pp. 187-210.

Jin, Y. and Koyama, T. (1990) Multi-agent planning through expectation based negotiation, In *Proceedings of the 10th International Workshop on DAI*, Texas.

Klein, M. (1995) iDCSS: Integrating Workflow, Conflict and Rationale-based Concurrent Engineering Coordination Technologies, *Concurrent Engineering: Research and Application,* 3(1):21-27.

Kwok A. and D.H. Norrie (1993) Intelligent Agent Systems for Manufacturing Applications, *Journal of Intelligent Manufacturing*, 4:285-293.

Labrou, Y. and Finin, T. (1997) A Proposal for a new KQML Specification. Technical Report, TR CS-97-03, Computer Science and Electrical Engineering Department, University of Maryland Baltimore.

Lander, S. and Lesser, V. (1993) Understanding the Role of Negotiation in Distributed Search among Heterogeneous Agents, In *Proceedings of the International Joint Conference on Artificial Intelligence*, Chambery, France, pp. 438-444.

Lesser, V. and Corkill, D. (1981) Functionally Accurate, Cooperative Distributed Systems, *IEEE Transactions on Systems, Man and Cybernetics*, 11(1):81-96.

Lesser, V. and Corkill, D. (1983) The Distributed Vehicle Monitoring Testbed: A Tool for Investigating Distributed Problem-Solving Networks, *AI Magazine*, 4(3):15-33.

Lesser, V.R. and Corkill, D.D. (1987) Distributed problem solving, In S.C. Shapiro, ed., *Encyclopedia of Artificial Intelligence*, Wiley, New York, pp. 245-251.

Lin, F., Norrie, D.H., Shen, W. and Kremer R. (1999) Schema-based Approach to Specifying Conversation Policies, In *Proceedings of Agents'99 Workshop on Specifying and Implementing Coversation Policies*, Seattle, WA.

Maturana F., Shen W. and Norrie, D.H. (1999) MetaMorph: an Adaptive Agent-Based Architecture for Intelligent Manufacturing, *International Journal of Production Research*, 37(10):2159-2174.

Mazer, M.S. (1991) Reasoning about knowledge to understand distributed AI systems, *IEEE Trans. Syst. Man Cybernet.*, 21(6):1333-1346.

Moulin, B. and Chaib-draa, B. (1996) An Overview of Distributed Artificial Intelligence, In O'Hare, G.M.P. and Jennings, N.R., eds., *Foundations of DAI*, Wiley & Sons, pp. 1-56.

Nwana, H., Lee, L. and Jennings, N. (1996) Coordination in Software Agent Systems, *BT Technology Journal*, 14(4).

Ossowski, S. (1999) *Coordination in Artificial Agent Societies*, Lecture Notes in Artificial Intelligence 1535, Springer.

Parunak, H.V.D. (1996) Visualizing Agent Conversations: Using Enhanced Dooley Graphs for Agent Design and Analysis, In *Proceedings of the Second International Conference on Multi-Agent Systems*, The AAAI Press, pp. 275-282.

Rosenschein, J.S. (1986) *Rational Interactions: Cooperation among Intelligent Agents*, Ph.D. Thesis, Computer Science Department, Stanford University, Stanford, CA.

Rosenschein, J.S. and Breese, J.S. (1989) Communication-Free Interactions among Rational Agents. In L. Gasser and M.N. Huhns, eds., *Distributed Artificial Intelligence 2*, Morgan Kaufmann, Los Altos, CA/Pitman, London, pp. 99-118.

Sandholm, T. and Lesser, V. (1996) Advantages of a Leveled Commitment Contracting Protocol, In *Proceedings of the Thirteenth National Conference on Artificial Intelligence (AAAI-96)*, Portland, OR, pp. 126-133.

Scalabrin, E. (1996) *Conception et Réalisation d'environnement de développement de systèmes d'agents cognitifs*, PhD Thesis, Université de Technologie de Compiègne, France.

Searle, J.R. (1969) *Speech Acts*, Cambridge University Press.

Searle, J.R., Kiefer, F. and Bierwisch, M. (1980) *Speech Act Theory and Pragmatics*, Dordrecht, Holland.

Shen, W. and Barthès, J.P. (1996) An Experimental Environment for Exchanging Design Knowledge by Cognitive Agents, In M, Mantyla, S. Finger and T. Tomiyama, eds., *Knowledge Intensive CAD*, Volume 2, Chapman & Hall, pp. 19-38.

Shen W. and Norrie, D.H. (1998) A Hybrid Agent-Oriented Infrastructure for Modeling Manufacturing Enterprises, In *Proceedings of the Eleventh Workshop on Knowledge Acquisition, Modeling and Management*, Banff, Alberta, Canada, available online at http://ksi.cpsc.ucalgary.ca/KAW/KAW98/shen/

Smith, R.G. (1980) The Contract Net Protocol: High-Level Communication and Control in a Distributed Problem Solver, *IEEE Transaction on Computers*, 29(12):1104-1113.

Walker, A. and Wooldridge, M. (1995) Understanding the Emergence of Conventions in Multi-Agent Systems, In *Proceedings of the First International Conference on Multi-Agent Systems*, San Francisco, CA, The MIT Press/The AAAI Press, pp. 384–389.

Washington, R. (1998) Markov tracking for agent coordination. In Proceedings of the second international conference on Autonomous agents, The ACM Press, pp. 70-77.

Winograd, T. and Flores, F. (1986) *Understanding Computers and Cognition*, Norwood, N.J.: Ablex.

Wong, A. and Sriram, D. (1993) SHARED: An Information Model for Cooperative Product Development, *Research in Engineering Design,* 5(1):21-39.

9

Collaboration, Task Decomposition and Allocation

9.1 INTRODUCTION

As noted in Chapter 8, collaboration is a most important form of cooperation in multi-agent systems. Collaboration means that a group of agents work together on a common project or a task. Under the term collaboration, we can also consider all the techniques used by these agents for task decomposition and resource allocation. Collaboration can pose difficult problems because of the range of issues involved; the agent's capabilities and their competence for the various tasks; the nature of the tasks; the social structure of the agents; and cost and efficiency; (Ferber, 1995).

In this chapter, we introduce techniques for task decomposition and allocation and, as an example, outline mechanisms developed for the MetaMorph project (Maturana and Norrie, 1996).

9.2 TASK DECOMPOSITION AND ALLOCATION

9.2.1 Definition

What is Task Decomposition?

Task decomposition is the process of decomposing a task into sub-tasks. For example, the process of task decomposition in a concurrent mechanical design system includes decomposing a product design (a high-level design task) into a number of assembly designs (sub-tasks) which may be further decomposed into a number of part designs (sub-sub-tasks). In manufacturing systems, high-level production tasks focus on the manufacture of products, each having a number of parts requiring various manufacturing operations, including subassembly. High-level production tasks can therefore be decomposed into several levels of sub-tasks (e.g., parts fabrication by different shop floors) and sub-sub-tasks (e.g., manufacturing operations of a part by different machines). There may even be further levels of decomposition.

What is Task Allocation?

Task allocation is the process of distributing tasks and sub-tasks (or sub-sub-tasks) among the available resources. In manufacturing systems, it is an essential element of production planning and scheduling.

9.2.2 Relationship between Task Decomposition and Allocation

A task that requires more know-how, resources, or operations than any one agent can supply should first be decomposed into sub-tasks. These can then be allocated to appropriate agents. These two steps (task decomposition and allocation) are closely related because task decomposition should take the skills of the different agents into account in a way that also facilitates the following task allocation.

9.2.3 Criteria for Task Decomposition

Researchers in DAI have proposed numerous criteria for task or problem decomposition (e.g., Bond and Gasser, 1988; Gasser and Hill, 1990). The following are commonly the most important criteria:

- the problems to be resolved should be naturally analyzable by levels of abstraction (from more general to more detailed);
- constraints (especially those involving control, data or resources) should also be taken into account;
- decomposed tasks (or sub-tasks) should be as independent as possible, to minimize the need for coordination among tasks;
- the amount of information to be transmitted between tasks should be minimized;
- decomposed tasks (or sub-tasks) should be able to use their local resources maximally, to reduce conflicts related to resources.

Although task decomposition is as important as task allocation, it is still very often done by human beings. It is, in fact, not easy to develop a software system for totally automatic task decomposition in manufacturing. In some agent-based concurrent design and manufacturing systems, task decomposition is carried out by special agents using product modeling databases, design/manufacturing knowledge bases, and/or rule bases. Examples of such agents are the mediators in MetaMorph I & II (Maturana and Norrie, 1996; Maturana et al., 1999; Shen et al., 1998, also see Chapters 14 and 15), and the project manager agent in DIDE (Shen and Barthès, 1996).

9.2.4 Different Techniques for Task Allocation

Task allocation techniques can be classified in two main categories: centralized approaches (coordinated allocation or allocation with some degree of centralized control) and distributed approaches (distributed allocation or task allocation without centralized control) (Ferber, 1995)[1].

There are two typical cases of centralized approach (coordinated allocation or allocation with some degree of centralized control):

- If the structure of subordination is hierarchical, it is the supervisor that decomposes the task or problem and assigns sub-tasks to suitable service or resource agents. Such task allocation is typically characterized by a procedure call in classic programming. It is a totally centralized approach (i.e. task allocation with totally centralized control). Although this system structure still often exists in agent-based manufacturing systems (see Chapter 7 Section 7.4.1), it does not essentially differ from traditional centralized approaches. There is thus no need to explore this situation further in this book.

- In contrast, if the system structure is non-hierarchical (e.g., heterarchical), task allocation will be realized through special functional agents, the so-called middle agents (such as mediators, brokers, and matchmakers, see Chapter 7 Section 7.4.2), that coordinate task allocation by *dynamically* centralizing the demands of clients and the offers of servers (service providers or service agents). This still allows using some techniques of the centralized approaches to resolve distributed allocation problems. This approach has been widely adopted for agent-based industrial applications (see Chapter 7 for descriptions of various organizational structures and of applications). We will explore this approach further in Section 9.3.

In the second category of distributed approaches (distributed allocation or task allocation without centralized control), each agent individually finds service providers (service agents) suitable for its objectives, without any degree of centralized system control. The distributed approach is often used for autonomous agent-based systems (see Chapter 7 Section 7.4.3 for the definition of autonomous agent systems). It is particularly suitable for open and evaluative agent-based systems where agents appear and disappear dynamically, and the relationships among agents are also dynamic. Clearly, this approach can be very useful for developing realistic agent-based concurrent design and manufacturing systems.

In the techniques described in the literature for distributed task allocation (task allocation without centralized control), three mechanisms can be distinguished:

- Direct allocation (see Section 9.4.1 for details)
- Allocation through delegation (see Section 9.4.2 for details)
- Allocation through bidding (see Section 9.4.3 for details)

Figure 9.1 shows a topology of techniques for task allocation, derived from (Ferber, 1995) with major modification. Although most techniques described in this

---

[1] Ferber (1995) called these centralized allocation and distributed allocation, while we prefer to describe the first category as "coordinated allocation and task allocation with some degree of centralized control," and the second as "distributed allocation or task allocation without centralized control."

chapter are for off-line (advance) allocation, they can be extended to resolve emergent (dynamic) allocation problems (see dotted lines in Figure 9.1).

There are also other techniques for task allocation described in the literature, including hybrid approaches combining techniques shown in Figure 9.1 (see Section 9.4.4). The mechanisms for task decomposition and allocation developed for MetaMorph I & II (Maturana and Norrie, 1996; Maturana et al., 1999; Shen et al., 1998), described in Section 9.5, are examples of hybrid approaches.

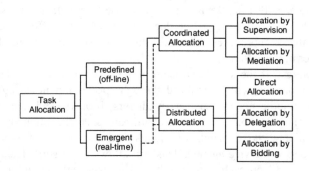

Figure 9.1 Topology of techniques for task allocation.

9.3 COORDINATED TASK ALLOCATION BY MEDIATION

As mentioned in the previous section, coordinated task allocation by mediation through middle agents uses *dynamically* centralized task allocation techniques for resolving distributed task allocation problems, to ensure reliability while maintaining efficiency and reducing complexity.

Middle agents described in Chapter 7 may be used to implement mediation mechanisms for task allocation in agent-based systems. Task allocation mechanisms in this category range from the very simple to the very complex. The simplest is a system composed of only one mediator (or other type of middle agent) coordinating a number of agents, each capable of carrying out one or more particular task(s) or sub-task(s). In this case, the mediator should have the necessary knowledge of all the agents it is coordinating, including their competences and their availability. This simple system is, in fact, a traditional centralized system, and task allocation here is purely centralized. The complex case is a system with a number of middle agents, even including dynamic middle agents (which are created dynamically as needed and destroyed after completing their coordination tasks). In this case, task allocation mechanisms can be very complex, with sometimes, task allocation being realized through multiple levels of the system, and other techniques or mechanisms being combined to resolve specific difficult problems. Most agent-based systems, however, are between these two extremes.

In this section, we consider the simplest example described in the previous paragraph, namely, a system composed of only one mediator coordinating a number of agents. This can be considered as being a subsystem of a complex system. This simple approach is widely used in agent-based concurrent design and manufacturing systems for subsystems and even whole systems, even though it uses a centralized approach. Figure 9.2 shows a mediator receiving a task request and allocating this task to a suitable agent through its knowledge about the agents it coordinates. It is assumed that the mediator has complete and consistent knowledge about all agents it coordinates. When the mediator receives a request (from a high-level mediator, or another agent in the system, or even from outside of the system), it sends an appropriate request to a relevant agent(s) according to its knowledge of the task capabilities of its agents. For example, suppose there are two agents capable of doing this task and they are each sent the same request. There are three possibilities:

- If only one of these two agents accepts the task, the mediator can simply allocate it to this agent.
- If both agents accept the task, the mediator needs to select one according to whatever are the relevant criteria.
- If both agents refuse the task, the mediator can only inform the originator of the request that it could not find a suitable agent for this task (if it cannot find, through any other mechanism, a capable agent to provide the service).

The advantages of coordinated task allocation by one mediator are:

- the mediator possesses sufficient knowledge about its coordinated agents and their competences to ensure *consistency* (coherent and appropriate behaviour of its group of agents);
- the system is easily implemented and maintained;
- optimization is easily realized.

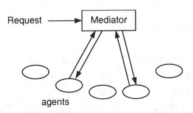

Figure 9.2 Task allocation by a mediator.

The main problem with this approach is its bottlenecks, especially for large numbers of agents and high volume of tasks. Moreover, this system with its centralized mediator is not fault tolerant (because failure of the mediator causes failure of that subsystem).

9.4 DISTRIBUTED TASK ALLOCATION

9.4.1 Direct Allocation

In the autonomous agent system architecture described in Chapter 7 (Section 7.4.3), each agent has knowledge about all other agents in the system, that is, it knows the skills or the competences of all the other agents. There are three possible situations:

1) An agent A knows that only one agent B in the system is capable of carrying out a requested task. Agent A can then simply assign this task directly to agent B without notifying or negotiating with other agents.

2) If more than one agent is capable of carrying out the requested task, two solutions exist:

   (a) If agent A knows not only the competence of all the other generally suitable agents but also their availability and performance, and it decides agent C is best for the requested task, it can then assign the task directly to agent C.

   (b) If agent A is not able to decide which agent is the most suitable for the requested task, the bidding approach (see Section 9.4.3) or a hybrid approach may be utilized.

3) If no agent in the system is able to carry out the requested task, the only thing agent A can do is to so notify the original requester of this task.

It might appear that direct task allocation is similar to the centralized approach (allocation by supervisor) described previously in this chapter. But, in direct allocation, any agent can become a supervisor (agent A in above example) when it has a task to be carried out by other agents, and an agent can be a supervisor and a service or resource agent at the same time, being involved simultaneously in several tasks.

9.4.2 Allocation through Delegation

The advantage of direct allocation as described above obviously is its simplicity. It simply assumes that agents know enough specialists for the tasks that they cannot carry out themselves. However, it is very limited and not optimized at the system level, because it cannot take into account agents whose skills are not known to the agent which has a task to be allocated (agent A in above example). This suggests it would be valuable to have a mechanism that allows agents to make indirect contact with each other.

In most real autonomous agent systems, it is usually not practical to have every agent having knowledge about all other agents. In fact, it would be very "expensive" to do so when the system has a large number of agents (as noted in Shen and Barthès, 1996). This situation is quite common in systems with acquaintance networks (Ferber, 1995).

In systems with acquaintance networks, it is assumed that each agent has a skills table in the form of a dictionary for the agents it knows. However, it is not assumed that each agent knows all the skills of the other agents. If each agent were to know all the skills of the other agents, the situation would be the same as that described in the previous section (Section 9.4.1) and the direct allocation approach could be used. For the present case, however, indirect allocation can be used. A typical indirect allocation mechanism is allocation through delegation (Ferber, 1995).

Allocation through delegation makes it possible for tasks to be transferred between agents that do not know each other directly. The situation is similar to that in federated agent-based systems, but it does not require a federated architecture and related middle agents (e.g., broker agents, etc.) for its implementation. Allocation allows an agent to forward a request to others when it is not able itself to carry out a task request from some agent.

Although it is not necessary that each agent know all the skills of the other agents, when using this approach, it is imperative that each agent can contact any other agent in the system and that each agent's knowledge about other agents should be consistent (for example, agent A's knowledge about agent B should correspond to the real competence of the agent B). The requirement for complete communication is the major limitation of this approach, if it is fully implemented.

For example, in a system of 10 agents (A1 – A10), assume that agent A1 knows only 3 other agents (A2, A3, and A4). Suppose agent A1 has a task T1 to be carried out by a specific skill S1, but knows no one (A2, A3 and A4) capable of carrying out that task. Using allocation through delegation, it will ask all the agents it knows (A2, A3 and A4) to ask all of their acquaintances who have skill S1 to carry out the requested task T1. This process has to be repeated until an agent is found who is capable and accepts the task T1, or until the entire network is covered and no one is found to carry out the task. This process is essentially a search through the network.

The key issue in implementing allocation through delegation is how to adopt or develop sufficiently powerful search mechanisms. Particularly suitable are the distributed search mechanisms described in the Artificial Intelligence and Distributed Artificial Intelligence literature. In (Ferber, 1995), several parallel graph search algorithms are described in detail, for allocation through delegation in agent systems with acquaintance networks.

9.4.3 Allocation through Bidding

Bidding may well be the most common mechanism for task allocation in agent-based systems. The contract net (Smith, 1980) and its variant versions are essentially collaboration protocols based on the notion of bidding, and these are widely used. The contract net has been introduced as an important concept in Distributed Artificial intelligence in Chapter 3 (Section 3.3.2), and then as a widely used coordination technique in Chapter 8 (Section 8.3.4.2). More detailed information on the contract net can be found in (Ferber, 1995) and in the literature.

In Section 9.4.1, if allocation by bidding is being used for situation 2.b, agent A sends a call for bids to all capable agents, and waits for bids. Each agent able to do the requested task generates and sends a bid. Agent A then makes a decision based on all the bids received, and awards the task to one agent. Alternatively, agent A could decompose the task into subtasks and call for bids for these. Or, a more sophisticated multi-level contract net approach could be used.

This approach is very suited to dynamic environments, and very easy to implement. In fact, it is widely used in agent-based manufacturing systems, particularly for (dynamic) manufacturing planning and scheduling (See Chapter 14 for more details).

9.4.4 Hybrid Approaches

The techniques and mechanisms described in the foregoing are "pure" approaches corresponding to those presented in Figure 9.1. Many researchers have proposed hybrid approaches, by either creating variants of one of above "pure" approaches or by merging several to combine their advantages. Some well-known hybrid task allocation approaches are:

- Variations of the contract net as described in Chapter 8 (Section 8.2.3)
- Combination of the contract net and acquaintance network approaches including direct allocation and allocation through delegation (Ferber, 1995)
- Combination of the contract net and the mediation approach (Section 9.5.2)

To illustrate how several techniques can be advantageously combined, we describe in the following section, in some detail, the hybrid approach developed for the MetaMorph projects.

9.5 TASK DECOMPOSITION AND ALLOCATION IN METAMORPH

9.5.1 MetaMorph I – Using Dynamic Mediators

In MetaMorph I (Maturana and Norrie, 1996), task decomposition and allocation are realized by using a clustering mechanism and dynamic mediators. A detailed description of the static mediators and dynamic mediators, including DAMs (Data Agent Managers) and AMs (Agent Managers) can be found in Chapter 14 (Section 14.2 MetaMorph). The following describes the use of the mediators.

At the resource community level, the shop floors are provided with high-level static mediators, also called mediator agents (for example, resource mediator, machine mediator, tool mediator), with each having knowledge of enterprise-level model of the system. A high-level task initially passes through the appropriate static mediator agent for recognition and decomposition. At this level, task decomposition is implemented by the static mediator agent through a centralized approach. According to the local resource capability (of the manufacturing resources registered with the static mediator agent), sub-tasks are assigned by the mediator to separate coordination clusters of agents, which are dynamically formed as needed. Each coordination cluster also includes a set of clone agents (see Chapter 14, Section 14.2 for description of clone agents) derived from active manufacturing resource agents. This originates a concurrent coordination framework as shown in Figure 9.3, where C1-C2 are clone agents1-C2 are clone agents.

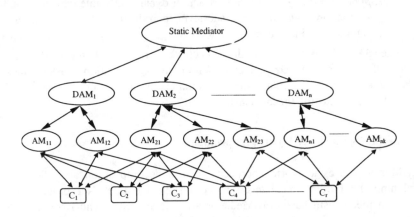

Figure 9.3 Concurrent Coordination Framework (Maturana and Norrie, 1996).

Subsequent second level task decomposition occurs in the coordination clusters, but at this level, the sub-tasks are assigned to various Agent Managers (AMS). As well, the virtual organization's meta-level activities allow the creation of sub-level clusters within a coordination cluster. This recursive property of the coordination activity creates partial dynamic hierarchies for multi-agent planning.

The following algorithm describes the process for task decomposition and clustering activities. Here, high-level tasks are represented by typical manufacturing products, which are presented as many parts requiring various manufacturing operations, including subassembly.

1) *Initially decompose to first-level sub-tasks (S_i) by product type*: The initial task decomposition is carried out at the static mediator agent level.
2) *Assign each sub-task to a Data Agent Manager (DAM)*: The static mediator assigns each new sub-task to an individual coordination cluster. The static mediator creates the coordination cluster, assigns the sub-tasks, and steps out of the scene. Within the coordination clusters, each DAM assumes the role of high-level agent to further distribute the sub-tasks.
3) *Decompose to second-level sub-tasks (S_{ij})*: This second level task decomposition may be required, for example, when manufacturing

operations need both material removal and assembly operations. The material removal operations are needed before the assembly operations.

4) *Assign each second-level sub-task to an AM*: The coordination clusters are provided with clone agents representing the manufacturing resources, which receive the second-level sub-tasks and initiate the negotiation process. The clone agents evaluate the technological requirements and schedules for material removal and assembly operations.

5) *Further decompose any relevant second-level sub-task that requires the support of other resources*: This action dynamically interconnects multiple coordination clusters. The sub-level coordination cluster applies task assignment and task decomposition strategies.

In a generalized sense, each coordination cluster is responsible for the solution of a sub-task through distributed planning. Within the coordination clusters, message exchange is kept to a minimum, since communication among clone agents is based on selective broadcasting mechanisms.

9.5.2 MetaMorph II – A Hybrid Approach

MetaMorph II uses a hybrid approach for task decomposition and allocation by combining the bidding mechanism based on the Contract Net protocol with a mediation mechanism based on Mediator architecture (Shen and Norrie, 1998a,b). More detailed information about MetaMorph II and its system architecture can be found in Chapter 15 (Section 15.5).

9.5.2.1 Organization of Resource Agents

In MetaMorph II, manufacturing resource agents are coordinated by appropriate mediators at all levels of the system. For example, as shown in Figure 9.4, a Machine Mediator is used to coordinate all the machines in a shop floor; a Tool Mediator is used to coordinate all tools, and so on. A high level Resource Mediator coordinates lower-level mediators such as Machine Mediators, Tool Mediators, Worker Mediators, and Transportation Mediators. In this type of architecture, the system level organization of mediators may appear to be hierarchical, but there is neither a hierarchical control structure nor a hierarchical coordination mechanism. For example, a machine agent can also communicate and negotiate directly with a Worker Mediator, worker agents, a Tool Mediator and tool agents, and so on. In other words, the communication between two different types of resource agents (such as between a machine agent and a worker agent) can be direct, rather than always through facilitators or mediators as in some other multi-agent systems. This approach can significantly reduce communication overhead and the risk of communication bottlenecks.

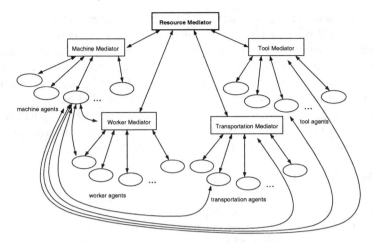

Figure 9.4 Organization of resource agents.

9.5.2.2 A Hybrid Approach for Task Decomposition and Allocation

The cooperative negotiation among resource agents in MetaMorph II is realized by combining the mediation mechanism based on hierarchical mediators and the bidding mechanism based on the Contract Net protocol (Smith, 1980) for dynamically generating and maintaining production schedules. Such cooperative negotiation may be achieved through two scheduling mechanisms according to manufacturing enterprise conditions: Machine-Centered Scheduling and Worker-Centered Scheduling.

In the system with Machine-Centered Scheduling, the Resource Mediator sends a request message to the Machine Mediator first. It is the machine agents that then take the responsibility to negotiate with worker agents and tools agents. In the system with Worker-Centered Scheduling, the Resource Mediator sends a request message to the Worker Mediator first. It is then the worker agents that take the responsibility to negotiate with machine agents that in turn negotiate with tool agents. The scheduling system can be shifted between these two mechanisms according to the manufacturing enterprise situation. Here, we discuss only the Machine-Centered Scheduling mechanism.

In MetaMorph II, parts are also modeled as agents, i.e. as part agents. In the prototype implementation, part agents do not communicate directly with resource agents. All manufacturing task requests are sent to the Resource Mediators.

The Contract Net protocol used in our implementation has been extended to meet our special requirements. For example, in the bidding and mediation process for selecting a machine for a manufacturing task, after receiving the different propositions (bids) from different machine agents, the Machine Mediator will select a machine agent to perform the task according to its criteria (mainly manufacturing cost criteria) and award a contract to it. The other machines that are not selected are noted as alternatives that may be contacted (negotiated with) in the future in case of unforeseen situations such as machine breakdown. This greatly reduces the rescheduling time when such unforeseen situations take place. Information about the alternative resources (machines, workers and tools) will be sent to the part agents along with the information about the selected resources. The part agents save these two sorts of information separately in their knowledge bases. When a selected resource cannot perform scheduled tasks due to unforeseen situations, the part agent may negotiate directly with alternative resource agents.

9.5.2.3 Bidding and Mediation Processes

❖   At the Machine Mediator Level

As shown in Figure 9.5, when the Machine Mediator receives a task *request* from the Resource Mediator (which itself had received the request from a part agent), it sends out multicast *announce* messages to relevant machine agents according to its knowledge of the capabilities of machine agents registered with it. After receiving an *announce* message, a machine agent calculates processing time and cost, verifies its availability in a reservation plan, and negotiates with worker agents through the Worker Mediator and with tool agents through the Tool Mediator. After receiving propositions (bids) from the Worker Mediator and the Tool Mediator, the machine agent sends a *bid* with information about the processing time and cost, proposed start time and end time, other free time slots, and information about the associated worker agent and tool agents to the Machine Mediator.

Waiting for a predefined period after sending out the *announce* messages to machine agents, the Machine Mediator then analyses the propositions (bids) from the various machine agents, and selects one machine agent and its associated worker agent and tool agent(s) to perform the task according to its selection criteria (based on a cost model as detailed in Maturana et al., 1996). As soon as the decision is made, the Machine Mediator sends an award message (*notice* type) to the selected machine agent which then locks a time slot and reserves it for the requested production task. This time slot now becomes unavailable for other tasks. At the same time, the Machine Mediator sends an *inform* message with information about both the selected resources and the alternative resources directly to the part agent (not through the Resource Mediator). After receiving the award message, the winning machine agent sends an award message to the winning worker agent and the winning tool agent(s). A more complex mediation mechanism is to be developed to optimize scheduling for a group of parts or for a given period.

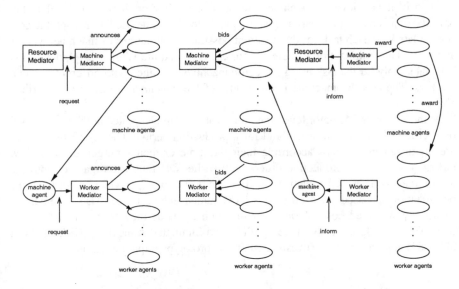

Figure 9.5 Mediation and bidding mechanism (Machine-Centered Scheduling).

❖ At the Worker Mediator Level
The bidding process between the Worker Mediator and the worker agents is basically similar to that at the Machine Mediator level (see Figure 9.5). But the Worker Mediator does not send an award message to the selected worker agent until it receives an award message from the winning machine agent who sent the *request* message earlier. Otherwise, it does not make an award to any worker agent.

❖ At the Tool Mediator Level
The bidding process between the Tool Mediator and the tool agents is similar to that at the Worker Mediator level. The main difference is that at the Worker Mediator level, only one worker agent can be awarded a task, whereas at the Tool Mediator level, several tool agents may be selected for a requested task.

❖ At the Transportation Mediator Level
When the Transportation Mediator receives a transportation task request from the Resource Mediator, it sends out multicast *announce* messages to all concerned transportation agents (AGVs) which then negotiate to find a suitable agent (AGV) to execute the requested task.

9.5.2.4 Optimization

Scheduling optimization is carried out at three levels: the Resource Mediator level, the Machine Mediator level, and the Worker Mediator level (also the Tool Mediator level).

At two lower levels (Machine Mediator and Worker Mediator levels), the optimization is based on a partial cost model. At the Machine Mediator level, the cost is calculated for each task for the different possible combinations of machine, worker and tools. The optimization is therefore for each task. At the Worker level, the cost is calculated only for the task using a specific machine combined with different possible workers. The specific machine is the sender of the *announce* message as described above.

At the Resource Mediator level, the optimization is implemented at the global level on a group of production orders and for a production period. The production orders are organized into groups according their due time or their emergence. Generally, emergent orders and production orders with earlier due time are organized into the group that will be scheduled first. For each group of production orders, the optimization is based on a cost model (Maturana et al., 1996). Task tardiness is treated by weighted cost. Various optimization methods may be employed at this level. Genetic Algorithm (Goldberg, 1989) and Simulated Annealing (Kirkpatrick et al., 1983) approaches were investigated during prototype implementation.

REFERENCES

Bond, A.H. and Gasser, L. (1988) An Analysis of Problems and Research in Distributed Artificial Intelligence. In A.H. Bond and L. Gasser, (eds.), *Readings in Distributed Artificial Intelligence*, Morgan Kayfmann.

Ferber, J. (1995) *Les systèmes multi-agents*, InterEditions, Paris.

Gasser, L. and Hill, R.W. (1990) Coordinated Problem Solvers. *Annual Reviews of Computer Science*, **4**, 203-253.

Goldberg, D. (1989) *Genetic Algorithms in Search, Optimization and Machine Learning*. Addison-Wesley.

Kirkpatrick, S., Gelatt, C.D. and Vecchi, M.P. (1983) Optimization by Simulated Annealing, *Science*, 220:671-680.

Maturana F. and Norrie D.H. (1996) Multi-Agent Mediator Architecture for Distributed Manufacturing. *Journal of Intelligent Manufacturing*, **7**, 257-270.

Maturana, F., Balasubramanian, S. and Norrie, D.H. (1996) A Multi-Agent Approach to Integrated Planning and Scheduling for Concurrent Engineering, In *Proceedings of the International Conference on Concurrent Engineering: Research and Applications*, Toronto, Ontario, pp. 272-279.

Maturana, F., Shen, W. and Norrie, D.H. (1999) MetaMorph: An Adaptive Agent-Based Architecture for Intelligent Manufacturing, *International Journal of Production Research*, 37(10):2159-2174.

Shen, W. and Barthès, J.P. (1996) An Experimental Multi-Agent Environment for Engineering Design, *International Journal of Cooperative Information Systems*, 5(2-3):131-151.

Shen W. and Norrie, D.H. (1998a) Combining Mediation and Bidding Mechanisms for Agent-Based Manufacturing Scheduling. In *Proceedings of Autonomous Agents '98*, Minneapolis/St Paul, MN, pp. 469-470.

Shen W. and Norrie, D.H. (1998b) An Agent-Based Approach for Dynamic Manufacturing Scheduling. In *Working Notes of Autonomous Agents'98 Workshop on Agent-Based Manufacturing*, Minneapolis/St. Paul, MN, pp. 117-128.

Shen, W., Xue, D. and Norrie, D.H. (1998) An Agent-Based Manufacturing Enterprise Infrastructure for Distributed Integrated Intelligent Manufacturing Systems, In *Proceedings of PAAM'98*, London, UK, pp. 533-548.

Smith, R.G. (1980) The contract net protocol: high-level communication and control in a distributed problem solver. *IEEE Transaction on Computers*, **29**(12), pp. 1104-1113.

10

Negotiation and Conflict Resolution

10.1 INTRODUCTION

Negotiation is an ambiguous term in the agent world. It ranges from situations where resources must be allocated to agents, to situations involving agent-to-agent bargaining. In both cases, however, negotiation is intended to improve the global state of affairs, or as stated by Lesser:

> Negotiation, the process of arriving at a state that is mutually agreeable to a set of agents, is intimately related to coordination. The negotiation process can be used as part of a multi-agent coordination algorithm that implements, for instance, a contracting mechanism for getting one agent to commit to solving a subproblem for another agent.

> Lesser, 1998

Clearly, to be able to negotiate, agents must be able to reason. Thus, negotiation is restricted to cognitive agents (see Chapter 3).

Since negotiating situations occur when there is a conflict of interest, the first step will be to detect such a conflict. Agents will then use communication channels and try to eliminate the conflicts. Conflicts may be about limited available resources, or may be a conflict between the beliefs of some agents, or else may be a disagreement on something such as the price of some object. In the first case, we are faced with an optimization problem, in the second case, one of the agents will have to change its beliefs, and in the third case, we have a bargaining situation. It is often difficult to see what exactly one wants to achieve when resolving the situation. Among possible choices are: (1) minimizing the time to find a solution; (2) minimizing the total resource usage in doing so (e.g., CPU time); (3) maximizing the quality of the result (in some cases this relates to the "social welfare" of the agents and corresponds to optimizing a global objective function). Another approach is to try to achieve Pareto optimality, meaning that the outcome maximizes the product of the agents' utilities, or to try to reach a Nash equilibrium[1].

Negotiation is done by exchanging messages among agents (often only two). Since the process involves several messages, a discussion will take place in which each agent's attitude will be an important factor. This attitude is governed by an agent's beliefs and goals and by the global situation. The negotiating process follows tactical

---

[1] A Nash equilibrium is a profile of strategies, one for each player, such that each player's strategy is a best reply to the other players' strategies in the profile.

rules, which together implement a strategy. Time may or may not be important in limiting the length of the negotiation.

In this chapter, we address the nature and use of negotiation protocols in Section 10.2 and of negotiation strategies in Section 10.3. We then look at a number of possible conflict situations in the area of concurrent design and manufacturing, in Section 10.4, giving additional examples in Section 10.5. As multi-agent negotiation and conflict resolution is a recent research topic, we discuss the research literature in the related sections and provide additional references in Section 10.6.

10.2 NEGOTIATION PROTOCOLS

Negotiation protocols set the stage for the negotiation process. If we follow Alonso (1998): "Negotiation is the process through which in each temporal point, an agent, say x, proposes an agreement from the negotiation set NS, and the other agent y, either accepts the offer or does not. If the offer is accepted, then the negotiation ends with the implementation of the agreement. Otherwise, the second agent then has to make a counter offer, or reject x's offer and abandon the process. So the protocol says when and how to exchange offers." Thus, the protocol contains the basic rules for the negotiation process and the communication. In addition to using a protocol, each agent will develop and use a negotiation strategy appropriate to the problem to be solved. Clearly, negotiation protocols and negotiation strategies will be quite different for different categories of negotiation. Since negotiation involves exchanges of messages, protocols structure what are called *conversations*, defining classes of dialogue. The simplest dialogues are found in contract-net approaches where they are limited to exchanges involving offers, bids, and grants of contract. More complex dialogs are found in human types of negotiations, when trying to change other agent's beliefs.

Reeds (1998), borrowing from Walton and Krabbe (1995) distinguishes five types of dialogue: persuasion, negotiation, inquiry, deliberation, and information seeking. Three of them are here of interest: (1) persuasion; (2) negotiation; and (3) deliberation.

Persuasion covers the case of conflicting beliefs. A basic case studied by Barbuceanu and Fox (1995) and Ekenberg et al. (1995) is when agent 1 tells agent 3 that it believes P, and at the same time agent 2 tells the same agent 3 that it believes Q. More complex situations were studied by Sycara (1990) and Chu-Carroll and Carberry (1996), as has already been mentioned.

Negotiation according to Reeds (1998) "differs procedurally from persuasion in an important respect, in that coherence between beliefs is not demanded, and the relevant beliefs of the participants may well remain at odds after negotiation." Negotiation dialogues arise when there are conflicts of interest among selfish agents over scarce resources.

Deliberation occurs when agents want to reach agreement over a plan for action. Situations described by Lesser (1998) fall into this category.

In all cases, conversations or dialogues are used to define the negotiation protocol. Each exchanged message is typed, usually by one of the performatives of the speech act theory (see Chapter 8). It is convenient to assign states to agents when they receive messages, and the type of messages and the corresponding state changes are usually specified by the equivalent of a finite state machine or by a Petri net. A simple example of such a graph taken from Barbuceanu and Fox (1995) is shown in Figure 10.1 (see also Barbuceanu and Fox, 1996). Other examples can be found in (Nodine and Unruh, 1998) in the InfoSleuth system, for a model of conversation policy which includes: (1) a common set of speech acts; (2) a common service ontology; and (3) a common set of prescriptive conversation policies, i.e., protocols. Sierra et al. (1998) developed more formal negotiation models using a general rhetorical language. Singh (1998) used Dooley graphs, and Parunak (1996) proposed using Enhanced Dooley graphs.

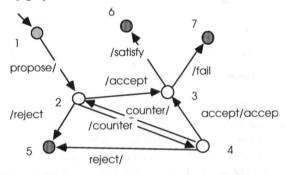

Figure 10.1 State transitions for negotiation (Barbuceanu and Fox, 1995) (©1995 AAAI).

In argumentative negotiation there will be arguments accompanying each proposal. This situation was considered by Chu-Carroll and Carberry (1996) and by Sierra et al. (1998).

In deliberative approaches, plans have to be exchanged. Lee and Durfee (1998) proposed new models and a language based on Grafcet to express such plans.

In auction type negotiation, offers are made in terms of monetary quantities, and the dialogue reduces to a simple exchange of offers until an agreement is reached or allocated time is exhausted. Finally, for contract-net interactions, Andersson and Sandholm (1998) have studied various algorithms, and in particular, a leveled commitment algorithm in which agents can de-commit by paying a penalty the amount of which increases with time.

10.3 NEGOTIATION STRATEGIES

If negotiation protocols govern the exchange of proposals (and perhaps arguments) among agents, negotiation strategies decide the position of a particular agent during the negotiation process. As noted by Alonso (1998): "A strategy is a function from the history of the negotiation to the current offer that is consistent with the protocol." Of course, each agent strategy will depend strongly on the type of application that each agent is involved in. Agents, in most applications, have a limited perception of the global problem and thus are restricted to acting according to their beliefs, desires or intentions. When one can define a utility function, then agents act to maximize their utility function. Early approaches involved fixed strategies, while recently more flexible strategies utilizing learning have been considered.

We classify negotiation strategies into the following categories:

- Contract based negotiation;
- Market based negotiation;
- Game theory based negotiation;
- Plan based negotiation;
- AI based negotiation;
- Other approaches.

10.3.1 Contract Based Negotiation

Contracting was first proposed in the Contract-Net (Smith, 1980) and then demonstrated on a distributed sensing system (see Section 3.3.2 in Chapter 3). To summarize: each agent (manager) having some work to subcontract broadcasts an offer and waits for other agents (contractors) to send bids. After some delay, the best offers are retained and contracts are allocated to one or more contractors who process their subtasks. The contract-net protocol provides for coordination in task allocation, with dynamic allocation and natural load balancing. The approach is quite simple and can be efficient. However, when the number of nodes is large, the number of messages on the network increases, which can lead to a situation where agents spend more time processing messages than doing the actual work, or worse, the system stops through being flooded by messages. Thus, various improvements to the basic contract-net approach have been proposed, such as:

- sending offers to a limited number of nodes, instead of broadcasting them;
- anticipating offers, i.e., contractors send bids in advance;
- varying the time when commitment is decided;
- allowing de-commitment (breaking commitments);
- allowing several agents to answer as a group (coalition formation);
- introducing priorities for solving tasks.

In the most basic approach, the choice of a contractor is done by comparing bids corresponding to a particular offer, using whatever mechanisms are relevant to the problem. In more complex cases, such as those involving de-commitment (Andersson and Sandholm, 1998; Jennings et al., 1996), one has to introduce penalties, thus bringing the approach nearer to a market one (see below).

10.3.2 Market Based Approaches

In a market-based approach, the goal is to solve a distributed resource allocation problem. Agents are classified as producers and consumers of goods and services (resources). Equilibrium is reached when the prices of goods is such that all resources are being used up. A particular agent wants to acquire goods but is limited by a budget. Thus, it will make offers based on the current price of goods and its own preferences. It has an internal utility function and its goal is to increase utility, which corresponds to the hypothesis of rational behavior. Producers have a specific production technology and seek to maximize their profits. Given a set of prices, the trading process involves a sequence of offers in which each consumer states how much of each resource it wants to purchase. If the demand differs from the supply, then prices will have to be adjusted by the producers. A good description of the approach is available in Mullen and Wellman (1996).

Under some conditions, there will be convergence towards a state of equilibrium which is Pareto optimal, i.e., there exists no solution that makes some agent better off yet does not make some other worse off (Petrie et al., 1995). A small example is provided by Mullen and Wellman (1996). The market-based approach was used by Huberman and Clearwater (1995) to study the control of room temperature at the Xerox research center, demonstrating its feasibility. Matsubara and Yokoo (1998) proposed using advising in the case of inaccurate payoffs.

One of the problems of the market-based approach using prices as a primary controlling mechanism is that the convergence process may be slow, involving a large number of offers (and computations). A new approach focusing on resources rather than prices (a dual of the first approach) was introduced by Ygge and Akkermans (1998). It appears to be complex but efficient. One appealing feature of the new approach is that, at each iteration, the computed scheme is feasible although not optimal. Finally, the Kasbah project at MIT (Chavez et al., 1997) is a simulation environment for web-based trading. In Kasbah, however, negotiation is done between one buyer and one seller in turn, there are no globally maintained prices and the information that each agent can have is limited, which is more in line with actual web applications. The market-based approach is appealing if one believes with Mullen and Wellman that "the market is the world-standard default interface for interacting entities."

Negotiation through network-based protocols (Section 10.3.1) in market approaches can use quantified models, which has some evident advantages. Network-based contract protocols can be associated with various strategies depending on the type of contract an agent can consider. Contracts for which agents once commited would keep their promises include Standard or Original-contract, Cluster-contracts, Swap-contracts, and Multi-agent-contracts. More recently, decommitment was considered by Sandholm and Lesser (1995) and tested in the ADEPT system (Jennings et al., 1998). A good discussion can be found in (Andersson and Sandholm, 1998). Various types of strategies exist, some of which were studied by Matos et al. (1998): time dependent, resource dependent, or behavior dependent (such as relative

Tit-For-Tat, Random Absolute Tit-For-Tat, Average Tit-For-Tat). Many variations have been proposed, as by Ito and Yano (1995), who make public the history of past negotiation, and by Matsubara and Yokoo (1998) who introduce inaccurate payoffs.

Market approaches have led to different types of bidding policies being introduced (Walsh and Wellman, 1998). Ygge and Akkermans (1998) propose that the agent strategy be determined by resources, rather than by prices. Lee (1998) proposes a strategy of task redistribution, trying to minimize the amount of computation in each agent. Other approaches based on probability theory have been proposed by Ekenberg and coworkers (1995) and by Giménez-Funes and coworkers (1998).

10.3.3 Game Theory Based Negotiation

Negotiation among self-interested agents has been studied from the perspective of game theory. Early research work in this area was carried out by Rosenshein and Zlotkin (1994).

In game theory, the global outcome for the system is given in a table showing the results of combined decisions. Each player, however, makes decisions independently. A typical example is the prisoner's dilemma game. The matrix for this is given in Figure 10.2, which shows that each player has a maximum return when it defects if the other cooperates, but a low return if both players defect, and an average return if both cooperate. The maximum return for the system as a whole occurs when both players cooperate. The question for each player is to select a strategy when the game is played repeatedly. Globally, the goal is to find strategies that lead to the maximum return for the system as a whole, i.e., the situation in which both agents cooperate.

	C	D
C	3 3	0 5
D	5 0	1 1

Figure 10.2 Gain table for the prisoner's dilemma game
(C=cooperate, D=defect)

Many improvements have been put forward for the prisoner's dilemma model. Mor and Rosenschein (1995) proposed introducing time or letting a player opt out. Ito and Yano (1995) proposed that the contract histories of the agents be made public and Carmel and Markovitch (1998) proposed a look-ahead exploration strategy. Other approaches deal with bluff (Myerson, 1989), or with assigning credibility to the statements of other agents using statistical theory (Ekenberg et al., 1995).

Game-theory based negotiation techniques have been widely used in agent systems. Here we take Rosenschein's game-theory approach (Rosenschein and Zlotkin, 1994), as an example. The key concepts in this game-theory approach to negotiation are:

- utility functions;
- a space of deals;
- strategies and negotiation protocols.

Utility is defined as the difference between the worth of achieving a goal and the price paid in achieving it. The *utilities* are usually given as decision matrices, where an agent looks up a value for a certain action. Corresponding to general strategy, it will perform the action with the lowest or highest value. In the context of negotiation, *utility functions* represent the prices or costs for activities.

A *deal* is an action an agent can take which has an attached utility. A *negotiation protocol* defines the rules that govern the negotiation, including how and when it ends (e.g. by agreement or no deal). A complex agent-based system usually has several *negotiation protocols*.

The actual negotiation proceeds as follows. Utility values, for each outcome of some interaction for each agent, are built into a payoff matrix which is typically common knowledge to both the parties involved in the negotiation. The negotiation process involves an interactive process of offers and counter-offers in which each agent chooses a deal which maximizes its expected utility value. There is an implicit assumption that each agent in the negotiation is an expected utility maximizer. At each step in the negotiation, an agent evaluates the other's offer in terms of its own negotiation strategy.

The negotiation process depends heavily on the agents' (internal) goal of maximizing its utilities. The various prices, costs, and values might serve as negotiation objects, but the decisions are settled on the basis of utility optimization.

According to Nwana et al. (1996), game-theory based negotiation for multi-agent systems fails to address some crucial issues:

- Agents are presumed to be fully rational and acting as utility maximizers using predefined strategies.
- Each has knowledge of its payoff matrix, and therefore full knowledge of the other agent's preferences; this is certainly unlike the real world where agents only have partial or incomplete knowledge of their own domains, let alone those of others. Therefore, this is unrealistic for truly non-benevolent and loosely coupled agents. Further, the payoff matrix can become very large and intractable for a negotiation involving many agents and outcomes.
- Agents only consider the current state when deciding on their deal; past interactions and future implications are simply ignored (this can be improved by learning, see Chapter 5).
- Agents are considered to have identical internal models and capabilities.

- Much of the work presumes two agents negotiating, though some later work is addressing n-agent negotiation.

It is unlikely that game-theory based negotiation will suit real-life industrial agent-based applications for the reasons already offered. In brief, its assumptions are untenable in these real applications.

10.3.4 Plan Based Negotiation

Plan based negotiation is another important negotiation technique. It is based on cooperation strategies for resolving conflicts among plans of a group of agents. Alder et al. (1989) used this approach for conflict detection and resolution in the domain of telephone network traffic control. They strongly maintain that negotiation and planning are very tightly intertwined. Agents need information from others to be able to plan and function effectively and efficiently.

A major contribution was made by Durfee and Lesser (1991) whose Generic Partial Global Planning (GPGP) requires agents to exchange descriptions of intermediate situations and results, enabling them to check for potential task overlaps and decide which agent should do what work. A good discussion of this approach can be found in (Lesser, 1998).

Kreifelt and von Martial (1991) proposed a negotiation strategy for autonomous agents. They view negotiation as a two-stage process: first, agents plan their activities separately, and then secondly, coordinate their plans. A separate coordination agent does coordination of all agents' plans, though they note that this role may be played by any of the agents. They then describe a negotiation protocol in terms of agents' states, message types and conversation rules. This proposal, however, also has some limitations. As Bussman and Müller (1992) rightly point out, it does not actually present a negotiation model but just prescribes one, and it is really left to the agents to achieve consensus. The protocol itself also needs some further clarification.

Conry et al. (1986) developed a negotiation protocol called *multistage negotiation* for cooperatively resolving resource allocation conflicts. They were specifically concerned with negotiation strategies for distributed constraint satisfaction problems, where a group of agents have a goal but each node or agent has only limited resources. The local constraints give rise to a complex set of global and inter-dependent constraints. This investigation was done in the context of the monitoring and control of a complex communication system. Their implementation involved developing algorithms for multi-agent planning taking the inevitable conflicts into consideration. This work is interesting because it investigates whether specific generic tasks can be linked with specific negotiation strategies. Identifying such links would significantly assist with implementation.

Subsequently, an extended protocol for multistage negotiation was reported (Kuwabara and Lesser, 1989). Durfee and Montgomery's hierarchical protocol for coordinating multi-agent behavior (Durfee and Montgomery, 1990) uses a similar approach. Naturally, protocols of this nature implicitly implement some negotiation strategy, and if it is suitable for some particular task then it may as well be reused.

Speaking generally, plan based negotiation inevitably suffers from limitations inherent in centralized or distributed multi-agent planning (see Chapter 8, Section 8.3.4.3).

10.3.5 AI Based Negotiation Approaches

It appears that almost every form of human interaction requires some degree of explicit or implicit negotiation. Hence, it is not surprising that many researchers draw from human negotiation strategies, which often leads to using AI techniques such as logic, case-based reasoning (CBR), and constraint-directed search. Detailed descriptions of these AI based approaches are outside of the scope of this book. Interested readers may refer to available books on AI, e.g., (Russell and Norvig, 1995). In the following paragraphs, we introduce some examples of AI based negotiation in agent-based systems.

10.3.5.1 Negotiation

Sycara (1989, 1990, 1991) considered negotiation as an iterative activity and proposed a general negotiation model to handle multi-agent, multiple-issue, single and repeated type negotiations. She exploited case-based reasoning (CBR) and multi-attribute utility theory. She argued for a case-based approach, since human negotiators draw from past negotiation experiences to guide present and future ones. Where there is an absence of past cases, she resorts to preference analysis based on multi-attribute theory. In this situation, issues involved in negotiation are represented by utility curves. By combining these curves in additive and multiplicative fashions, a proposal can be chosen which maximizes the utility. A prototype system called PERSUADER was constructed to resolve adversarial conflicts in labour relations with the aid of two practicing negotiators. In this system, agents can modify other agent's beliefs, behaviors and intentions via persuasion.

10.3.5.2 Negotiation Based on Constraint-Directed Search

Sathi and Fox (1989) view negotiation as a constraint-directed search of a problem space using negotiation operators. In their approach, negotiation involves two phases: a communication phase where all information is communicated to participating agents, and a bargaining phase where deals are made between individuals or within a group. Agents negotiate via the relaxing of conflicts and constraints until agreement is reached. Alternatively, solutions may be modified until acceptable. Initially, negotiation preferences are modeled as constraints. The negotiation operators are drawn from human negotiation studies (Pruitt, 1981). These operators are for relaxation, reconfiguration and composition, and are used to generate new constraints. This approach has been used to build a system for resource allocation. The main limitation of this iterative approach is this: selecting relaxation to achieve a compromise can cause a major problem, since no criteria are provided, hence agents easily get caught in an infinite loop of exchanging offers (i.e. deadlock).

Lander and Lesser (1991) proposed a negotiated search paradigm for conflict resolution among heterogeneous and reusable expert agents, in their TEAM framework. In this approach, *extended search* is carried out by an agent recognizing a conflict, in order to find another solution in its local search space that avoids the conflict, while relaxation is used to expand the local solution space if the search is constrained or expensive. Conflict resolution combines search and solution composition. Local search by an agent produces a solution to a subproblem. These solutions have to be integrated into an overall solution in the composite space given by the interaction of the local solution spaces. A selected solution is acceptable if it satisfies the requirements imposed by each agent and by the user.

10.3.5.3 Knowledge-Based DFI Model

Werkman (1990) proposed a knowledge-based model of incremental negotiation and his Designer Fabricator Interpreter (DFI) is based largely on various human models of negotiation. This scheme uses a shared knowledge representation called shareable agent perspectives, which allows agents to negotiate in a similar manner to cooperating (or competing) experts with a common background of domain knowledge. Essentially, it exploits a blackboard having partitions for requested proposals, rejected proposals, accepted proposals, communications, and shared knowledge. Such rich detail and knowledge of the perspectives of others is invaluable information, allowing agents to make better proposals in future. In the DFI model, negotiation has a three-phase cycle:

- The first phase involves an agent announcing a proposal which is evaluated by the receiving agent;
- The second phase involves generating a counter proposal if the latter is not happy with the initial proposal, or the original proposal may be simply accepted;
- The third phase involves submitting the counter proposal for review by other agents.

An arbitrator agent assists whenever two agents deadlock, by reviewing their negotiation dialogue and using their mutual information network to generate alternative proposals. This is done using issue relaxation techniques or some intelligent proposal generator approach. This process may fail to achieve resolution in which case the arbitrator may set time limits or use other techniques. Arbitration can be very useful for avoiding or resolving deadlocks.

10.3.6 Other Approaches

Negotiation strategies are more difficult to model in situations where conflicts arise between different sets of beliefs, and exchanged arguments then have to be taken into account. Extensive symbolic reasoning can be required at each stage and may depend closely on the context of the particular application. For example, the simple situation of belief conflict described by Barbuceanu and Fox (1995) was solved by introducing roles, credibility and deniability.

It is clear that time plays an important role in the strategy of each agent. It does so at two levels: (1) in time-limited negotiations; (2) in changing strategies after experiments (learning). Time-limited situations were studied by Kraus and coworkers (1995). Learning was introduced by Zeng and Sycara (1997) in the BAZAAR system, which incorporated Bayesian learning.

Matos and coworkers (1998) applied genetic algorithms to negotiation strategies involving two populations of agents: buyers and sellers. A new strategy was formed as a weighted combination of a number of existing strategies. The conclusion from their experiments was that no strategy continues to dominate and that combining strategies constitutes a good approach.

Bussmann and Müller's negotiation framework for cooperating agents is one of the most interesting approaches in the literature (Bussmann and Müller, 1992). Drawing from socio-psychological theories of negotiation, particularly Gulliver's eight phases of the negotiation process (Gulliver, 1979), they evolve a cyclic negotiation model that is both general and simple. They attempt to address the limitations of other negotiation proposals and models. The cyclic nature of the model addresses the thorny issue of conflict resolution. The general strategy is that negotiation begins with one, some or every agent making a proposal. Next, the agents evaluate and check the proposals against their preferences, and criticize them by listing any of their preferences violated by the proposals. The agents then update their knowledge about the other agents' preferences and the negotiation cycle resumes with a new proposal or proposals in the light of this newly gleaned information. Conflicts between agents are handled in a concurrent conflict resolution cycle. Bussmann and Müller's approach is untested but it does seem to address the limitations of some other approaches. In this respect, it is an interesting and promising proposal.

A framework for argumentation-based negotiation, derived from work in the domain of business process management, was proposed by Sierra et al. (1998). A formal model was put forward by Alonso (1998), who introduced the idea of reasonable equilibrium theory. In this area, one should also mention the work of Chu-Carroll and Carberry (1996) on detecting and resolving conflicts in beliefs during collaborative planning, which involves beliefs and justification chains, as well as the work of Malheiro and Oliveira (1996) dealing with dynamic and conflicting agent beliefs.

The last approach to be mentioned is that of Polat et al. (1993) who developed a negotiation platform, including a flowchart that describes the general conflict resolution process between agents, to support multi-agent conflict detection and resolution.

10.4 NEGOTIATION FOR CONFLICT RESOLUTION

10.4.1 Conflicts in Agent-Based Design and Manufacturing Systems

The use of agents in concurrent design and manufacturing systems brings the possibility of conflicts between agents. It may in theory be possible to build a system where all potential conflicts are considered during development and measures taken to ensure they will not arise during run time, but in practice this either cannot be done or would be extremely difficult. Moreover, in many cases, conflict resolution of this kind is not even desirable since there are many advantages to run time conflict resolution. Run time interactions among agents can support concurrent engineering activities. Negotiation results in emerging behavior as the system responds to the necessities of its current problems. Such a system is more flexible and does not suffer from the brittleness problem of traditional expert systems. Run time conflict resolution, however, requires that conflict resolution knowledge be treated as first class knowledge in the agents just like domain knowledge and control knowledge (Klein, 1991; Klein and Lu, 1990). This requires that separation and explicit representation of different knowledge types be enforced.

Conflicts in agent-based concurrent design and manufacturing systems occur for a variety of reasons:

- conflicts from agents having different points-of-view;
- conflicts from agents having different knowledge;
- conflicts from agents having different beliefs, desires and intentions;
- conflicts from agents not providing the right information (e.g., less accurate than required);
- conflicts from agents having different local plans;
- conflicts resulting from manufacturability analysis;
- conflicts resulting from assembleability/disassembleability analysis;
- conflicts resulting from manufacturing (shop floor) resource capability and capacity analysis.

10.4.2 Conflict Resolution through Negotiation

As mentioned in Chapter 8, arbitration, mediation and negotiation are important techniques widely used for conflict resolution in multi-agent systems and for ensuring the stability and performance of these systems. Arbitration is usually a decision-making mechanism based on predefined rules for all agents in the system and for the society of agents. Mediation is usually used in federated multi-agent systems. But arbitration is not well suited for deliberative (cognitive) agent system, whereas negotiation techniques are. Both arbitration and mediation, however, are still useful when a deadlock arises in a multi-agent system using negotiation techniques for conflict resolution (Werkman, 1990; Werkman and Barone, 1992).

The negotiation process is very flexible and encompasses all other conflict resolution methods. For example, methods where conflicts are resolved by a table lookup mechanism, as in the I3D+ system (Victor and Brown, 1994) where the outcome of every conflict is stored, may be seen as a restricted case of negotiation.

Almost all the negotiation techniques (strategies) described in Section 10.3 can be applied to conflict resolution in multi-agent systems. In particular, the following negotiation techniques have been widely used:

-   compromise negotiation (a solution is iteratively revised by sliding a value, or set of values along some dimension until a middle point that is mutually acceptable is found);
-   integrative negotiation (identify the most important goals of each agent, and find a solution which fulfills all these goals);
-   constraint relaxation and adjustment;
-   case-based and utility reasoning methods;
-   multi-agent truth maintenance;
-   market-based approaches.

10.5 EXAMPLES IN CONCURRENT DESIGN AND MANUFACTURING

In this section, we provide two examples to illustrate the types of conflict that can exist in agent-based concurrent design and manufacturing systems, and how these can be resolved through negotiation.

10.5.1 Conflict Resolution Using Single Function Agents

The primary objective of the SiFAs project (see Chapter 13 for a description of this project) is to investigate conflicts in concurrent engineering design using single function agents, that is, where each agent performs a single function in the design process and therefore contains knowledge that is as specialized as possible. The approach to conflicts resolution is negotiation between the single function agents. To illustrate conflict resolution in agent-based concurrent design systems, we briefly survey the different types of conflict investigated in SiFAs and the techniques or mechanisms that can be used to resolve these conflicts. Interested readers may refer to (Dunskus et al., 1995; Becker, 1995) for details.

10.5.1.1 Types

Possible types of conflict between single function agents in this system fall into the following categories:

- *Not enough information in output.* This conflict occurs when one agent needs information from another agent and the information provided is less comprehensive than the first agent needs. The conflict occurs with all agents that use the output of another agent, especially the pairs Evaluator-Estimator, Evaluator-Selector, Selector-Estimator, Selector-Evaluator, and Critic-Selector.

- *Information quality not sufficient.* Among those situations where an agent uses another agent's output, the agent that uses the information might have quality requirements, such as reliability or error tolerance values. Use of high quality information will help produce high quality designs but provision of lower quality information can constitute a conflict. Agent pairs for which this type of conflict can exist are Evaluator-Estimator, Selector-Evaluator, Critic-Estimator, and Critic-Evaluator.

- *Poor processing model.* This conflict can only occur when an agent has knowledge about another agent's different processing models. An agent might have both a rough and a detailed model for deriving a value. Knowledge about these processes can be acquired through communication and is mostly used by critics, suggestors and selectors. The selector, for example, might request that the evaluator use the detailed model, so the evaluations be more usable for selection purposes, and would be dissatisfied with the more approximate value.

- *Differing preferences.* When several agents compete in assigning a value to the same design parameter, they can have differing local optimizing processes or criteria. Two selectors can have different first choices, and also two suggestors might suggest different solution approaches.

- *Design constraint violation.* Critics most commonly discover violations of constraints relating to parameter values, though selectors and advisors produce them because they are the agents making the design decisions. A suggestion can also violate an agent's constraints and selectors can have built-in constraints that fail.

10.5.1.2 Resolution

Conflicts in this system can be solved in various ways, depending on the conflict type. What the type is will influence the nature of the negotiation between the conflicting agents. These types can form the basis for taxonomy of conflict resolution, as introduced by Klein (1991):

- *Add needed information.* When an agent needs more information as input, the agent that can output this information can try to fill in with the kinds of data that have been requested. Often, the agent will not be able to produce precisely what is required. In those cases, either the producing agent will have to supply default values or the information-using agent will have to make default assumptions.

- *Improve information quality.* Information quality is often dependent on the amount of processing done. For estimators especially, statistical information can sometimes be improved by performing a more thorough analysis of the data. This will produce a higher reliability and/or less error margin in the estimate.

- *Change to a better processing model.* In many cases, only a poor or shallow model has been used, to allow rough design. In the detailed design phase, these models should be abandoned in favor of better models that use the additional information available at that point.

- *Agree on common value.* Depending on the strategy, an agreement on a common value may be reached through different means. The agents are assumed to be cooperative and the primary criteria for searching for the common value will relate to the generated design.

10.5.2 Plan Based Negotiation for Conflict Resolution in MetaMorph

There are many different types of conflicts in agent-based manufacturing systems, including those detected during the analysis of manufacturability, assembleability, manufacturing resource capability and capacity. In this example, we address only conflicts resulting from manufacturing resource capacity. Conflict resolution in MetaMorph (see Chapter 14 for description of this project and the organization of mediators and resource agents) is through plan-based negotiation, combining a mediation mechanism based on layered mediators with a bidding mechanism based on the Contract Net protocol (Smith, 1980). A detailed description of the negotiation process can be found in Chapter 9 (Section 9.5.2), and further in (Shen and Norrie, 1998).

10.6 ADDITIONAL REFERENCES

Multi-agent negotiation and conflict resolution is an exciting research topic, and we have surveyed its research literature in the related sections of this chapter. In this section, we provide some additional references.

Note that Muller (1996) provides a good discussion of negotiation principles and the classification of negotiation categories, from a different point of view not cited in above sections. Further information on negotiation and conflict resolution can be found in the following resources:

Books, Special Issues of Journals and Workshops

- Special Issue on Conflict Management, Journal of *Artificial Intelligence for Engineering Design and Manufacturing* (AI-EDAM), I. Smith (ed.), Volume 9, Number 4, 1995.
- Special Issue of the Journal of Concurrent Engineering: Research and Applications, Volume 2, Number 2, 1994.
- *CSCW: Cooperation or Conflict?* S. Easterbrook (ed.), Springer-Verlag, London, 1993.
- AAAI'99 Workshop on "Negotiation: Settling conflicts and identifying opportunities" (http://www.mcs.utulsa.edu/~sandip/wshop/aaai99/)
- AAAI'99 Workshop on "Agents' Conflicts" (http://www.cert.fr/fr/dcsd/PUB/AAAI99/conflicts.html)
- ECAI'98 Workshop on "Conflicts among agents: avoid or use them?" (http://www.cert.fr/fr/dcsd/DCSDPUB/ECAI98/cfp.html)
- IJCAI'97 Workshop on "Collaboration, Cooperation and Conflict in Dialogue Systems" (http://www.cs.umd.edu/~traum/CCCinDS/workshop.html)
- IJCAI'97 workshop on Collaboration, Cooperation and Conflict in Dialogue Systems (http://www.cs.umd.edu/~traum/CCCinDS/Papers/papers.html)
- ECAI'96 Workshop on "Modeling Conflicts in AI"
- AID'94 Workshop on Conflict Resolution

Some other relevant web sites:

- Computational Models of Conflict Management (MIT – Mark Klein) (http://ccs.mit.edu/conflicts/)
- Negotiation and Learning for Conflict Resolution in Multi-Agent Design (WPI – David Brown) Systems (http://www.cs.wpi.edu/Research/aidg/)
- Negotiation by Dialectic Argumentation (QMW – Nick Jennings) (http://www.elec.qmw.ac.uk/dai/projects/negot_via_arg.html)
- Conflict Detection and Resolution Project at Univ. of Oregon (Steve Fickas) (http://www.cs.uoregon.edu/~fickas/negotiation.html)
- Conflict Problems in Knowledge Management (INRIA – Rose Dieng) (http://www.inria.fr/Equipes/ACACIA-eng.html)
- Multi-agent Negotiation Under Time Constraints (UMD – Sarit Kraus) (http://www.umiacs.umd.edu/users/sarit/)
- Managing Conflict in Business – an Internet Class (Roger Moore) (http://home1.gte.net/rnsmoore/index.htm)
- Multi-Agent Negotiation at University of Massachusetts (Victor Lesser) (http://mas.cs.umass.edu/index.shtml)
- Market-Based Mechanisms for Collaborative Design Conflict Management (ERIM - Van Dyke Parunak) (http://www.erim.org/~van/)

- Resolving Design Conflicts using Agents (Stanford CDR) (http://cdr.stanford.edu/)
- Conflict Mitigation in Concurrent Engineering (MIT – Feniosky Pena-Mora) (http://ganesh.mit.edu/feniosky/)
- Automated Negotiation (CMU – Katia Sycara) (http://www.cs.cmu.edu/~softagents/pleiades.html)

REFERENCES

Alder, M.R., Davis, A.B., Weihmayer, R. and Forest, R.W. (1989) Conflict-resolution strategies for non-hierarchical distributed agents, In L. Gasser and M.N. Huhns, eds., *Distributed Artificial Intelligence 2*, Morgan Kaufmann, pp. 139-161.

Alonso, E. (1998) How Individuals negotiate Societies, In *Proceedings of the Third International Conference on Multi-Agent Systems*, IEEE Computer Society, pp. 18-25.

Andersson, M.R. and Sandholm, T.W. (1998) Leveled Commitment Contracting among Myopic Individually Rational Agents, In *Proceedings of the Third International Conference on Multi-Agent Systems*, IEEE Computer Society, pp. 26-33.

Barbuceanu, M. and Fox, M.S. (1995) COOL: A Language for Describing Coordination in Multi Agent Systems, In *Proceedings of the First International Conference on Multi-Agent Systems*, AAAI Press / The MIT Press, pp. 17-24.

Barbuceanu, M. and Fox, M.S. (1996) The Architecture of an Agent Building Shell, In M. Wooldridge, J.P. Müller, M. Tambe, eds., *Intelligent agents II: Agent Theories, Architectures, and Languages*, Lecture Notes in Artificial Intelligence 1037, Springer, pp. 235-250.

Becker, I. (1995) *Conflicts and Negotiations in SiFA Based Design Systems*, MS Thesis, Dept. of Computer Science, Worcester Polytechnic Institute, USA, available through the Web at: http://www.cs.wpi.edu/Research/aidg/SiFA/ilan-thesis/thesis.html

Bussmann, S. and Müller, H.J. (1992) A Negotiation Framework for Cooperating Agents, In *Proceedings of CKBS-SIG*, Dake Centre, University of Keele, pp. 1-17.

Carmel, D. and Markovitch, S. (1998) How to Explore Your Opponent's Strategy (almost) Optimally, In *Proceedings of the Third International Conference on Multi-Agent Systems*, IEEE Computer Society, pp. 64-71.

Chavez, A. Dreilinger, D., Guttman, R. and Maes, P. (1997) A Real-Life Experiment in Creating an Agent Marketplace, In *Proceedings of the First International Conference on the Practical Application of Intelligent Agents and Multi-Agent Technology*, London, UK.

Chu-Carroll, J. and Carberry, S. (1996) Conflict Detection and Resolution in Collaborative Planning, In M. Wooldridge, J.P. Müller, M. Tambe, eds., *Intelligent agents II: Agent Theories, Architectures, and Languages*, Lecture Notes in Artificial Intelligence 1037, Springer, pp. 111-126.

Conry, S.E., Meyer, R.A. and Lesser, V.R. (1986) Multistage Negotiation in Distributed Planning, COINS Technical Report, University of Massachusetts, Amherst, Boston.

Dunskus, B., Grecu, D.L., Brown, D.C. and Berker, I. (1995) Using Single Function Agents to investigate Negotiation, *Artificial Intelligence for Engineering Design, Analysis and Manufacturing (AI EDAM)*, 9(4):299-312.

Durfee, E.H. and Lesser, V.R. (1991) Partial Global Planning: A Coordination Framework for Distributed Hypothesis Formation, *IEEE Transactions on Systems, Man and Cybernetics.* 21(5):1167-1183.

Durfee, E.H. and Montgomery, T.A., (1990) A hierarchical protocol for coordinating multi-agent behaviour, In *Proceedings. of the 8th National Conference on Artificial Intelligence, Boston,* MA, pp. 86-93.

Ekenberg, L., Boman, M. and Danielson, M. (1995) A Tool for Coordinating Autonomous Agents with Conflicting Goals, In *Proceedings of the First International Conference on Multi-Agent Systems,* AAAI Press / The MIT Press, pp. 89-93.

Giménez-Funes, E., Godo, L., Rodriguez-Aguilar, J.A. and Garcia-Calvés, P. (1998) Designing Bidding Strategies for Trading Agents in Electronic Auctions, In *Proceedings of the Third International Conference on Multi-Agent Systems,* IEEE Computer Society, pp. 136-143.

Gulliver, P. (1979) *Disputes and Negotiations – A cross-cultural perspective,* Academic Press.

Huberman, B.A. and Clearwater, S.H. (1995) A Multi-Agent System for Controlling Building Environments, In *Proceedings of the First International Conference on Multi-Agent Systems,* AAAI Press/The MIT Press, pp. 171-176.

Ito, A. and Yano, H. (1995) The Emergence of Cooperation in a Society of Autonomous Agents – The Prisoner's Dilemma Game under the Disclosure of Contract Histories, In *Proceedings of the 6th International Joint Conference on Artificial Intelligence,* pp. 201-208.

Jennings, N.R., Faratin, P., Johnson, M.J., Norman, T.J., O'Brien, P. and Wiegand, M.E. (1996) Agent-Based Business Processing, *International Journal of Cooperative Information Systems,* 5(2-3):105-130.

Jennings, N.R., Norman, T.J. and Faratin, P. (1998) ADEPT: An Agent-based Approach to Business Process Management, *ACM SIGMOD Record,* 27 (4):32-39.

Klein, M. (1991) Supporting Conflict Resolution in Cooperative Design Systems, *IEEE Transactions on Systems, Man, and Cybernetics,* 21(6):1379-1390.

Klein, M. and Lu, S.C.-Y. (1990) Conflict Resolution in Cooperative Design, *Artificial Intelligence in Engineering,* 4(4):168-180.

Kraus, S., Wilkenfeld, J., Zlotkin, G. (1995) Multiagent Negotiation under Time Constraints, *Artificial Intelligence,* 75(2):297-345.

Kreifelt, T. and von Martial, F. (1991) A negotiation framework for autonomous agents, In Y. Demazeau and J. P. Müller, eds., *Decentralized AI 2,* Elsevier Science.

Kuwabara, K. and Lesser, V.R. (1989) Extended Protocol for Multistage Negotiation, In *Proceedings of the 9th Workshop on Distributed Artificial Intelligence.*

Lander, S.E. and Lesser, V.R. (1991) Customizing Distributed Search Among Agents with Heterogeneous Knowledge, In *Proceedings of the 5th International Symposium on AI Applications in Manufacturing & Robotics,* Cancun, Mexico.

Lee, L.C. (1998) A Model of Progressive Multi-Agent Negotiation, In *Proceedings of the Third International Conference on Multi-Agent Systems,* IEEE Computer Society, pp. 448-449.

Lee and Durfee (1998) On Explicit Plan Languages for Coordinating Multiagent Plan Execution, In M.P. Singh, A. Rao and M.J. Wooldridge, eds., *Intelligent Agents IV: Agent Theories, Architectures, and Languages,* Lecture Notes in Artificial Intelligence 1365, Springer, pp. 113-126.

Lesser, V.R. (1998) Reflections on the Nature of Multi-Agent Coordination and Its Implications for an Agent Architecture, *Autonomous Agents and Multi-Agent Systems*. 1(1):89-111.

Malheiro, B. and Oliveira, E. (1996) Consistency and Context Management in a Multi-Agent Belief Revision Testbed, In M. Wooldridge, J.P. Müller, M. Tambe, eds., *Intelligent agents II: Agent Theories, Architectures, and Languages*, Lecture Notes in Artificial Intelligence 1037, Springer, pp. 361-375.

Matos, N., Sierra, C. and Jennings, N.R. (1998) Determining Successful Negotiation Strategies: An Evolutionary Approach, In *Proceedings of the Third International Conference on Multi-Agent Systems*, IEEE Computer Society, pp. 182-189.

Matsubara, S. and Yokoo, M. (1998) Negotiations with Inaccurate Payoff Values, In *Proceedings of the Third International Conference on Multi-Agent Systems*, IEEE Computer Society, pp. 449-450.

Mor, Y. and Rosenschein, J.S. (1995) Time and the Prisoner's Dilemma, In *Proceedings of the First International Conference on Multi-Agent Systems*, AAAI Press/The MIT Press, pp. 276-282.

Mullen, T. and Wellman, M.P. (1996) Some Issues in the Design of Market-Oriented Agents, In M. Wooldridge, J.P. Müller, M. Tambe, eds., *Intelligent agents II: Agent Theories, Architectures, and Languages*, Lecture Notes in Artificial Intelligence 1037, Springer, pp. 283-298.

Müller, H.J. (1996) Negotiation Principles, In G.M.P. O'Hare and N.R. Jennings, eds., *Foundations of DAI*, Wiley & Sons, pp. 211-229.

Myerson, R.B. (1989) Credible Negotiation Statements and Coherent Plans, *Journal of Economic Theory*, 48:264-303.

Nodine, M. and Unruh, A. (1998) Facilitating Open Communication in Agent Systems: The InfoSleuth Infrastructure, In M.P. Singh, A. Rao and M.J. Wooldridge, eds., *Intelligent Agents IV: Agent Theories, Architectures, and Languages*, Lecture Notes in Artificial Intelligence 1365, Springer, pp. 281-295.

Nwana, H., Lee, L. and Jennings, N. (1996) Coordination in Software Agent Systems, *BT Technology Journal*, 14(4).

Parunak, H.V.D. (1996) Visualizing Agent Conversations: Using Enhanced Dooley Graphs for Agent Design and Analysis, In *Proceedings of the Second International Conference on Multi-Agent System*, The AAAI Press, Menlo Park, CA.

Petrie, C., Webster, T. and Cutkosky, M. (1995) Using Pareto Optimality to Coordinate Distributed Agents, *Artificial Intelligence for Engineering Design, Analysis and Manufacturing (AI EDAM)*, 9(4):269-282.

Polat, F., Shekar, S. and Guvenir, H.A. (1993) A Negotiation Platform for Cooperating Multi-Agent Systems, *Concurrent Engineering: Research and Applications*, 1(3):179-187.

Pruitt, D.G. (1981) *Negotiation Behaviour*, Academic Press.

Reeds, C. (1998) Dialogue Frames in Agent Communication, In *Proceedings of the Third International Conference on Multi-Agent Systems*, IEEE Computer Society, pp. 246-253.

Rosenschein, J.S. and Zlotkin, G. (1994) *Rules of Encounter: Designing Conventions for Automated Negotiation among Computers*, The MIT Press.

Russell, S. and Norvig, P. (1995) *Artificial Intelligence: A Modern Approach*, Prentice-Hall.

Sandholm, T. and Lesser, V. (1995) Issues in Automated Negotiation and Electronic Commerce: Extending the Contract Net Framework, In *Proceedings of the First International Conference on Multi-Agent Systems*. AAAI Press / MIT Press, pp. 328-335.

Sathi, A. and Fox, M.S. (1989) Constraint-directed negotiation of resource allocations, In L. Gasser and M.N. Huhns, eds., *Distributed Artificial Intelligence 2*, Morgan Kaufmann.

Shen W. and D.H. Norrie (1998) An Agent-Based Approach for Dynamic Manufacturing Scheduling, In *Working Notes of the Agent-Based Manufacturing Workshop*, Minneapolis, MN, pp. 117-128

Sierra, C., Jennings, N.R., Noriega, P. and Parsons, S. (1998) A Framework for Argumentation-Based Negotiation, In M.P. Singh, A. Rao, M.J. Wooldridge, eds., *Intelligent Agents IV: Agent Theories, Architectures, and Languages*, Lecture Notes in Artificial Intelligence 1365, Springer, pp. 177-192.

Singh, M.P. (1998) Developing Formal Specifications to Coordinate Heterogeneous Autonomous Agents, In *Proceedings of the Third International Conference on Multi-Agent Systems*, IEEE Computer Society, pp. 261-268.

Smith, R.G. (1980) The contract net protocol: high-level communication and control in a distributed problem solver, *IEEE Transaction on Computers*, 29(12):1104-1113.

Sycara, K. (1989) Multi-agent Compromise via Negotiation, In L. Gasser and M.N. Huhns, eds., *Distributed Artificial Intelligence 2*, Morgan Kaufmann.

Sycara, K. (1990) Persuasive Argumentation in Negotiation, *Theory and Decision*, 28:203-242.

Sycara, K. (1991) Cooperative Negotiation in Concurrent Engineering Design, In *Cooperative Engineering Design*, Springer Verlag Publications, pp. 269-297

Victor S.K. and Brown, D.C. (1994) Designing with Negotiation Using Single Function Agents, In *Proceedings of the 9th International AI in Engineering Conference*, Pennsylvania, USA. Computational Mechanics Publications, pp. 173-179.

Walsh, W.E. and Wellman, M.P. (1998) A Market Protocol for Decentralized Task Allocation, In *Proceedings of the Third International Conference on Multi-Agent Systems*, IEEE Computer Society, pp. 325-332.

Walton, D.N. and Krabbe, E.C.W. (1995) *Commitment in Dialogue: Basic Concepts of Interpersonal Reasoning*, State University of New York Press.

Werkman, K.J. (1990) Knowledge-based model of negotiation using shareable perspectives, In *Proceedings of the 10th International Workshop on DAI*, Texas.

Werkman, K.J. and Barone, M. (1992) Evaluating Alternative Connection Designs through Multiagent Negotiation, In D. Sriram, R. Logcher, S. Fukuda, eds., *Computer Aided Cooperative Product Development*, Lecture Notes Series, No. 492, Springer, pp. 298-333.

Ygge, F., and Akkermans, H. (1998) On Resource-Oriented Multi-Commodity Market Computations, In *Proceedings of the Third International Conference on Multi-Agent Systems*, IEEE Computer Society, pp. 365-371.

Zeng, D. and Sycara, K. (1997) How Can Agents Learn to Negotiate? In J.P. Müller, M.J. Wooldridge and N.R. Jennings, eds., *Intelligent Agents III: Agent Theories, Architectures, and Languages,* Lecture Notes in Artificial Intelligence 1193, Springer, pp. 233-244.

11

Ontologies

11.1 INTRODUCTION

Chapter 4 considered the need for different types of representation ranging from those modeling the product to those that model the user. Chapter 8 looked at how agents communicate and cooperate and Chapter 10 discussed how agents negotiate. To communicate, cooperate, and negotiate, agents need to be able to handle descriptions of the world, usually based on systems of symbols. In Section 4.3.3 we mentioned ontologies as a possible mechanism for agents to share the meaning of exchanged symbols. Concurrent intelligent design and manufacturing projects are complex, involve many different actors, and require sharing knowledge independently from the use of agent technology. Ontological engineering is a new discipline, which allows descriptions of the world to be handled within human–human, human–machine, and machine-machine communication, irrespective of the particular computing environments in which these occur. This chapter introduces and discusses the concept of ontology and ontological engineering in a broader perspective than simply focusing on agents.

11.2 WHAT IS AN ONTOLOGY?

The concept of ontology is difficult to explain because it has different meanings for different categories of people, in particular for philosophers, and for software engineers. Some authors even argue that the definition of ontologies is less important than the role they can play.

11.2.1 Philosophy

According to the Catholic Encyclopedia (1913) the term Ontology was introduced by von Wolff in the eighteen century to distinguish a branch of metaphysics, the general metaphysics, from the special metaphysics which included cosmology, psychology, and theodicy. The term is built from *on, ontos,* being and *logos,* science, and thus stands for the science or philosophy of the being. Aristotle is the first Western philosopher credited with constructing "a well defined and developed ontology by analyzing in his 'Metaphysics' the simplest elements to which the mind reduces the world reality."

Recently Guarino and Giaretta (1995) have given a more compact definition: "Ontology is the branch of philosophy which deals with the nature and the organization of reality [...] Ontology tries to answer the question *What is being?* or, in a meaningful reformulation; *What are the features common to all beings?* The domain of Ontology has further been refined into Formal Ontology and Material Ontology, where Formal Ontology is concerned not so much with the bare existence of certain objects, but rather the rigorous description of their *form of being,* i.e., their structural features." Finally Guarino and Giaretta indicate that "Ontology as such is usually contrasted with Epistemology [or theory of knowledge] which deals with the nature and the organization of reality."

Thus, Ontology is concerned with the description of entities and the relations among entities of the world (or categories used to model the world), independently of any particular state or situation in which they happen to be.

11.2.2 Computer Science

For computer scientists as for Guarino and Giaretta (1995), the concept of ontology has two meanings corresponding to two different levels: (1) at an abstract level it is a conceptualization denoting a set of structured concepts, corresponding to the nature of Formal Ontology; (2) at an operational level it is a model of entities or categories expressed in some representational language. The exact definition of ontology at that stage then depends on who is using it.

For Gruber and for those interested in sharing knowledge, "an ontology is an explicit specification of a conceptualization" or more practically: "An ontology is a set of definitions of content-specific knowledge representation primitives: classes, relations, functions, and object constants" (Gruber, 1992). For Chandrasekaran and coworkers (1999) "Ontologies are content theories about the sort of objects, properties of objects, and relations between objects that are possible in a specified domain of knowledge." For Mahesh and the researchers involved in the Mikrokosmos project, "an ontology is a computational entity, a resource containing knowledge about what 'concepts' exist in the world and how they relate to one another. A concept is a primitive symbol for meaning representation with well-defined attributes and with other concepts. An ontology is a network of such concepts forming a symbol system where there are no uninterpreted symbols (except for numbers and a small number of known literals)" (Mahesh and Nirenburg, 1996).

11.3 ONTOLOGIES AND KNOWLEDGE SHARING

An ontology is a repository of concepts structured by relations describing the world. As such, it can be used by people to communicate, provided they agree on the same concepts. Concepts may be simple like <color> or more complex like <gearbox>. Concepts must not be confused with the words or phrases which label them. In particular, they are independent of any language. Thus, the concept <gearbox> could be labeled as <#C-138AC25> and referred to as 'gearbox' in English or 'boîte de vitesse' in French. A difficult step in the development of an ontology is not only to have people agree on the nature of concepts but to be able to phrase them in an unambiguous equivalent fashion in different languages. However, an ontology defined and built through common understanding does allow the sharing of an unambiguous way to refer to the external world. Then, from the shared structure of concepts (Plato's ideals), it is possible to express rules governing such concepts, or state facts about instances of the concepts. The latter allows the description of specific situations, yielding an epistemic knowledge.

An interesting use of an ontology is in the building of artificial systems, like databases, knowledge-based systems, or natural language understanding systems. Indeed, if we are able to process an unambiguous agreed-upon structure of concepts, adding labels to define a representation language is an easy task. In addition, the ontology provides the means to translate between the internal machine representation and the external natural language. The ontology gives a semantic foundation to the representation language. If in any specific domain we do not have an ontology to start with, the first step will be to develop one through a careful analysis of the domain, extracting the main concepts and selecting the appropriate vocabulary (which happens also to be good software engineering practice).

In principle, knowledge sharing can be achieved through a common universal ontology. Unfortunately, it is not clear that such a thing will ever exist. Several difficulties interfere with building a common universal ontology: (1) its large size; (2) its lack of focus; (3) the definition of its granularity; (4) the cultural references; and (5) its maintenance. A single common universal ontology would be very large indeed, as was demonstrated by the Cyc project (Guha et al., 1990).

11.4 USES OF ONTOLOGIES

Ontologies are used in several domains: (1) knowledge management; (2) knowledge bases; (3) natural languages; (4) information access; and (5) agents.

11.4.1 Communications and Knowledge Management

Building an ontology for a particular domain provides powerful clarification and can be used to improve person to person communications. An ontology-oriented analysis yields, in addition to the basic structure of useful concepts, a normalized vocabulary which for use in the design of the application, and later when the application is run. Even if the application does not support a computer tool, and is for example purely organizational, a common understanding of the relevant concepts, leading to improved communications, can generate substantial savings. Ontologies developed for this purpose sometimes only contain definitions, in which case they are then called lexicons or glossaries. When they gather terms with relational links, they are called thesauri. When both definitions and links are available, they constitute terminological ontologies. Most often such ontologies, developed for knowledge management, are produced as paper documents or web pages.

11.4.2 Knowledge-Based Systems

As noted by Neches and coworkers (1991), ontologies have a central role in developing knowledge-based systems. They provide the necessary conceptual structures and corresponding terminology. There are two types of ontology that can be developed: (1) domain ontologies; and (2) generic ontologies.

Domain ontologies centered on a particular subject (e.g., medicine) can be large, but relations between concepts are simple (Swartout et al., 1997).

Generic ontologies involve higher level concepts like time or causality, in which case they contain a smaller number of terms, but relations among the different terms can be quite complex, allowing powerful reasoning mechanisms (Gruber, 1992).

11.4.3 Natural Language Understanding Systems

Natural language processing, for automatic translation or for accessing data or knowledge bases, uses lexicons or semantic dictionaries organized around words and phrases of a particular language. Taking into account deep semantics as well as pragmatics requires a conceptual modeling of the world. Two approaches exist: (1) general ontologies; (2) situated ontologies.

General ontologies are intended to be universal and independent of any type of application. The Cyc Project (Guha et al., 1990) was developed along such lines.

Situated ontologies are developed with a particular application in mind, for example translation in a specific domain, which was the case for Mikrokosmos (Mahesh and Nirenburg, 1996).

11.4.4 Organizing Web Information

The web environment needs ontologies to improve the efficiency of search engines. This can be done in two steps: (1) giving some semantic content to the web pages by using special markers; (2) defining reference ontologies. A user building a web site will thus have the additional task of declaring page structure with respect to a particular ontology. Moreover the reference ontology will need to be extended if its content is insufficient.

Current proposals for enhancing web information can be categorized as: (1) minimal; (2) more expressive; (3) within W3C; and (4) elsewhere. HTML 2.0 is an example of a minimal enhancement which does not significantly change the user's habits. Some proposals are based on an entity/relationship model, which makes them more expressive. The World Wide Web Consortium (W3C) is developing standards for special mechanisms like Resource Description Languages (RDF) coupled with the eXtensible Markup Language (XML). See, for example, the excellent paper by Lassila (1998). Other researchers propose different solutions, e.g., SHOE (Simple HTML Ontology Extensions) at the University of Maryland, or OML & CKML (Ontology Markup Language & Conceptual Knowledge Markup Language) at Washington State University. See also the excellent paper by Luke et al. (1997).

11.4.5 Agents

Multi-agent systems need ontologies to decipher the content of exchanged messages. Their use is, however, different from the web situation since: (1) agents are local independent programs which can solve particular classes of problems; (2) agents deal with much less information than webbots; (3) agents need to communicate more complex information than that used for extracting web pages; (4) agents work automatically; and (5) specialized agents like assistants must be able to model the user and to learn.

One of the first projects using ontologies for agents was the SHADE project (Gruber et al., 1992), developed at Stanford and discussed elsewhere in this book. The project continued with the Palo Alto Cooperative Testbed (PACT) discussed by Cutkosky and coworkers (1993). The Ontolingua environment was developed in part on this occasion (Gruber, 1992).

Some of the problems linked with the use of ontologies in agent systems are: (1) the transfer of knowledge between agents; (2) learning; and (3) interfacing agents and humans.

11.5 FORMALISMS AND TOOLS

In this section, we consider some of the formalisms used for representing ontologies and describe some of the tools developed for ontologies.

11.5.1 Formalisms

Taking an operational approach, we can cite Gruber (1992): "An ontology is a vocabulary of terms (names of relations, functions, individuals), defined in a form that is both human and machine readable. An ontology, together with a kernel syntax and semantics, provides the language by which knowledge-based systems can interoperate at the knowledge level: exchanging assertions, queries, and answers." If a machine is to process ontologies, a language must be chosen. This language must be able to represent lists of terms standing for concepts, to express relations among the terms, and constraints involving terms and relations.

Many existing languages can be used for representing ontologies. However, some are more convenient than others. Two approaches – both derived from research in artificial intelligence – are suitable: (1) logic-based languages; and (2) frame-based languages. Frame-based languages are good at expressing concepts and structures, while logic-based languages are good at expressing relations and constraints. Consequently, exchanging knowledge tends to be supported by logic formalisms, while managing concept libraries is more commonly frame-oriented.

The same information can be represented very differently as shown by the following example defining the concept or primary-color.

```
"A primary color is either red, or green, or else blue."

<primary-color> ::= RED | GREEN | BLUE

primary-color = {red, green, blue}
        ∀x.color(x) ⇒ red(x)∨green(x)∨blue(x)

λ(x) (member x (list "red" "green" "blue"))

(defclass primary-color ()
   ((value (red green blue))))

<THEORY type="PrimaryColor" genus="Color">
   <object type="red"/>
   <object type="green"/>
   <object type="blue"/>
</THEORY>

<rdf:Class rdf:ID="PrimaryColor">
<rdf:Property ID="color-value">
   <rdf:range rdf:resource="colorValue"/>
   <rdf:domain rdf:resource="PrimaryColor"/>
</rdf:Property>
<rdfs:Class rdf:ID="colorValue"/>
   <colorValue rdf:ID="Red"/>
   <colorValue rdf:ID="Green"/>
   <colorValue rdf:ID="Blue"/>
</rdf:RDF>

(define-class PRIMARY-COLOR (?color)
   "primary colors are defined extensionally"
   (member ?color (listof red green blue)))
```

Thus, many different syntactic variations are possible for expressing the same thing. The choice will be guided by convenience and by the need to rely on precise semantics.

In the domain of knowledge-based systems, two languages are used primarily: CycL and KIF. CycL was developed for the Cyc project and details can be found in (Lenat and Guha, 1990). Its syntax uses first order predicate logic with second order constructs. It looks like this:

```
(#%ForAll y #%PrimaryColor
  (#%And (#%Color y)
         (#%Member y (#%Red #%Green #%Blue))))
```

KIF was developed at Stanford and its syntax can be found in (ANS98). KIF also uses first order predicate logic plus a few second order extensions.

```
(defobject primary-color (?x) :=
  (item ?x (listof red green blue)))
```

Both languages were originally defined in a Lisp environment.

In the web context, most languages are less expressive than CycL or KIF, since they have to deal with very large number of pages, and the reasoning process cannot be too complex. The underlying syntax is usually XML. Among the available formalisms, one finds RDF (Brickley et al., 1998), SHOE (Luke et al., 1997), and OML/CKML (OML, 1998).

In a somewhat intermediate position, are languages used for developing ontologies for linguistic purposes (essentially automatic translation). As examples, one can cite the Mikrokosmos ontology (Mahesh and Nirenburg, 1996) and the CG proposal for representing conceptual graphs (Sowa, 1997).

A general syntax for expressing knowledge is normally sufficient for writing axioms which can then be bundled into logic theories, to express knowledge about a particular restricted domain. Since a frame-like construction is easier to manage when handling structured concepts, extensions to languages have been proposed that define the frame representation itself. For example, an extension to KIF written as a specific ontology, the Frame ontology, defines the semantics of standard frames. The corresponding language, named Ontolingua, can then be used to create and to maintain ontologies more easily. It is interesting to note that such languages are defined reflexively, following the tradition in frame representation languages begun by RLL in artificial intelligence (Greiner and Lenat, 1980).

11.5.2 Tools

Ontologies are extremely difficult to define and to construct, since, most of the time, they have to be hand-crafted. Moreover, the structures and constraints as written in the various languages are definitional and there are no guaranties for forming a consistent set of axioms. Hence, tools have been defined with two purposes: (1) simplifying the process of creating, browsing, and modifying ontologies; and (2) checking for the coherence of the theories being developed. Such tools are usually web-based and a side benefit is that sometimes they allow ontologies to be developed cooperatively. Among the available editors, one can mention the Ontolingua editor (Farquhar et al., 1997), the Mikrokosmos editor (Mahesh and Nirenburg, 1996), the JOE (Java Ontology Editor) by Mahalingam and Huhns (1997), and several others in projects being developed worldwide. Since the domain is changing rapidly, it is best to check on the web for new tools and languages when needed.

11.5.3 Existing Ontologies

Fridman Noy and Hafner (1997) carried out a survey in which they examined 10 ontologies and tried to define criteria for comparing them. In this section, we simply give some examples of existing ontologies to illustrate various formalisms coming from different approaches.

Cyc

Cyc is the largest ontology developed in the world (100,000 terms, not including hundreds of thousands of proper names, over a million hand-entered assertions, not including trillions of one-step deductions). It was begun during the 10 years of the Cyc project at MCC (Austin, Texas), spun out as a separate company, Cycorp, at the end of 1994, which by now has expended an even larger number of person-centuries expanding it and applying it. Although it is used for commercial purposes, the top of the ontology is freely available. It contains the following concepts: Fundamentals, Top Level, Time and Dates, Types of Predicates, Spatial Relations, Quantities, Mathematics, Contexts, Groups, Doing, Transformations, Change Of State, Transfer Of Possession, Movement, Parts of Objects, Composition of Substances, Agents, Organizations, Actors, Roles, Professions, Emotions, Propositions, Attitudes, Social, Biology, Chemistry.

The following (truncated) example details the concepts of a professional and relations between employer and employee.

```
#$Professional
        A set of agents. Elements of #$Professional are agents who
spend a significant part of their waking hours doing activities
that are characteristic of some occupation, skilled or unskilled.
However, elements of #$Professional need not be working the entire
duration of when they are a professional, such as a #$Professor on
summer break, or someone who is temporarily unemployed. The
elements of #$Professional are persons, most of whom belong to at
least one such collection during some portion of their lives.
Typically their actions are performed for pay, but not always
(e.g., #$Artist-Visual). What are colloquially considered
professions or occupations are subsets of #$Professional; for
example, #$LumberJack, #$Scientist, #$Lifeguard, #$StockBroker,
#$Technician, #$CraftWorker, #$Housekeeper, #$SportsCoach,
#$Athlete, #$LegalProfessional, #$Publicist, etc.
        isa: #$ExistingObjectType
        genls: #$Person
        some subsets: #$Researcher #$Athlete #$Executive
#$MilitaryPerson #$Employee #$DeskWorker #$PublicSectorEmployee
#$Consultant #$AcademicProfessional #$NonProfitEmployee
#$SalesPerson #$EntertainmentOrArtsProfessional
#$PrivateSectorEmployee #$Farmer #$SelfEmployedWorker (plus 31
more public subsets, 596 unpublished subsets)

        RELATIONS between employers and employees
#$hasPositionIn : <#$Person> <#$Organization> <#$PositionType>
```

The predicate #$hasPositionIn is used to relate a particular person to his or her position in a particular organization. (#$hasPositionIn PER ORG POS-TYPE) means that the #$Person PER works in the #$Organization ORG, in a position of type POS-TYPE. POS-TYPE may or may not specifically indicate PER's occupational field or training; that is, one individual occupying a position of #$Manager may be a #$MarketingPerson, while another is an #$ElectricalEngineer. In addition, POS-TYPE may or may not correspond to an official 'Job Title' (cf. #$hasTitle) or be the primary occupation of PER. Contrast, e.g., (#$hasPositionIn JerryLewis MarchOfDimes #$Spokesman) with (#$hasPositionIn DellaStreet PerryMasonsLawOffice #$Secretary). Note that assertions using #$hasPositionIn should be properly time-constrained, for example with #$holdsIn.
      isa: #$TernaryPredicate

Ontolingua

Ontolingua was developed by Gruber (1992) at the Knowledge Systems Laboratory at Stanford (which hosts libraries of ontologies, as well as tools to develop new ones). Ontolingua is not an ontology but is meant to be used for developing ontologies, and for translating them into target languages capable of complex inferences. It is thus a neutral language. The following examples are taken from the frame ontology (Olsen, 1994). They correspond to high-level ontologies.

```
(define-class BINARY-RELATION ($relation)
    "A binary relation is a relation of fixed arity = 2.
Binary relations go by many names.
In LOOM binary relations are called roles.
In CycL, binary relations are called slots.
In KEE, binary relations are called slots (but they are not
objects).
In Epikit, binary relations are not distinguished from other
relations."
    :iff-def (and (relation ?relation)
                  (not (empty ?relation))
                  (forall (?tuple)
                          (=> (member ?tuple ?relation)
                              (double ?tuple))))
    :constraints (= (arity ?relation) 2)
    :theory kif-relations)

(define-function INVERSE (?binary-relation) :-> ?inverse-relation
    "One binary relation is the inverse of another if they are
equivalent when their arguments are swapped."

    :lambda-body (if (binary-relation ?binary-relation)
                     (setofall (listof ?y ?x) (holds ?binary-
                     relation ?x ?y)))

    :constraints (and (binary-relation ?binary-relation)
                      (binary-relation ?inverse-relation))

    :axiom-def (<=> (holds ?binary-relation ?x ?y)
                    (holds (inverse ?binary-relation) ?y ?x))
```

```
:theory kif-relations
:issues
("Note that INVERSE is a function.  It is possible to have more
  than one relation constant naming the inverse of a relation,
  but they are all = to each other."))
```

RDF

RDF has the same goal as SHOE but aims at becoming a standard. The following is taken from a W3C Working Draft edited by Brickley, Guha, and Layman (1998) (Copyright: http://www.w3.org/Consortium/Legal/copyright-documents-19990405.html).

In this example, `Person` is a class with a corresponding human-readable description of "The class of people". A `Person` is a sub-class of `Animal`. A `Person` may have an `age` property. The value of `age` is an `integer`. A `Person` may also have an `ssn` ("Social Security Number") property. The value of `ssn` is an `integer`. A `Person`'s marital status is one of {Married, Divorced, Single, Widowed}. This is achieved through use of the range constraint: we define both a `maritalStatus` property and a class `MaritalStatus` (adopting the convention of using lower case letters to begin the names of properties, and capitals for classes). We then use `rdfs:range` to state that a `maritalStatus` property only 'makes sense' when it has a value which is an instance of the class `MaritalStatus`. The schema then defines a number of instances of this class. Whether resources declared to be of type `MaritalStatus` in another graph are trusted is an application level decision; such decisions will be aided by the provisions in RDF for digital signatures.

```
<rdf:RDF
    xmlns:rdf="http://www.w3.org/TR/WD-rdf-syntax#"
    xmlns:rdfs="http://www.w3.org/TR/WD-rdf-schema#">

  <rdfs:Class rdf:ID="Person">
    <rdfs:comment>The class of people.  correspond to a single
person.
    </rdfs:comment>
    <rdfs:subClassOf
rdf:resource="http://www.classtypes.org/useful_classes#Animal"/>
  </rdfs:Class>

  <rdf:Property ID="maritalStatus">
    <rdfs:range rdf:resource="#MaritalStatus"/>
    <rdfs:domain rdf:resource="#Person"/>
  </rdf:Property>

  <rdf:Property ID="ssn">
    <rdfs:comment>Social Security Number</rdfs:comment>
    <rdfs:range
rdf:resource="http://www.datatypes.org/useful_types#Integer"/>
    <rdfs:domain rdf:resource="#Person"/>
  </rdf:Property>

  <rdf:Property ID="age">
```

```
      <rdfs:range
 rdf:resource="http://www.datatypes.org/useful_types#Integer"/>
      <rdfs:domain rdf:resource="#Person"/>
    </rdf:Property>

    <rdfs:Class rdf:ID="MaritalStatus"/>

    <MaritalStatus rdf:ID="Married"/>
    <MaritalStatus rdf:ID="Divorced"/>
    <MaritalStatus rdf:ID="Single"/>
    <MaritalStatus rdf:ID="Widowed"/>
    </rdf:RDF>
```

FIPA

FIPA is a consortium formed to develop standards for agent systems. The following examples are extracted from the "FIPA SL ontology for MNA Memoir Agent" (Gibbins, 1998). It defines a person and the concept of document. It also shows how an agent can use the ontology for sending queries to other agents, or for providing information.

Object model:

```
(Person <email-address>
        <first-name>
        <last-name>
        <url>
        <interests>
        <phone-number>
        <fax-number>)
```

<email-address> is used to uniquely qualify a person.

Instantiation:

```
(person "wh@ecs.soton.ac.uk" "Wendy" "Hall"
        "http://www.ecs.soton.ac.uk/~wh/"
        ("multimedia" "open hypermedia" "hypertext")
        "4060" "9090")
```

Relations:

```
(has-interest <email-address> <interest>)
(bookmarked <email-address> <url>)
(seen <email-address> <url>)
```

Use:

Example of message for telling the MEMOIR agent that a user explored a given document (messages are written using the KQML protocol):

```
(inform :sender agentui.host
        :receiver memoir.snoopy
        :reply-with msg002
```

```
:ontology memoir-ontology
:content (seen "nmg97r@ecs.soton.ac.uk"
         "http://www.ecs.soton.ac.uk/~nmg97r/agent."))
```

Example of question: Who else has seen a document ?

```
(query-ref :sender agent-ui.host
           :receiver memoir.snoopy
           :reply-with msg001
           :ontology memoir-ontology
           :content (iota ?y (and (seen ?x "http://www.dai.ed.ac.uk/")
                             (has-email-address ?y ?x)))))
```

11.6 ONTOLOGIES IN CONCURRENT DESIGN AND MANUFACTURING

In the field of Concurrent Design and Manufacturing, ontologies are mainly used for the following purposes: (1) to improve communication among humans; (2) to improve data exchange among programs; (3) to improve the design or manufacturing process; and (4) to organize knowledge management.

Improving communications among humans involves standardizing the vocabulary and integrating new concepts. The goal is to increase mutual understanding among people from different departments, such as the design department and the production planning department. A typical example was reported by Genc and coworkers (1998) who tried to put some order into the domain of snap-fit assemblies.

Improving electronic data exchange requires compatible representation models. The ontology can be used to specify the concepts and vocabulary needed for developing exchange software (using frameworks like CORBA or STEP/EXPRESS), or in integrating legacy systems when implementing concurrent engineering. When cognitive agents are used in concurrent engineering, the use of ontologies is critical in sharing knowledge among the agents, as in the PACT project (Gruber et al., 1992)

Improving the design process using ontologies is linked either to the reuse of previous designs (for example, from a repository of such designs) or to guiding the design process. Note that whenever a design domain is decomposed into basic primitives, these can be thought of as elements of an ontology. For examples involving features, see (Shah and Rogers, 1993; Rosen, 1993; Brown et al., 1995). In the following, however, ontologies are used for guiding the design process. Oksala (1993) and Aksoy and Saglamer (1993) use grammars for generating housing, trying to automate the design process. Tomiyama and coworkers (1995) use a combination of reasoning methods for the design of mechanical assemblies. Kumar and Upadhyaya (1998) propose the use of component-based ontologies for reasoning about functions during the design process. Gershenson and Stauffer (1999) advocate using a taxonomy for design requirements that implement company knowledge, to guide the design process. Borst and Akkermans (1997) show how ontologies are useful for reasoning about product disassembly.

Organizing knowledge management also requires ontologies. A good example is provided by the ATOS-1 project from the European Space Agency, for organizing information associated with a particular satellite (Jones et al., 1994).

11.7 RESEARCH LITERATURE AND FURTHER REFERENCES

11.7.1 Research Literature

Ontology engineering is a relatively new domain. Despite widespread interest in developing ontologies, there are few tools or methods readily available to help do the work, so most existing ontologies have been crafted by hand. There has, however, been considerable research on classifying various types of ontologies, linking ontologies to problem solving techniques, and designing methods for developing new ontologies.

In 1997, Fridman Noy and Hafner (1997) carried out a survey in which they examined 10 ontologies and tried to define criteria for comparing them. Then, more recently, Uschold and Jasper (1999) developed a framework for classifying ontology applications, using the following criteria: intended purpose or benefit; role of the ontology; the actors using the ontology, the supporting technologies; the level of maturity; the degree of formalism for representing concepts; whether an ontology is shared or copied. The framework can be used to classify various approaches and projects in the domain.

Building ontologies has so far been an art. Mahesh and Nirenburg (1995), and Grüninger and Fox (1995) offer recommendations and guidelines for the process. Swartout and coworkers (1997) give some recommendations for building large ontologies. Lopez (1999) noted some methodologies, referring also to other methodologies in various domains suited, in particular, for developing software.

Connecting ontologies with problem solving is an important issue, since this leads to better programming of systems containing knowledge. A discussion of this can be found in Fensel et al. (1997). Also, Swartout and coworkers (1999) propose a framework, EXPECT, for automatically organizing and indexing problem solving methods with the use of ontologies. And recently, Gomez Pérez and Benjamins (1999) give an overview of ontologies and problem solving methods.

Finally, it is evident that more experimental work is needed, in particular, for designing ontologies cooperatively. The (KA)^2 initiative (Benjamins and Fensel, 1998) is one such project (http://www.aifb.uni-karlsruhe.de/WBS/broker/KA2.html).

11.7.2 Related Conferences and Workshops

Specialized workshops or sessions on ontologies have been held during the following conferences:

- IJCAI (International Joint Conferences on Artificial Intelligence) (http://www.ijcai.org/)
- AAAI (American Association of Artificial Intelligence) (http://www.aaai.org/)
- ECAI (European Conferences on Artificial Intelligence) (http://www.eccai.org/)
- KAW (Knowledge Acquisition Workshops) (http://ksi.cpsc.ucalgary.ca/KAW/)
- EKAW (European Knowledge Acquisition Workshops) (http://www.aifb.uni-karlsruhe.de/ekaw99/)

There have also been few specific meetings on Ontological Engineering:

- 1997 AAAI Spring Symposium on Ontological Engineering (http://www-ksl.stanford.edu/projects/htw/sss97/)
- International Workshop on Ontological Engineering on the Global Information Infrastructure. (http://www.swi.psy.uva.nl/usr/richard/workshops/ka2-99/ka2-cfp.html)

REFERENCES

Aksoy, M. and Saglamer, G. (1993) A Shape-Grammar for Courtyard and Houses, In M.R. Beheshti and K. Zreik, eds., *Advanced Technologies*, Elsevier Science Publishers, pp. 251-258.

Benjamins, V.R. and Fensel, D. (1998) The Ontological engineering Initiative (KA)^2, In N. Guarino, ed., *Formal Ontology in Information Systems*, IOS Press, The Netherlands, pp. 287-301.

Borst P. and Akkermans H. (1997) An Ontology Approach to Product Disassembly, In E. Plaza and R. Benjamins, eds., *Knowledge Acquisition, Modeling and Management*, Lecture Notes in Artificial Intelligence 1319. Springer, pp. 33-48.

Brickley, D., Guha R.V. and Layman, A., eds., (1998) Resource Description Framework (RDF) Schema Specification, WD-rdf-schema-19981030, http://www.w3.org/TR/1998/WD-rdf-schema-19981030/.

Brown K.N., McMahon C.A. and Sims Williams J.H. (1995) Features, a.k.a. The Semantics of a Formal Language of Manufacturing, *Research in Engineering Design*. 7(3):151-172.

Chandrasekaran, B. and Josephson, J.R. (1999) What Are Ontologies, and Why Do We Need Them? *IEEE Intelligent Systems*, 14(1):20-26.

Cutkosky, M.R., Engelmore, R.S., Fikes, R.E., Genesereth, M.R., Gruber, T.R., Mark, W.S., Tenenbaum, J.M. and Weber, J.C. (1993) PACT: An Experiment in Integrating Concurrent Engineering Systems, *IEEE Computer*, 26(1):28-37.

Farquhar, A., Fikes, R. and Rice, J. (1997) The Ontolingua Server: a Tool for Collaborative Ontology Construction, *International Journal of Human-Computer Studies*, 46(6):707-728.

Fensel D., Motta E., Decker S. and Zdrahal Z. (1997) Using Ontologies for Defining Tasks, Problem-Solving methods and their Mappings, In E. Plaza and R. Benjamins, eds., *Knowledge Acquisition, Modeling and Management*, Lecture Notes in Artificial Intelligence 1319. Springer, pp. 113-128.

Fridman Noy, N. and Hafner, C.D. (1997) The State of the Art in Ontology Design, A Survey and Comparative Review, *AI Magazine*, 18(3):53-74.

Genc, S., Messler, R.W.Jr. and Gabriele, C.A. (1998) A Hierarchical Classification Scheme to Define and Order the Design Space for Integral Snap-Fit Assembly, *Research in Engineering Design*, 10(2):94-106.

Gershenson, J.K., and Stauffer, L.A. (1999) A Taxonomy for Design Requirements from Corporate Customers, *Research in Engineering Design*. 11(2):103-115.

Gibbins, N. (1998) FIPA SL ontology for MNA Memoir Agent, Department of Electronics and Computer Science, University of Southampton,
http://www.staff.ecs.soton.ac.uk/~nmg97r/agent/fipa-mna.html

Gomez Pérez A. and Benjamins R. (1999) Overview of Knowledge Sharing and Reuse Components: Ontologies and Problem-Solving Methods, In *Proceedings of the IJCAI-99 Workshop on Ontologies and Problem Solving Methods: Lessons Learned and Future Trends*, CEUR Publication, Vol 18, The Netherlands, pp. 1-1 to 1-15.

Greiner, R. and D.B. Lenat (1980) A Representation Language Language, In *Proceedings AAAI-80*, Stanford, pp. 165-169.

Gruber, R.T. (1992) Ontolingua: A Mechanism to Support Portable Ontologies (version 3.0), Technical Report KSL 91-66, Stanford University.

Gruber, R.T., Tenenbaum, J.M. and Weber, J.C. (1992) Toward a Knowledge Medium for Collaborative Product Development, In *Proceedings of the Second International Conference on Artificial Intelligence in Design*, Kluwer Academic Publishers, pp. 413-432.

Grüninger M. and Fox M. (1995) Methodology for the Design and Evaluation of Ontologies, In *Proceedings of the Workshop on Basic Ontological Issues in Knowledge Sharing*, Montreal, Canada.

Guarino, N. and Giaretta, P. (1995) Ontologies and Knowledge Bases, Towards a Terminological Clarification, In N. Mars, ed., *Towards Very Large Knowledge Base Building and Knowledge Sharing*, IOS Press, Amsterdam, pp. 25-32.

Guha, R.V., Lenat, D.B., Pittman, K., Pratt, D. and Shepherd, M. (1990) Cyc: A Midterm Report, *Communications of the ACM,* 33(8).

Jones, M., Wheadon, J., O'Mullane, W., Whitgift, W., Poulter, K., Niezette, N., Timmermans, R., Rodriguez, I. and Romero, R. (1994) ATOS-1: A Basis for Knowledge Sharing in Spacecraft Mission Operations, In *Proceedings of ISMICK'94*, IIIA. France, pp. 21-33.

Kumar, A.N. and Upadhyaya, S.J. (1998) Component-Ontological Representation of Function for Reasoning about Devices, *Artificial Intelligence in Design*, 12:399-415.

Lassila, O. (1998) Web Metadata: A Matter of Semantics, *IEEE Internet Computing*, 2(4):30-37.

Lenat, D.B. and Guha, R.V. (1990) *Building Large Knowledge Based Systems: Representation and Inference in the Cyc Project*, Addison-Wesley Publishing, Massachusetts.

López, F, (1999) Overview of Methodologies for Building Ontologies, In *Proceedings of the KRR-5 Workshop Ontologies and Problem-Solving Methods: Lessons Learned, and Future trends*, IJCAI'99, CEUR Publications, Vol 18, pp. 4-1 to 4-13.

Luke, S., Spector, L., Roger, D. and Hendler, J. (1997) Ontology-Based Web Agents, In *Proceedings of the First International Conference on Autonomous Agents*, ACM Press, New York.

Mahalingam, K. and Huhns, M.N. (1997) A Tool for Organizing Web Information, *Computer*, 30(6):80-83.

Mahesh, K. and Nirenburg, S. (1995) A Situated Ontology for Practical NLP, In *Proceedings of IJCAI-95 Workshop on Basic Ontological Issues*, Montréal, Canada.

Mahesh, K. and Nirenburg, S. (1996) Meaning Representation for Knowledge Sharing in Practical Machine Translation, In *Proceedings of the. Florida Artificial Intelligence Research Symposium,* FLAIRS-96, Special Track on Information Interchange.

Neches, R., Fikes, R., Finin, T., Gruber, T., Patil, R., Senator, T. and Sartout, W.R. (1991) Enabling Technology for Knowledge Sharing, AI Magazine 3(12):36-56.

Oksala, T. (1993) Typology versus Change: Studies in the Generation of Apartment Types in Housing Production, In M.R. Beheshti and K. Zreik, eds., *Advanced Technologies,* Elsevier Science Publishers, pp. 435-442.

Olsen, G. R. (1994) Theory Frame Ontology, Department of Mechanical Engineering, Stanford University, http://piano.stanford.edu/concur/examples/html-lib/frame-ontology/.

OML. (1998) Ontology Markup Language, WSU. http://www.ontologos.org/

Rosen, D.W. (1993) Feature-Based Design: Four Hypotheses for Future CAD Systems, *Research in Engineering Design.* 5:125-139.

Shah, J.J. and Rogers, M.T. (1993) Assembly Modeling as an extension to Feature-Based Design, *Research in Engineering Design.* 5:218-237.

Sowa, S. (1997) *Standard for Conceptual Graphs,* a draft for the ANSI X3H32 group ANSI.

Swartout B., Patil R., Knoght K. and Russ T. (1997) Toward Distributed Use of Large Scale Ontologies, In *Proceedings of the AAAI Spring Symposium Series: Workshop on Ontological Engineering.*

Swartout B., Gil Y. and Valente A. (1999) Representing Capabilities of Problem Solving Methods, In *Proceedings of the IJCAI-99 Workshop on Ontologies and Problem Solving Methods: Lessons Learned and Future Trends,* CEUR Publication, Vol 18, The Netherlands, pp. 8-1 to 8-8.

Tomiyama, T., Umeda, Y., Ishii, M., Yoshioka, M. and Kiriyama, T. (1995) Knowledge Systematization for a Knowledge Intensive Engineering Framework, In T. Tomiyama, M. Mäntylä and S. Finger eds., *Knowledge Intensive CAD-1, Preprints of the First IFIP WG 5.2 Workshop,* Espoo, Finland, pp. 55-80.

Uschold M. and Jasper R. (1999) A Framework for Understanding and Classifying Ontology Applications, In *Proceedings of the IJCAI-99 Workshop on Ontologies and Problem Solving Methods: Lessons Learned and Future Trends,* CEUR Publication, Vol 18, The Netherlands, pp. 11-1 to 11-12.

12

Other Issues

12.1 INTRODUCTION

Chapters 4–11, considered key issues in developing agent-based concurrent design and manufacturing systems, including

- knowledge representation
- learning
- agent architectures
- agent system architectures
- communication, coordination and cooperation
- collaboration and task decomposition
- negotiation and conflict resolution
- ontology

In this Chapter, we consider some other important issues relating to development of agent-based concurrent design and manufacturing systems, including:

- agent encapsulation
- human-machine interaction
- system dynamics
- design and manufacturability assessment
- integration of manufacturing planning, scheduling and execution
- distributed dynamic scheduling
- enterprise integration and supply chain management
- legacy problems
- other mature technologies and tools

12.2 AGENT ENCAPSULATION

There are two distinctly different approaches to agent encapsulation in agent-based design and manufacturing systems: functional decomposition and physical decomposition.

In functional decomposition, agents are used to encapsulate modules assigned to functions such as order acquisition, product design, process planning, scheduling, material handling, transportation management, and product distribution. There is no explicit relationship between agents and physical entities. Agent-based concurrent engineering design systems that are not directly connected to real-time manufacturing (see Chapter 13 for examples) usually use this approach; however, RAPPID (Parunak et al., 1997; see also Chapter 13) is an exception. Examples of other agent-based systems that use this approach for enterprise integration, manufacturing planning and scheduling are ISCM (Fox et al., 1993), CIIMPLEX (Peng et al., 1998), ABACUS (McEleney et al., 1998), and LMS (Fordyce and Sullivan, 1994).

In physical decomposition, agents are used to represent entities in the physical world, such as machines, tools, fixtures, products, parts, features, operators and operations. There exists an explicit relationship between an agent and a physical entity. Examples can be found in MetaMorph I & II (Maturana and Norrie, 1996; Shen et al., 1998), ADDYMS (Butler and Ohtsubo, 1992), AIMS (Park et al., 1993), AARIA (Parunak et al., 1998), YAMS (Parunak, 1987).

The functional approach tends to share many state variables across different functions. Separate agents thus share state variables, leading to problems of consistency and unintended interaction. The physical approach naturally defines distinct sets of state variables that can be managed efficiently by individual agents with limited interactions.

However, the functional approach is very useful for integrating existing systems (e.g., CAD tools, MRP systems, etc.) and resolving legacy problems (see Section 12.9). Even in an agent-based manufacturing system using primarily the physical approach, the functional approach may still be useful. Agents encapsulating special functions may provide such services at the system level, such as the facilitators in PACT (Cutkosky et al., 1993), the broker agents in CIIMPLEX (Peng et al., 1998) and AIMS (Park et al., 1993), the mediators in MetaMorph I & II (Maturana and Norrie, 1996; Shen et al., 1998), and the system monitors in DIDE (Shen and Barthès, 1995; 1997) and in IFCF (Lin and Solberg, 1992).

12.3 HUMAN-MACHINE INTEGRATION

Agent-based design and manufacturing systems usually involve both human and artificial agents. As in any software application, appropriate provision for human–machine interaction in an agent-based system is key to effective user operation.

Two areas of computer science have been developing techniques to implement partnership between people and machines:

-   Classical work in multi-agent systems builds systems of artificial agents, into which we can introduce people.
-   Computer-supported collaborative work (CSCW) which uses computer technology to integrate communities of humans, and opens the door to including artificial agents.

Traditional CSCW develops computer mechanisms (such as electronic mail, computer conferencing, interactive video conferencing, and shared databases) to help networks of humans interact more efficiently. Such systems provide each human with an "interface" that provides connectivity to other humans. Once such a network is in place, non-human entities can be incorporated by giving them an "interface" compatible with the others in the system.

While multi-agent systems open the door to a partnership between humans and machines, they do not guarantee it. Many multi-agent systems still model people as operators, for example, ARCHON (Cockburn and Jennings, 1996) and PACT (Cutkosky et al., 1993). But several others seek closer integration.

In theory, there are two ways to incorporate people into a network of agents. Either artificial agents can be made so intelligent that people come to view them as peers, or people can be represented in the network by artificial agents that make them appear to other agents as computers or software entities. The first approach is the objective of classical AI's Turing test, but most agent-based concurrent design and manufacturing systems take the second approach.

Chang (1987) defined participant systems as computer systems that support collaborative intellectual tasks among a number of persons, possibly distributed in different locations. The system proposed in this research provides mechanisms conducive to the establishment of social cohesiveness. Croft and Lefkowitz (1988) used the expert system shell KEE to build POLYMER, a system that supports users involved in cooperative activities that can be performed either by human or artificial agents.

Pan and Tenenbaum (1991) described a software Intelligent Agent framework for integrating people and computer systems in large, geographically dispersed manufacturing enterprises. This framework was based on the vision of a very large number (e.g. 10000) computerized assistants, known as Intelligent Agents (IAs). Human participants are encapsulated as Personal Assistants, a special type of IA that knows how to communicate with humans (through multimedia interfaces), and with other IAs and a shared knowledge base (called the knowledge service).

DCSS (Klein, 1991) was developed to support collaboration among designers of local area networks by providing them with a protocol for representing and justifying design decisions, mechanisms for detecting conflicts among proposals, and resources for suggesting alternative solutions. The system offers valuable support for purely human teams of designers, but once the protocols are defined, it can be augmented by the inclusion of automated design agents as well.

12.4 SYSTEM DYNAMICS

Agent-based concurrent design and manufacturing systems may have long life times. However, real world manufacturing environments are highly dynamic because of diverse and frequently changing situations: bank rates change overnight; political situations change; materials do not arrive on time; power supplies breakdown; production facilities fail; workers are absent; new orders arrive and existing orders are changed or canceled. Hence, it is necessary that the architecture not only facilitate rapid response to change but also allow agents to join or to quit the overall system without disturbing its overall operation. Systems that allow this kind of behavior are called open systems. In most existing design and manufacturing systems, however, whenever a change occurs, everything must be brought to a halt, code must be recompiled, and the system has to be reinitialized. This process is not acceptable for large engineering design projects which must continue to work even when some subsystems have a problem, or when new agents or new modules are plugged into the system. Traditional approaches, in which a group of powerful tools may be tightly integrated into a large, efficient, decision support system, typically are not based on open architectures

In the DIDE system (Shen and Barthès, 1995; 1997), which does have an open architecture, agents may appear or disappear with minimal disturbance to the whole system. New agents, whether slaves (such as encapsulated existing tools) or more active entities, are created from generic agents as needed. They possess communication capabilities and some expertise (which may result from programs, rules, databases, etc.). They are not required to know anything about the existing agents or the ongoing tasks to be solved. When a new agent connects to the group, it can immediately receive broadcast messages and respond to them. Other agents do not need to be modified because of the introduction of a new agent, so the system keeps running. Once a new agent is connected, it must build its own representation of the task being solved as well as acquire information about the other agents. Currently, two strategies can be used: (i) either a "tutor" agent helps the new agent to do it; or (ii) models will be built as needed through exchanged messages. Two things could be done in addition: (i) an agent could advertise its skills as soon as it is connected; (ii) an agent could actively inquire once it is connected, by broadcasting requests.

In AIMS (Park et al., 1993), by using the Internet to connect agents, services can be added or taken out of service at any time, with incremental impact on capacity, simply by informing the appropriate directories and brokers. MetaMorph I & II (Maturana and Norrie, 1996; Shen et al, 1998) also provide this feature by using a registration mechanism at both the system level and the mediator level. AARIA (Parunak et al., 1998) provides such a feature through its Autonomous Agent approach, and ATP (NIST, 1998) through its plug-and-play framework.

In conclusion, next generation concurrent design and manufacturing systems require dynamic reconfiguration. Agents, if properly implemented, can offer such features. Of course, some organizational effort has to be deployed to prevent users introducing agents anarchically.

12.5 DESIGN AND MANUFACTURABILITY ASSESSMENT

Geometric and functional specifications, the availability of raw materials, and the capability and availability of shop-floor resources each have a major influence on manufacturability. A design may be manufacturable under one combination of product requirements and shop-floor resources, but not under another. An agent-based concurrent design and manufacturing system should be able to provide dynamic manufacturability assessments during the product design process.

The MetaMorph I architecture provides a mechanism for immediate manufacturability assessments (Maturana and Norrie, 1996). As a product part is progressively designed by repeated instantiation of features, manufacturability is evaluated by resource agents for every instantiation. Design Mediators and Resource Mediators ensure coordination among design parts and resource agents. Design subsystems (or design agents) interact with resource agents via Resource Mediators to obtain manufacturability assessments during the product design process. This process not only ensures the manufacturability of a product, but also results in incremental identification of general process plans.

12.6 INTEGRATION OF MANUFACTURING PLANNING, SCHEDULING AND EXECUTION

12.6.1 Manufacturing Planning, Scheduling and Execution

Manufacturing planning is the process of selecting and sequencing activities to achieve one or more goals and satisfy a set of domain constraints. Manufacturing scheduling is the process of selecting among alternative plans and assigning resources and times to the set of activities in the plan. These assignments must obey a set of rules or constraints that reflect the temporal relationships between activities and the capacity limitations of a set of shared resources. The assignments also affect the optimality of a schedule with respect to criteria such as cost, tardiness, or throughput. In summary, scheduling is an optimization process where limited resources are allocated over time among both parallel and sequential activities (Zweben and Fox, 1994). Execution is the process of schedule realization. A description of several agent-based systems for manufacturing planning, scheduling, and execution can be found in Chapter 14.

12.6.2 Integration of Planning and Scheduling

Traditional approaches to planning and scheduling do not consider the constraints of both domains simultaneously. In spite of being sub-optimal these approaches have been in vogue due to the non-availability of a unified framework. Agent-based approaches provide a possible way to integrate planning and scheduling activities through enterprise-level coordination between the product design system and the factory resource scheduling system. MetaMorph I implemented such a mechanism through enterprise level coordination between Design Mediators and Resource Mediators who in turn coordinate resource agents at the shop floor level (Maturana et al., 1996).

12.6.3 Concurrent Scheduling and Execution

Traditional systems alternate scheduling and execution. For example, a firm develops a schedule each night for its manufacturing operations the next day. The real world tends to change in ways that invalidate advance schedules. In fact, natural systems do not plan in advance, but adjust their operations on a time scale comparable to that in which their environment changes. MetaMorph II (Shen et al., 1998) proposed using Execution Mediators to coordinate the execution of the machines, AGVs, and workers as necessary. Using such an approach, manufacturing resources (e.g., machines, interfaces for operators, etc.) can be connected directly to the manufacturing system. As manufacturing is in progress, information about the execution process (progress of the schedule) is captured by the Execution Mediators which in turn send this information to Resource Mediators for adjusting the schedule as necessary.

12.7 DISTRIBUTED DYNAMIC SCHEDULING

Corresponding to the two distinct approaches for agent encapsulation described in Section 12.1, two types of distributed manufacturing scheduling systems can be distinguished:

- Systems in which scheduling is an incremental search process that can involve backtracking. Agents, responsible for scheduling orders, perform local incremental searches for their orders and may consider multiple resources. The global schedule is obtained through the merging of local schedules. This is very similar to centralized scheduling.
- Systems in which each agent represents a single resource (e.g., a work cell, a machine, a tool, a fixture, or an operator) and is responsible for scheduling this resource. One agent may negotiate with other agents to carry out overall scheduling.

Examples of the first type of scheduling system can be found in (Sadeh and Fox, 1989; Sycara et al., 1994; Burke and Prosser, 1991; Fordyce and Sullivan, 1994; Murthy et al., 1997; McEleney et al., 1998).

In the second approach, the scheduling mechanism is generally realized through negotiation among agents. Common protocols for negotiation include the voting mechanism (Fordyce and Sullivan, 1994), and Smith's Contract net (Smith, 1980) or its modified versions, such as the Extended Contract Net Protocol (ECNP) proposed by Fischer et al. (1995), the Market-Driven Contract Net by Baker (1991), the B-Contract-Net by Scalabrin (1996), and the Leveled Commitment Contracting Protocol by Sandholm and Lesser (1996). Other examples can be found in (Duffie and Piper, 1986; Parunak, 1987; Ow and Smith, 1988; Shaw, 1988; Butler and Ohtsubo, 1992; Lin and Solberg, 1992; Saad et al., 1995; Shen and Norrie, 1998; Ouelhadj et al., 1998). The bidding mechanism can be part-oriented (Duffie and

Piper, 1986; Ow and Smith, 1988; Lin and Solberg, 1992), resource-oriented (Butler and Ohtsubo, 1992; Baker, 1991; Shen and Norrie, 1998), or bi-directional (Saad et al., 1995).

We use "dynamic scheduling" here to indicate that a real-time manufacturing scheduling system can update its schedule to adapt to changing situations such as new order insertion, machine failures, or job tardiness. ADDYMS developed a dynamic scheduling mechanism for local resource allocation at the local work cell level (Butler and Ohtsubo, 1992). In MetaMorph II, several mechanisms were developed for dynamic scheduling and rescheduling by combining a bidding mechanism based on Contract Net protocol with a mediation mechanism based on the Mediator architecture (Shen and Norrie, 1998). Also, Sousa and Ramos (1997) have proposed a dynamic scheduling architecture using holons.

12.8 ENTERPRISE INTEGRATION AND SUPPLY CHAIN MANAGEMENT

To support its global competitiveness and rapid market responsiveness, an individual manufacturing enterprise has to integrate its activities (e.g., purchasing, order acquisition, product design, production process planning and scheduling, execution control, product delivery, and personnel, material, and quality control), with those of its partners, suppliers and customers. This integration is via both information networks (local networks, the Internet or Intranet), and material networks which are, in general, heterogeneous software and hardware environments.

The supply chain of a manufacturing enterprise is a worldwide network of suppliers, factories, warehouses, distribution centers, and retailers through which raw materials are acquired, transformed and delivered to customers (Fox et al., 1993). This network, in general, involves heterogeneous environments.

Agent based approaches provide a natural way to design and implement manufacturing enterprise integration and supply chain management within heterogeneous environments. A description of several such systems can be found in Chapter 15.

Research has shown that agent based approaches have the following advantages for enterprise integration and supply chain management:

- increase the responsiveness of the enterprise to the market requirements;
- involve customers in total supply chain optimization;
- realize supply chain optimization through effective resource allocation;
- achieve dynamic optimization of materials and inventory management;
- realize total supply chain optimization including all linked enterprises;
- increase the effectiveness of the information exchange and feedback.

However, security problems resulting from the open architecture of agent-based systems, particularly when using Internet and the mobile agent technology, have been recognized by both manufacturing enterprises and the researchers in this area. These are not unique to agent-based systems and may be mitigated through further research.

12.9 LEGACY PROBLEMS

Product design and production management systems used by today's manufacturing enterprises are separate application software packages, each for a different part of the product design, process planning, scheduling and execution control processes. For example, Capacity Analysis software determines a Master Production Schedule that sets long-term production targets. Enterprise Resource Planning (ERP) software generates material and resource plans. Scheduling software determines the sequence in which shop floor resources (people, machines, tools, materials, etc.) are used in producing different products. A Manufacturing Execution System (MES) tracks the real-time status of work in progress, enforces routing integrity, and reports labor or material claims. Most of these application software packages are legacy systems developed over years. Existing legacy architectures provide little interoperability among the individual application software packages. Modern manufacturing enterprises are searching for solutions to integrate these legacy systems within an open environment. Traditional technologies do not yet have a satisfactory solution to this problem. Agent technology, however, provides a possible way to reuse legacy systems within an open environment, provided that they possess sufficiently good programmable interfaces (APIs). In this case, they can be encapsulated into agents, and integrated into more complex systems.

A number of techniques have been proposed to resolve legacy problems in agent-based industrial applications. EXPORT demonstrated how commercial tools could be encapsulated and become part of a larger system using a blackboard architecture (Monceyron and Barthès, 1992). Other projects like ANAXAGORE (Trousse, 1993) and KAD (Gupta et al., 1996) used a similar approach.

One of the major objectives of the ARCHON project (Cockburn and Jennings, 1996) was to address the challenge of legacy systems. ARCHON's solution is its "head-body" agent architecture. Each agent consists of a "head" – an ARCHON Layer (AL), a "body" – an application program known as an Intelligent System (IS), and an AL-IS interface. Any pre-existing (legacy) system can be taken as an IS and then easily encapsulated into an ARCHON agent by adding an AL and an AL-IS interface. PACT (Cutkosky et al., 1993) and DIDE (Shen and Barthès, 1995; 1997) used a wrapper, which is very similar to the AL in ARCHON, to encapsulate a legacy system as an agent. Jha et al. (1998) also used wrapper agents to encapsulate legacy systems including tools for geometry manipulation and optimization. One wrapper agent can be used to connect more than one application to the system.

CIIMPLEX (Peng et al., 1998) uses a special service agent, called the Gateway Agent, to provide an interface between the agent world and the application world (including MES, ERP, Capacity Analysis, Scheduling, and Customer Response). The Gateway Agent's functions include making connections between the two transport mechanisms (TCP/IP and IBM's MQ Series) and converting messages between the two different formats (KQML and BOD – Business Object Document). Usually only one Gateway Agent is needed for the system to connect all integrated legacy systems.

Barbuceanu and Fox (1996) employ a conversation mechanism using rules to activate the legacy application (rather than checking input messages and sending out responses), to communicate with it and reason about its operation. The integration can be realized in several ways, ranging from batch execution of an application (by preparing input data, spawning its process, reading the produced outputs) to interacting with the application through its API functions.

12.10 USING OTHER MATURE TECHNOLOGIES AND TOOLS

To implement next generation concurrent design and manufacturing systems that can meet all of the fundamental requirements stated in Chapter 2 (Section 2.4), may require more than a single technology. Thus, Agent Technology can be used with with Object Technology, Internet Technology, CSCW and Groupware Technology, Knowledge Engineering Methodologies, and Holonic Manufacturing concepts, to obtain the following advantages:

- Object Technology is a mature technology widely used in industry. Internationally accepted object-oriented standards (e.g., CORBA by OMG and COM/DCOM by Microsoft) and object-oriented analysis and design methods (with UML) can be used in system analysis, design, and prototype development to facilitate system development and deployment, and reduce development time and cost.
- Internet Technology is a rapidly expanding technology that is being recognized as the future for network computing. Since it is widely implemented, it is well suited for distributed information (data and knowledge) sharing. Earlier attempts at such sharing were reported by Kumar et al. (1994) and Hong et al. (1995). However, the Internet uses a client/server architecture which is quite different from truly open multi-agent system architecture. One possible solution is to implement HTTP servers as "active" servers that can work together with other agents in a real agent-based system.
- Researchers in CSCW and Groupware Technology have developed widely used software, such as Microsoft Netmeeting™, to facilitate synchronous discussions using powerful functions like application sharing, and Audio and Video communications. These capabilities could be used to improve cooperation among product designers, production managers, and other staff.
- Sophisticated knowledge engineering methodologies are available that can provide a good basis for multi-agent system modeling, which will help in implementing more intelligent agent-based industrial applications.
- Some HMS concepts (like "holon" and "holarchy", see Chapter 15, Section 15.6 for details) as well as the holonic architecture are very useful for implementing intelligent manufacturing systems, particularly at lower physical level for manufacturing control.

REFERENCES

Baker, A.D. (1991) *Manufacturing Control with a Market-Driven Contract Net*, PhD thesis, Rensselaer Polytechnic Institute, NY.

Barbuceanu, M. and Fox, M.S. (1996) The Architecture of an Agent Building Shell, In M. Wooldridge, J. P. Muller and M. Tambe (eds.), *Intelligent Agents II*, LNAI 1037, Springer, pp. 235-250.

Burke, P. and Prosser, P. (1994) The Distributed Asynchronous Scheduler, In M. Zweben and M.S. Fox, eds., *Intelligent Scheduling*, Morgan Kaufman Publishers, San Francisco, CA, pp. 309-339.

Butler, J. and Ohtsubo, H. (1992) ADDYMS: Architecture for Distributed Dynamic Manufacturing Scheduling, In A. Famili, D.S. Nau, and S.H. Kim, eds., *Artificial Intelligence Applications in Manufacturing*, The AAAI Press, pp. 199-214.

Chang, E. (1987) Participant Systems for Cooperative Work, In M.N. Huhns, ed., *Distributed Artificial Intelligence*, Morgan Kaufmann, San Mateo, CA/Pitman, London, pp. 311-339.

Cockburn, D. and Jennings, N.R. (1996) ARCHON: A DAI System for Industrial Applications, In G.M.P. O'Hare, and N.R. Jennings, eds., *Foundations of DAI*, Wiley Interscience, pp. 319-344.

Croft, W.B. and Lefkowitz, L.S. (1988) Knowledge-Based Support of Cooperative Activities, In *Proceedings of 21ʰ Annual International Conference on System Science*, Volume 3, pp. 312-318.

Cutkosky, M.R., Engelmore, R.S., Fikes, R.E., Genesereth, M.R., Gruber, T.R., Mark, W.S., Tenenbaum, J.M. and Weber, J.C. (1993) PACT: An Experiment in Integrating Concurrent Engineering Systems, *IEEE Computer*, 26(1):28-37.

Duffie, N.A. and Piper, R.S. (1986) Non-Hierarchical Control of Manufacturing Systems, *Journal of Manufacturing Systems*, 5(2):137-139.

Fischer, K., Muller J.P, Pischel, M. and Schier, D. (1995) A Model for Cooperative Transportation Scheduling, In *Proceedings of ICMAS'95*, AAAI Press/MIT Press, San Francisco, CA, pp. 109-116.

Fordyce, K. and Sullivan, G.G. (1994) Logistics Management System (LMS): Integrating Decision Technologies for Dispatch Scheduling in Semiconductor Manufacturing, In M. Zweben, M. S. Fox, eds., *Intelligent Scheduling*, Morgan Kaufman Publishers, San Francisco, CA, pp. 473-516.

Fox, M.S., Chionglo, J.F. and Barbuceanu, M. (1993) The Integrated Supply Chain Management System, Internal Report, Dept. of Industrial Engineering, Univ. of Toronto.

Gupta, L., Chionnglo, J.F. and Fox, M.S. (1996) A Constraint Based Model of Coordination in Concurrent Design Projects, In *Proceeding of WET ICE'96*, Stanfrod, CA.

Hong, J., Toye, G. and Leifer L. (1995) Personal Electronic Notebook with Sharing, In *Proceeding of 4ʰ WET ICE*, West Virginia, IEEE Press, pp. 88-94.

Jha, K.N., Morris, A. Mytych, E. and Spering, J. (1998) MADEsmart: Agents for Design, Analysis, and Manufacturability, In *Working Notes of the Agent-Based Manufacturing Workshop*, Minneapolis, MN, pp. 57-63.

Klein, M. (1991) Supporting Conflict Resolution in Cooperative Design Systems, *IEEE Systems, Man and Cybernetics*, 21(6):1379-1390.

Kumar, V., Glicksman, J. and Kramer, G.A. (1994) A Shared Web To Support Design Teams, In *Proceeding of IEEE 3ʳᵈ Workshop on Enabling Technologies: Infrastructure for Collaborative Enterprises*, Morgantown, WV.

Lin, G.Y.-J. and Solberg, J.J. (1992) Integrated Shop Floor Control Using Autonomous Agents, *IIE Transactions: Design and Manufacturing*, 24(3):57-71.

Maturana, F. and Norrie, D. (1996) Multi-Agent Mediator Architecture for Distributed manufacturing, *Journal of Intelligent Manufacturing*, 7:257-270.

Maturana, F., Balasubramanian, S. and Norrie, D.H. (1996) A Multi-Agent Approach to Integrated Planning and Scheduling for Concurrent Engineering, In *Proceedings of the International Conference on Concurrent Engineering: Research and Applications*, Toronto, Ontario, pp. 272-279.

McEleney, B., O'Hare, G.M.P. and Sampson, J. (1998) An Agent Based System for Reducing Changeover Delays in a Job-Shop Factory Environment, In *Proceedings of PAAM'98*, London, pp. 591-613.

Monceyron, E. and Barthès, J.P. (1992) Architecture for ICAD Systems: an Example from Harbor Design, *Revue Sciences et Techniques de la Conception*, 1(1):49-68.

Murthy, S., Akkiraju, R., Rachlin, J. and Wu, F. (1997) Agent-Based Cooperative Scheduling, In *Proceedings of AAAI Workshop on Constrains and Agents*, AAAI Press, pp. 112-117.

NIST (1998) Advanced Technology Program, http://www.atp.nist.gov/.

Ouelhadj, D. and Bouzouia, B. (1998) Multi-Agent System for Dynamic Scheduling and Control in Manufacturing Cells, In *Working Notes of the Agent-Based Manufacturing Workshop*, Minneapolis, MN, pp. 96-105.

Ow, P.S. and Smith, S.F. (1988) A Cooperative Scheduling System, In *Proceedings of the 2nd International Conference on Expert Systems and the Leading Edge in Production Planning and Control*, pp. 43-56.

Pan, J.Y.C. and Tenenbaum, M.J. (1991) An Intelligent Agent Framework for Enterprise Integration, *IEEE Transactions on Systems, Man, and Cybernetics*, 21(6):1391-1408.

Park, H., Tenenbaum, J. and Dove, R. (1993) Agile Infrastructure for Manufacturing Systems: a Vision for Transforming the US Manufacturing Base, In *Proceedings of 1993 Defense Manufacturing Conference*.

Parunak, V.D. (1987) Manufacturing Experience with the Contract Net, In M.N. Huhns, ed., *Distributed Artificial Intelligence*, Pitman, pp. 285-310.

Parunak, V., Ward, A., Fleischer, M. and Sauter, J. (1997) A Marketplace of Design Agents for Distributed Concurrent Set-Based Design, In *Proceedings of 4th ISPE International Conference on Concurrent Engineering*, Troy, Michigan.

Parunak, V.D., Baker, A. and Clark, S. (1998) The AARIA Agent Architecture: From Manufacturing Requirements to Agent-Based System Design, In *Working Notes of the Agent-Based Manufacturing Workshop*, Minneapolis, MN, pp. 136-145.

Peng, Y., Finin, T., Labrou, Y., Chu, B., Long, J., Tolone, W.J. and Boughannam, A. (1998) A Multi-Agent System for Enterprise Integration, In *Proceedings of PAAM'98*, London, pp. 155-169.

Saad, A., Biswas, G., Kawamura, K., Johnson, M.E. and Salama, A. (1995) Evaluation of Contract Net-Based Heterarchical Scheduling for Flexible Manufacturing Systems, In *Proceedings of Intelligent Manufacturing Workshop at IJCAI'95*, Montreal, pp. 310-321.

Sadeh, N. and Fox, M.S. (1989) CORTES: An Exploration into Micro-Opportunistic Job-Shop Scheduling, In *Proceedings of Workshop on Manufacturing Production Scheduling at IJCAI-89*, Detroit.

Sandholm, T. and Lesser, V. (1996) Advantages of a Leveled Commitment Contracting Protocol, In *Proceedings of the Thirteenth National Conference on Artificial Intelligence (AAAI-96)*, Portland, OR, pp. 126-133.

Scalabrin, E. (1996) *Conception et Réalisation d'environnement de développement de systèmes d'agents cognitifs*, Thèse de Doctorat, Université de Technologie de Compiègne, France.

Shaw, M.J. (1988) Dynamic Scheduling in Cellular Manufacturing Systems: A Framework for Networked Decision Making, *Journal of Manufacturing Systems*, 7(2):83-94.

Shen, W. and Barthès, J.P. (1995) DIDE: A Multi-Agent Environment for Engineering Design, In *Proceeding of The First International Conference on Multi-Agent Systems (ICMAS)*, San Francisco, USA, The AAAI press / The MIT Press, pp. 344-351.

Shen, W. and Barthès, J.P. (1997) An Experimental Environment for Exchanging Design Knowledge by Cognitive Agents, In M, Mantyla, S. Finger and T. Tomiyama eds., *Knowledge Intensive CAD*, Volume 2, Chapman & Hall, pp. 19-38.

Shen, W. and Norrie, D.H. (1998) An Agent-Based Approach for Dynamic Manufacturing Scheduling, In *Working Notes of the Agent-Based Manufacturing Workshop*, Minneapolis, MN, pp. 117-128.

Shen, W., Xue, D. and Norrie, D.H. (1998) An Agent-Based Manufacturing Enterprise Infrastructure for Distributed Integrated Intelligent Manufacturing Systems, In *Proceedings of PAAM'98*, London, UK, pp. 533-548.

Smith, R.G. (1980) The Contract Net Protocol: High-Level Communication and Control in a Distributed Problem Solver, *IEEE Transaction on Computers*, 29(12):1104-1113.

Sousa, P. and Ramos, C. (1997) A Dynamic Scheduling Holon for Manufacturing Orders, *Journal of Intelligent manufacturing*, 9(2):107-112.

Sycara, K.P., Roth, S.F., Sadeh, N. and Fox, M.S. (1994) Resource Allocation in Distributed Factory Scheduling, *IEEE Expert*, 6(1):29-40.

Trousse, B. (1993) Towards a multi-agent approach for cooperative distributed design assistants, In M.R. Beheshi and K. Zreik, eds., *Advanced Technologies*, pp. 451-460.

Zweben, M. and Fox, M. eds., (1994) *Intelligent Scheduling*, Morgan Kaufman Publishers, San Francisco, CA.

Part III: Agent-Based Concurrent Engineering Systems

13

Agent-Based Engineering Design Systems

13.1 INTRODUCTION

Real world concurrent engineering design projects require the cooperation of multidisciplinary design teams using sophisticated and powerful engineering tools such as commercial CAD tools, engineering database systems, and knowledge-based systems. Individuals and multidisciplinary design teams work in parallel and independently with various engineering tools are located on different sites, often for quite a long time. At any moment, individual members may be working on different versions of a design or viewing the design from various perspectives (e.g., profitability, manufacturability, resource capability or capacity), at various levels of detail. To coordinate the design activities of the various groups and guarantee good cooperation among the different engineering tools, an efficient distributed intelligent design environment is required. This environment should not only automate individual tasks, in the manner of traditional computer-aided engineering tools, but also help individual members share information and coordinate their actions as they explore alternatives in search of a globally optimal or near-optimal solution.

Engineering design has been using some techniques from artificial intelligence (AI) for a number of years. However, the recent development of multi-agent systems within the domain of distributed artificial intelligence (DAI) brings new and interesting possibilities.

Some projects have already been undertaken using agent-based approaches. SHADE (McGuire et al., 1993) was primarily concerned with the information sharing aspect of the concurrent engineering problem. It demonstrated a flexible infrastructure for knowledge-based, machine-mediated collaboration between various engineering tools. PACT (Cutkosky et al., 1993) was a landmark demonstration of both collaborative research efforts and agent-based technology. SHARE (Toye et al., 1993) is concerned with developing open, heterogeneous, network-oriented environments for concurrent engineering. FIRST-LINK (Park et al., 1994) is a system of semi-autonomous agents helping specialists work on one aspect of the design problem. NEXT-LINK (Petrie et al., 1994a, 1994b) is a continuation of the previous project for testing agent coordination. All the previous systems were developed at Stanford University.

ANARCHY (Quadrel et al., 1993) was a working prototype of an asynchronous design environment for generating, evaluating and modifying designs of medium- and high-rise office buildings. The simulated annealing method was applied in ANARCHY as a control strategy. ITX (Lee et al., 1993) addresses the issue of interaction control. SiFA (Brown et al., 1995; Grecu and Brown, 1995), developed at Worcester Polytechnic, was intended to address the issues of interaction, communication, and conflict resolution. Interesting experiments were also conducted at CMU and Stanford University, in particular by Finger, in class projects for industrial problems (Reddy et al., 1995; Toye et al., 1993). DIDE (Shen and Barthès, 1996) was developed to study system openness, legacy problems, and long distance collaboration.

In this chapter, we consider some of the projects mentioned above, giving for each project:

- its origin, domain of application, and research objectives;
- its main technical features including overall approach and system architecture, agent structure, inter-agent communication, and dynamic or global behavior;
- implementation details when available;
- its most significant results;
- its position with respect to other projects in the same group or its further development.

13.2 PACT (PACE)

13.2.1 Origin, Domain of Application, and Research Objectives

PACT (Palo Alto Collaborative Testbed) was a project for joint experimentation in computer-aided concurrent engineering, pursued by research groups at Stanford University, Lockheed, Hewlett-Packard, and Enterprise Integration Technologies (Cutkosky et al., 1993). Its objective was to explore a new methodology for cooperatively solving engineering problems based on knowledge sharing. It was a landmark demonstration of both collaborative research efforts and agent-based technology. Its application domain was electro-mechanical design and robotics.

13.2.2 Technical Features

13.2.2.1 Overall Approach and System Architecture

PACT was an infrastructure for concurrent engineering encompassing multiple sites, subsystems, and disciplines. PACT experiments explored issues in building a framework along three dimensions: (i) cooperative development of interfaces, protocols and architecture; (ii) sharing of knowledge among systems that maintain their own specialized knowledge bases and reasoning mechanisms; and (iii) computer-aided support for the negotiation and decision making that characterize concurrent engineering.

Conventional approaches to integrating engineering tools depend on standardized data structures and a unified design model, both of which require substantial commitments from tool designers. PACT departs from such approaches in two fundamental ways. First, tool data and models are encapsulated rather than standardized and unified. Each tool is thus free to use the most appropriate internal representations and models for its task. Second, the encapsulating agents help tools communicate by translating their internal concepts into a shared language (grammar, vocabulary and meanings) of engineering.

The PACT architecture is based on interacting agents (programs that encapsulate engineering tools). The agent interaction in turn relies on shared concepts and terminology for communicating knowledge across disciplines, an *interlingua* for transferring knowledge between agents, and a communication and control language that enables agents to request information and services. This technology allows agents working on different aspects of a design to interact at the knowledge level, sharing and exchanging information about the design independent of the format in which the information is encoded internally.

The PACT experiments have shown the feasibility of design through the interaction of distributed knowledge-based reasoners. Each reasoner has its own local knowledge base and reasoning mechanisms. As shown in Figure 13.1, agents interact through (initially very simple) facilitators that translate tool-specific knowledge into and out of a standard knowledge interchange language. Each agent can therefore reason in its own terms, asking other agents for information and providing other agents with information as needed through the facilitators.

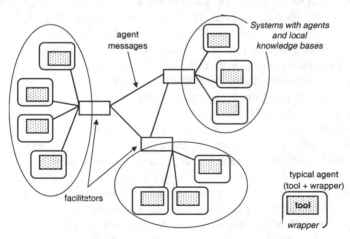

Figure 13.1 Schematic diagram of the PACT architecture
(Cutkosky et al., 1993) (©1993 IEEE).

The PACT architecture was designed to conveniently and flexibly integrate the diverse operating systems, control mechanisms, data/knowledge formats, and environmental assumptions of software not originally written for wide-area interoperation. It is an extension of the Open Distributed Processing (ODP) architectures appearing on mainstream computing platforms. On a larger scale, the Object Management Group's Common Object Request Broker Architecture (OMG/CORBA) (OMG, 1991) defines more sophisticated service naming and parameter type translation.

The software interoperation architecture of PACT extends ODP architectures by further standardizing the style of program interaction. This is done by defining message content at three levels:

1) Messages are expressions in a common agent communication language (ACL) that supports knowledge-base operations (assertions, queries, etc.)

2) The message contents of the knowledge-base operations are expressions in a common knowledge interchange format that provides an implementation-independent encoding of information.

3) The expressions stated in the interchange format use a standard vocabulary from common ontologies that are defined for the shared application domain.

At the first level, an agent communication language called Knowledge Query and Management Language (KQML) (Finin et al., 1993) was proposed and developed. KQML specifies a relatively small set of *performatives* that categorize the services that agents may request of one another. For example, one agent may request to assert a fact to another's local data/knowledge, retract a previous assertion, or obtain the answer to some query.

At the second level, is a specification of the knowledge interchange format called KIF (Knowledge Interchange Format) (Genesereth and Fikes, 1992). KIF can be used as a format for KQML arguments (KQML allows multiple formats). KIF is a prefix version of the language of first order predicate calculus, with various extensions to enhance its expressiveness. KIF provides for the communication of constraints, negations, disjunctions, rules, quantified expressions, and so forth. With the application of AI technology to practical problems, more programs are able to manipulate information of this sort.

At the third level, is the development of engineering ontologies (Gruber, 1993). This effort is focusing on defining formal vocabularies for representing knowledge about engineering artifacts and processes, specifying the assumptions underlying the common views of this information. Given the application-specific nature of such ontologies, this activity is not producing a single specification, but is addressing various domains of importance to knowledge sharing such as device behavior modeling. This effort is complementary to the work of PDES/STEP vocabulary committees, which have been most successful in specifying data models for domains such as solid modeling and finite-element geometry.

Plugging into the PACT architecture requires a substantial commitment to supporting the above-described message content. To ease this burden, the use of *facilitators* was introduced. Each *facilitator* is responsible for providing an interface between a local collection of agents and remote agents, by serving four purposes:

(1) providing a layer of reliable message-passing, (2) routing outgoing messages to the appropriate destination(s), (3) translating incoming messages for consumption by its agents, and (4) initializing and monitoring the execution of its agents. Communication and coordination, therefore, occur between agents and facilitators and among facilitators, but not directly between agents. This arrangement of agents and facilitators is called a *federation architecture*. Messages from agents to facilitators are undirected, i.e., they have content but no address. It is the responsibility of the facilitators to route such messages to agents able to handle them. In performing this task, facilitators can go beyond simple pattern matching – they can translate messages, decompose problems into sub-problems, and schedule the work on those sub-problems. In some cases, this can be done interpretively (with messages going through the facilitator); in other cases, it can be done in one-shot fashion (with the facilitator setting up specialized links between individual agents and then stepping out of the picture).

13.2.2.2 Agent Structure

In PACT, an agent is a program that encapsulates an engineering tool. As shown in Figure 13.1, a PACT agent is composed of a tool and a wrapper that enable it to communicate with other agents using KQML and KIF.

13.2.2.3 Inter-Agent Communication

The agents communicate among themselves using KQML (Knowledge Query and Manipulation Language) and KIF (Knowledge Interchange Format). Most KQML messages consist of queries and assertions in KIF (as well as basic control functions such as reset). Agents also use KIF expressions to inform each other of their interests and to request notification of informational changes that may affect them.

The PACT demonstration involved 31 agent-based programs executing on 15 workstations and microcomputers. When grouped by engineering discipline, these programs compose into the following six "top-level" agents: the digital circuitry agent, the control software agent, the power system agent, the physical plant agent, the sensor agent, and the parts catalog agent. All but the latter two existed before the PACT system was built.

Each top-level PACT agent worked through a facilitator to coordinate its interactions with other agents. Each PACT facilitator consisted of two parts: a connection associate and an agent manager. A connection associate implemented a layer of reliable message-passing above the widely available TCP/IP transport protocol, and was nearly identical in implementation across facilitators. The connection associate supplied message strings for the agent manager. If the message was a control message from another facilitator, the agent manager would interpret the message and react accordingly; if the message was from an agent (via a facilitator), the agent manager would perform any necessary translations before forwarding the message to its agent.

13.2.2.4 Dynamics / Global Behavior

To create the illusion of a shared design model, all interactions between agents are mediated by *facilitators* as described in Section 13.2.2.1. Information is passed among agents and facilitators in an *interlingua* KIF, based on first order logic, with agents translating between the *interlingua* and the internal representations of their clients. Knowledge sharing across disciplines is possible because of *a priori* ontological agreements among the agents about the meanings of terms. Agents and facilitators coordinate their activities using the communication and control language KQML.

13.2.3 Implementation Details

The first prototype integrated four pre-existing concurrent engineering systems into a common framework: Next-Cut1, a mechanical design and process planning system from the Stanford Center for Design Research (Cutkosky et al., 1991), Designworld, a digital electronics design, simulation, assembly, and testing system from the Stanford Computer Science Department and Hewlett-Packard (Genesereth, 1991), NVisage, a distributed knowledge-based integration environment for design tools developed by the Lockheed Information and Computing Sciences group (Weber et al., 1992), and DME (Device Modeling Environment), a model formulation and simulation environment from the Stanford Knowledge Systems Laboratory (Iwasaki and Low, 1991). In the initial experiments, each system modeled different aspects of a small robotics device and reasoned about them from the standpoint of a different engineering discipline (software, digital and analog electronics, and dynamics). The systems then cooperated to produce a distributed device simulation, and to synchronize on a subsequent design modification. The systems interacted via knowledge-based communication languages and services.

The subject of the PACT experiments was the design of a robotics manipulator, chosen as a convenient example of a system that combined mechanics, electronics, and software, with extensive design documentation. The experiment prototype had six top-level agents (facilitators) with four pre-existing engineering tools: a *digital circuit* using the Designworld by Logic group of Stanford University and Hewlett-Packard; *control software* using NVisage by Lockheed; *power system* and *sensor* using DME (Device Modeling Environment) by the Knowledge Sharing Laboratory of Stanford University; and *physical plan* and *parts catalog* using Next-Cut by the Center for Design Research of Stanford University and Enterprise Integration Technology.

---

[1] Next-Cut (Cutkosky et al., 1991) is a prototype system for concurrent product and process design of mechanical assemblies developed at the Center for Design Research of Stanford. It consists of several modules that surround representations of design artifacts, process plans and tooling. It has been applied to an industrial project on design and fabrication of cable harnesses.

Each participant had his or her own hardware and software platforms, including two Macintoshes, a DEC 3100, a TI Explorer LISP machine, and HP and SUN Workstations—all linked through the Internet.

13.2.4 Most Significant Results

PACT actually demonstrated a mechanism for distributing reasoning, not a mechanism for automatically building and sharing a design model. The model sharing in PACT, as in other efforts, is still implicit, i.e. not given in a formal specification enforced in software. The ontology for PACT was documented informally in email messages among developers of the interacting tools. The tools in PACT are able to interact coherently because the commitments to common concepts and vocabulary were "wired" into their interactions.

PACT provides the framework for specifying shared content. It enables tools that should interoperate to do so without committing to common data formats. It also makes possible routing and notification based on message content, not just simple syntactic patterns. Such an approach is clearly advantageous in situations where installing the software locally would require expensive system reconfiguration or where experts are required to run and maintain the code.

13.2.5 Position with Respect to Other Projects or Future Work

PACT is the first practical engineering application of the SHADE effort (McGuire et al., 1993), and may also be the first real multi-agent system for cooperative engineering design in the world.

Recently, PACT has been replaced by PACE (Palo Alto Collaboration Environment) [http://cdr.stanford.edu/PACE/] which involves Stanford's CDR (Center for Design Research), EIT (Enterprise Integration Technologies), Stanford's KSL (Knowledge Systems Laboratory) and Lockheed Artificial Intelligence Center (LAIC). PACE projects, including CDR's Design Space Colonization (DSC) (see the following section), EIT's Virtual Design Workspace (VDW) [http://www.madefast.org/VDW/], KSL's How Things Work project [http://www-ksl.stanford.edu/htw/] and LAIC's Context Integrated Design (CID) project, are part of the DARPA DSO RaDEO program.

13.3 SHARE

13.3.1 Origin, Domain of Application and Research Objectives

SHARE was developed in the Center for Design Research of Stanford University in collaboration with Enterprise Integration Technologies (EIT), and supported by

ARPA (Advanced Research Projects Agency) (Toye et al., 1993).

The objective of SHARE project was to help teams of engineers achieve a shared understanding of their designs and design processes, using agent-based computational tools and services for communication, collaboration, analysis, and synthesis. The approach was based on developing:

- design representations that encompass decisions and rationale linked to the design artifact;
- a distributed architecture that enables agents (human and computational) to communicate and cooperate in solving engineering problems;
- incremental interactive concurrent engineering tools for analysis and synthesis.

Implementation of the SHARE collaborative engineering environment was intimately tied to supporting real industrial design efforts. Its application domain was ME-210, a 9-month long engineering design course in which teams of designers conceived, designed and prototyped substantial electromechanical systems for industrial sponsors. The designers were typically first or second year graduate students with one to two years of industrial experience.

13.3.2 Technical Features

13.3.2.1 Overall Approach and System Architecture

According to the objective of developing an open, heterogeneous, network-oriented environment for concurrent engineering, thus enabling engineers to participate in a distributed team using their own tools and data bases, the environment should provide:

- familiar displays that put information at engineers' fingertips, including on-line notebooks, handbooks, requirements documents, and design libraries;
- collaboration services, including multimedia mail and desktop video conferencing, that enable team members to communicate and share tools and data;
- on-line catalog ordering and fabrication services, with information about pricing and shipping and bid solicitations, leading to delivery of components without numerous phone calls to clarify the designers' intent;
- specialized services for simulation, analysis and planning, (e.g. cost estimation, dynamics simulation) and shared engineering knowledge bases;
- a distributed product data management service that accepts postings from on-line tools and services, and maintains dependencies so that when changes occur, the right people are notified, tools invoked and sources consulted;
- an integration infrastructure that enables heterogeneous design tools and databases to inter-operate transparently across platforms, creating a shared project environment.

The windows onto this world of networked resources and services are multimedia "notebooks" in which information about a project is captured and organized: CAD drawings and solid models, audio notes, sketches, spreadsheets, pages from handbooks and catalogs, animated simulations, mail, excerpts from video conferences, and so forth.

SHARE's tools and services complement existing engineering tools in three ways:

1) Facilitate the real-time capture, annotation and structuring of information associated with product development so that it can be preserved, retrieved and shared;

2) Support communication and collaboration among engineers separated by time and physical location; and

3) Enable the specialized tools used by various members of a development team to inter-operate and share data.

The top-level architecture of SHARE is a set of agents interacting as peers over the Internet (See Figure 13.2). Each agent can represent one or more of the following:

- a designer and his personal CAD tools,
- a database or other information service,
- a computational service that supports engineering or the engineering process.

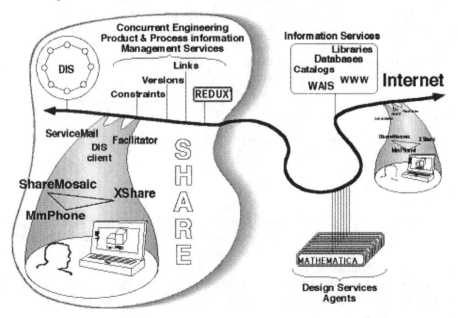

Figure 13.2 SHARE architecture (Toye et al., 1993) (©1993 IEEE).

The agents exchange information and services using KQML (Finin et al., 1993) and representation for multimedia information. The messages are sent using standard e-mail and TCP/IP transport services.

13.3.2.2 Agent Structure

SHARE agents are Internet-based ones. In this environment, agents are essentially human designers, or special interfaces or tools for human designers.

13.3.2.3 Inter-Agent Communication

E-mail serves as the primary medium for both human communication and tool integration. Engineers use e-mail routinely for discussions and information exchange. Computer agents can too. Using structured messages, they can request information and services from other agents, communicate results, notify agents of design changes, and so forth. E-mail has a combination of characteristics that make it well suited for integrating loosely coupled applications in wide-area networked environments. Most importantly, e-mail is already pervasive and familiar to large numbers of designers. It inter-operates seamlessly across thousands of networks and hence is demonstrably heterogeneous and scalable. Second, recent standards in multimedia e-mail make it easy to exchange compound documents containing text, images, audio, video, and programs. Third, it's relatively easy to wire existing applications up to e-mail. Finally, e-mail's explicit asynchrony matches how engineers prefer to work; publishing design changes and "checking in" for change notices at will allows them to feel in full control of the design process.

13.3.2.4 Dynamics / Global Behavior

Interaction among design students during the life cycle of the project.

13.3.3 Implementation Details

Implementation of the SHARE architecture focused on providing tools to help people capture and organize the multimedia e-mail messages, annotations, etc., that made up the bulk of typical design records. Several existing tools were combined to provide the necessary functions: NoteMail and DIS which help engineers capture and manage product information; MovieMail and X-Share which help them collaborate with colleagues; and ServiceMail and Facilitators which enable their tools to inter-operate and invoke remote services.

NoteMail is a tool for creating, viewing, and sharing multimedia engineering documents in a network environment. It combines the functions of an engineering notebook, hypermedia browser and authoring environment, mail tool, and file application manager. The engineers used NoteMail to construct multimedia messages documenting their thoughts. These messages could contain information (either copied or referenced) from any on-line source, such as handbooks, catalogs, CAD tools, simulations, e-mail and video conferences. Messages could also contain programs so that, for example, recipients could redo analyses and simulations with their own parameters.

Messages are inserted in chronological order as pages in an electronic "design notebook". These pages can be marked up and annotated; items of information can be linked to related items on other pages. The result is a personal hyper-document that captures and structures an engineer's knowledge about a project. Selected information can be shared by e-mailing pages to other engineers or to a central project repository, complete with embedded reference pointers and hyperlinks. What emerges is an Internet-wide information web that documents and organizes the shared understanding of an entire engineering team.

NoteMail messages are formatted in MIME (Multi-purpose Internet Mail Extension), the Internet standard for multimedia mail so that it is easy to send and read.

Unlike most other engineering notebooks and multimedia authoring environments, any application that displays through an X-server can insert its output (audio, video or graphics) dynamically onto a notebook page through an embedded "virtual window". When a data object or file is selected for inclusion in the notebook, the system will automatically invoke the appropriate application for displaying that item in the notebook. If the needed application is not locally resident (a likely occurrence in the case of MIME external body references), it will be located and run remotely over the network.

Multimedia engineering documents containing raw text, encoded images, audio clips, video clips, etc. can get quite large. Sending such documents via e-mail to everyone on a large design team can be costly in terms of both time and storage. Instead of transferring full copies to everyone, it is more efficient to store the components of the message in one place and just transmit a set of reference pointers. NoteMail uses an object-oriented knowledge base, known as DIS (Distributed Information Service), for this repository function.

DIS provides a centralized information storage and management service for all the data associated with a design: CAD files, e-mail messages, specifications, simulation results, and so forth. DIS uses a commercial object-oriented database as a persistent object store. The store is accessed via e-mail or TCP/IP through a service layer that provides predefined object classes and APIs for common design applications (simulators, CAD modelers, etc.). Service enhancements such as notification, constraint management, access control and version management also reside conceptually at this level.

DIS evolved into a fully distributed, OMG (Object Management Group) – compatible service for messaging and information sharing. What DIS provides, above and beyond OMG's basic publish and subscribe services are an e-mail CORBA API, and the predefined class hierarchies and APIs for value-added engineering services and legacy applications.

MovieMail and Xshare were used for communicating and collaborating in the SHARE environment. MovieMail is a groupware tool that captures a workstation session as a series of screen dumps and video clips, narrated with mouse (cursor pointer) gestures and spoken comments. The "movie" can then be e-mailed for automatic replay at receiving workstations.

MovieMail builds on the spatial layout and hypertext extensions to MIME that were added to support NoteMail. In particular, MovieMail also requires temporal constraints for synchronizing separate streams such as audio and gesturing. Additionally, MovieMail requires a new video type to capture the cursor movements that direct a viewers' attention to relevant portions of the window. By making MovieMail compatible with MIME, MovieMail "documents" can be sent as e-mail messages in the same way as NoteMail, and then inserted directly into NoteMail notebooks.

Xshare is a set of programs that provide real-time conferencing and application sharing over the Internet. Specifically they provide interactive audio, video, text, and graphics connections among participants as well as the ability to jointly run applications.

Application sharing relies on a single instance of the application driving multiple display servers. With this approach, applications that have machine, license, or other restrictions can still be included in shared sessions. Also, there are none of the synchronization problems that can occur when linking replicated applications.

Xshare conferences have some significant advantages: they can occur spontaneously at the designers' desks; participants can share their data and tools while they converse, with very low communication overhead. In particular, users can share NoteMail and MovieMail "documents" with colleagues during Xshare conferences, and conversely, preserve portions of Xshare conferences as MovieMail messages in their personal notebooks or the DIS repository. In SHARE, all project data is connected in a giant distributed web, and can be navigated in a uniform manner, regardless of the source, format or location.

The KQML toolkit (Finin et al., 1993 – see also Chapter 16) provides mid-level network services and application interfaces that enable these tools to inter-operate and share data.

ServiceMail toolkit was used to facilitate the treatment of the incoming messages, decompose the messages, and route them to the appropriate person or program. ServiceMail was tightly integrated with NoteMail to produce an "active" record of the design process.

Using e-mail, a design engineer is able to retrieve CAD models from a library, send them off to a supercomputer for analysis and simulation, make design changes and have automatic e-mail change notifications sent to interested engineers. Later, this thread of e-mail messages can be replayed with minor editing, to explore design alternatives.

13.3.4 Most Significant Results

SHARE is a particularly important project focusing on collaboration, from the Center of Design Research (CDR) of Stanford University.

It is a good example of collaborative engineering design using Internet and e-mail tools, for capturing and organizing engineering design data and knowledge. It was based on the view of design as a process of gathering, organizing and exchanging information. This process begins with notes, sketches, and evolves to encompass more formal representation as models are generated and communicated to others.

It is also an interesting example of applying agent-based technology to collaborative engineering design, through graduate design courses.

13.3.5 Position with Respect to Other Projects or Future Work

Subsequently, several prototypes and tools were developed through related projects. The software prototype PENS (Personal Electronic Notebook with Sharing) was developed at CDR (Hong et al., 1995). It is an off-line authoring client for the Web, which simplifies posting of Web information for both students and staffs. It was used by the teaching staff for curriculum development notes and for posting FAQs to the ME210 Web. It was slated for student usage in team mini-projects. Technically, PENS is a HyperCard-based notebook database that sends MIME-encoded messages via SMTP to the ME210 Web, where the message is received by ServiceMail and placed in a predetermined HTML directory. PENS eliminates the need for direct Web server write access to edit HTML documents, and eliminates the need for HTML markup by incorporating limited style parsing with pre-structured document organization schemes. The user only needs to pay attention to note content, as with any conventional word processor, and click the "SEND" button to post contents to the ME210 Web.

A mature edition of the SHARE environment, including the PENS was developed and tested by engineering design teams in ME210. The SHARE environment was used by all ME210 teams, two of which had team members that were remotely located: one in Washington and one in Pennsylvania. In 1996, all teams had remote participants, three being international: one in Spain, one in the Netherlands, and one in Sweden. Supporting files for each project were captured on laptop computers and published on CDROMs and the WWW, and analyzed to study design processes and design re-use.

SHARE also collaborated with Sandia National Laboratories to develop an agile, Internet-based service for design and manufacturing during the Sandia Intelligent Agents for Manufacturing (SIAM) project. Demonstration agents were constructed using JAVA.

The SHARE team joined with other MADE (Manufacturing Automation and Design Engineering) [http://dso.sysplan.com/DSO/programs/mfgdes/made.html] projects in MADEFAST (Cutkosky et al., 1996) to collaboratively design an infrared seeker prototype. The design team used SHARE collaboration tools to facilitate communications and sharing of information over the Internet. The result was not only a demonstration that a seeker could be designed and built collaboratively by distributed participants with no management hierarchy in six months, but also an extensive Web of design and process data useful for future experiments.

In cooperation with the SHADE project (McGuire et al., 1993), agents were constructed for controls, dynamics and structural analysis in the SHARE STRAND (STRuctural ANalysis and Design agent) project. These agents incorporate commercial software (e.g., Mathlab/Simulink), knowledge-based systems shells and solid modelers. In the SHARE CONCUR project (Olsen et al., 94), the agents exploit ontology for knowledge sharing and the KQML API for communicating via multiple protocols and transport mechanisms. In particular, this last project added the mechanical components ontology.

A project called GCDK (Generation and Conservation of Design Knowledge) focused on research tools and methodologies for modeling, indexing and retrieval of design information during the design process (Baudin et al., 1993). These tools and methods were integrated with the Engineering Design Notebook developed in the SHARE project.

The SHARE group and Lockheed developed and demonstrated a working prototype of ACaPS: an Internet-based service for design and manufacture of wire harness assemblies.

In total, 17 different industrial organizations including 3M, AT&T, Boeing, DEC, GM, NASA, and Nikon participated in the SHARE ME210 testbed environment. Electronic documents were delivered to industrial sponsors for redesign and published as a design library on the WWW.

Since 1996, the SHARE project has been replaced by a DARPA DSO sponsored project DSC (Design Space Colonization) [http://cdr.stanford.edu/DSC/] which is part of the Palo Alto Collaboration Environment (PACE) (see Section 13.2.)

13.4 FIRST-LINK, NEXT-LINK AND PROCESS LINK

13.4.1 Origin, Domain of Application, Research Objectives

Starting at the beginning of the 90s, a series of projects called First-Link, Next-Link, and Process-Link were developed at Stanford in the domain of concurrent design and manufacturing. The first project, First-Link, itself a successor of a previous research project Next-Cut, was supervised by Cutkosky and undertaken as a joint project between Lockheed and the Center for Design Research at Stanford University. It was intended to test a specific software architecture in which a central product model was accessed by independent software modules, somewhat like a blackboard architecture. The First-Link project was considered promising enough to be followed by a second one, appropriately called Next-Link. Next Link was integrated within the SHARE framework described in the previous section. The Next Link project was developed to coordinate the independent agents of First Link, and a mechanism called the *Redux'* services, developed by Petrie (1993) was added for this purpose. Next-Link was applied to the problem of the configuration and routing of aircraft electrical cables using a novel theory of design decisions. A third project, Process-Link, was initially developed at Kaiserslautern University in cooperation with the Center for Design Research of Stanford University, and was then regarded as a successor of Next-Link. Its goal was to provide tools and methodologies for the integration, coordination, and project management of distributed interacting CAD tools and services in large engineering projects.

13.4.2 Technical Features

The three projects First-Link, Next-Link, and Process-Link, belong to the same family. We present their technical features focusing first on First-Link and then noting the improvements introduced in the following projects.

13.4.2.1 Overall Approach and System Architecture

The main approach in the three projects derives from the global architecture, in which independent software modules (agents) exchange information to carry out tasks, and access a central complex product model. Hence, the key issues are:

- the representation (product model);
- the global architecture;
- the decomposition and coordination of tasks.

First-Link

First-Link was developed to support the cooperative work of human designers, and to regulate their access to a common central representation of the developed artifact.

Product Model

The target problem of designing cable harnesses for aircraft was decomposed into top-level tasks, each one addressed by an agent. The problem domain was decomposed into multiple layers of features and represented as a hierarchy of objects, and divided into three main levels of abstraction: (1) aspects; (2) domain features; and (3) parameters.

- *aspects* constitute the top of the hierarchy; for the example of the cable harness design, they correspond to the basic attributes of a cable harness: electrical properties, configuration, and geometry. The electrical properties embody the primary function of the harness; the configuration defines the harness as an assembly of elements, and is represented by a tree; and the geometry defines the space in which the harness resides, and the paths taken by the cables (see Figure 13.3);

- *domain-features* are objects associated with each of the aspects that correspond to the main features in the domain, e.g., the wiring list for the electrical properties aspect; bundle properties and cable assembly for the configuration aspect; and geometry model and path model for the geometry aspect;

- *parameters* are objects that constitute "slots" in the domain-features, e.g., Schematics and Line Specification for the wiring list feature; Parts, Weight and Cost for the cable assembly feature; and Packages, Structure and Free Space for the geometry model feature.

The product model as defined here can adequately represent the designed artifact.

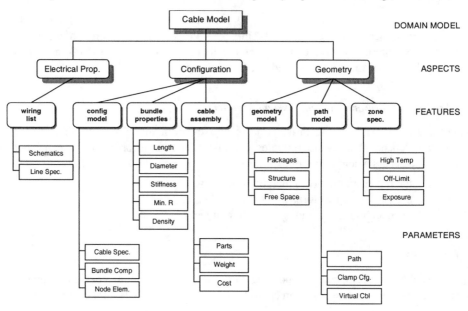

Figure 13.3 Abstraction layers of the domain model (Park et al., 1994)
(Reprinted with the permission of Cambridge University Press).

System Architecture

The First-Link system architecture is shown in Figure 13.4. A *Central Node* maintains a list of input and output features (cf. Section 13.4.2.3) for each agent, and manages the flow of messages whenever an agent publishes or requests information. Introducing a new agent only requires the update of the Central Node. The Central Node can thus be regarded as an administrative agent providing a directory service.

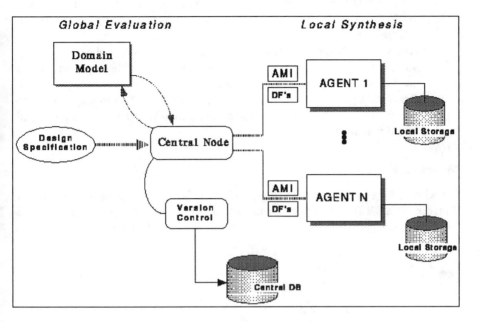

Figure 13.4 First-Link system architecture (Park et al., 1994)
(Reprinted with the permission of Cambridge University Press).

Next-Link

Next-Link was developed from First-Link by adding the *Redux'* services (Petrie, 1993) and other additional functionalities like *Subtasking, Decision Revision* and *Constraint Violations* (cf. Section 13.4.2.4). *Redux'* services are a partial implementation of a general model of design called *Redux* developed by the Center for Design Research at Stanford. In the *Redux* model of design, the design begins with goals that are to be satisfied (alternatively, tasks to be accomplished). In the scenario of the cable harness design, the goals are to route cables. In order to satisfy a goal, or make progress in doing so, a design decision must be made. The results of a decision may be set as sub-goals and/or assignments. In the scenario, assignments are routed paths. In the Next-Link architecture, *Redux'* is used to capture the dependencies generated by any problem solver.

System Architecture

Before it was superceded by Process-Link, Next-Link consisted of a set of First-Link domain-specific agents along with a domain-independent *Redux* agent called *Redux'*, and a mail system based distributed object database called DIS (Glicksman et al., 1991). The First-Link domain-specific agents in Next-Link are of three types:

- Task agents that each know how to perform a set of services for some part of the design task, such as cable configuration (Cable Editor), parts selection (Part Selector), etc.
- The Central Node (CN) which embodies domain-specific message routing knowledge. It knows, for instance, that a message publishing a new cable path should go to the Cable Editor agent.
- The Design Manager that embodies a partly hierarchical model. It initiates the design by setting forth the top-level requirements. It also has a privileged position in resolving conflicts.

Process-Link

The primary Process-Link system (Figure 13.5) consists of four parts: (1) the Project Manager (PM); (2) the Agent Manager (AM); (3) the *Redux'* server and the Constraint Manager (CM); and (4) a set of Application Agents (AAs).

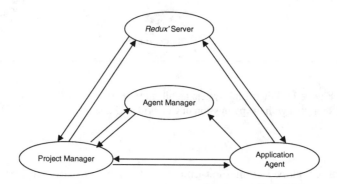

Figure 13.5 Initial Process-Link system architecture (Goldmann, 1996)
(©1996 IEEE).

- *Project Manager* is responsible for constructing the plan and schedule according to its hierarchical structure and the input/output relationships of the corresponding tasks. The scheduling decisions for all tasks are made by the Project Manager. To do this, the Project Manager requests information about the availability of agents and equipment, as well as their capabilities, from the Agent Manager. The Project Manager then constructs a schedule and calculates its critical path, and for each task, its earliest and latest start and end time. As the application agents request or refuse inputs to the tasks they are working on, the Project Manager has to update the plan and schedule and inform all concerned agents. It is also responsible for making necessary changes (and notify *Redux'*) as the plan gets more detailed and the higher levels of the plan have to be updated. During the execution of the plan, the Project Manager monitors the start and end times of all tasks, and sends warnings to the concerned agents (the agent assigned to the task and the agent responsible for its supertask[2]), if a task latest start time or end time has been violated.

- *Agent Manager* controls the utilization and availability of all agents and resources. For each agent, worker or piece of equipment, it provides a schedule that states its utilization and availability at each point in time. When an application agent comes on-line, it will send a message to the Agent Manager, stating its name, location, and a description of its skills and capabilities. On request, the Agent Manager makes this information available to the Project Manager or other agents. Should an agent or resource become unexpectedly unavailable, the Agent Manager will inform the Project Manager, which then will do re-scheduling and inform all concerned agents.

- *Redux'* agent assists the user with the frequently required replanning by keeping track of the decisions that have been made so far, and in determining the consequences of proposed changes to the plan and schedule. *Redux'* provides the notification mechanism for conflicts, decision rejections and necessary re-evaluation of rejected decisions. When a decision is rejected, all other decisions that are dependent on the rejected decision, are invalidated. *Redux'* sends notifications to the concerned application agents when: (1) a new task assignment occurs, (2) the input for one of an agent's tasks becomes available or changes, (3) a task or task assignment is invalidated, or (4) the timing for a task has been changed. *Redux'* also keeps track of committed tasks, and sends a warning if the user tries to reject a decision that would lead to the invalidation of an already committed task.

- *Application Agents* are responsible for solving special task(s). In the solution process, they make both *planning decisions*, which divide tasks into subtasks, and *doing decisions*, which make assignments to output parameters and features. They also state the level of commitment of a decision. For each decision, an application agent should, on request from the Project Manager, state the suggested inputs and outputs of the created subtasks. It should also be able to state the agent skills and equipment properties which are required for solving the task, and estimate the number of hours the task will take up from an agent with the stated skill, or from a piece of equipment with the required properties. This information will enable the Project Manager to make a scheduling decision for the task.

13.4.2.2 Agent Structure

In the three projects, agents have the same structure. An agent is a software module consisting of three primary elements (see Figure 13.6):

---

[2] Supertask: from which the current task is decomposed.

(1) a local model and associated operations;
(2) mechanisms for interaction;
(3) a user-interface.

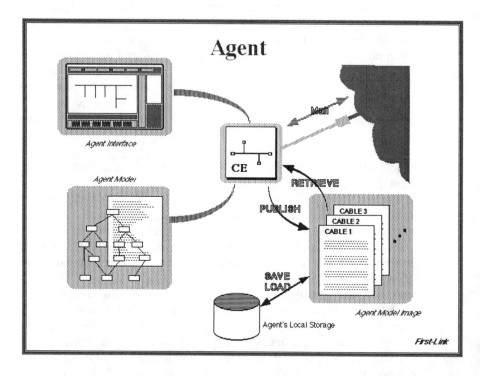

Figure 13.6 Schematic diagram of a First-Link agent [http://cdr.stanford.edu/FirstLink/].

Agents are used for storing/retrieving versions of design models, publishing versions of their design, and responding to requests from other agents. When an agent is not on-line, requests and notifications accumulate in its mail queue.

Each agent contains an Agent Model, which consists of objects, rules and procedures. An Agent Model for the Free Space Manager, for example, includes procedures and representations pertaining to the task of cable-routing. It also contains a set of graphical objects that display the resultant paths on the screen. Similarly, the Agent Model for the Cable-Editor contains information for re-computing, manipulating and displaying the configuration.

To exchange information among the agents, a special format is proposed. An agent builds Agent Model Images (AMIs), which are composed of Feature Images (FIs) corresponding to features in the Domain Model. As shown in Figure 13.7, a compact collection of data structures is extracted from the Agent Model for transmission to other agents and for archival by the Central Node. Each agent has enough knowledge to rebuild its Agent Model from the associated images.

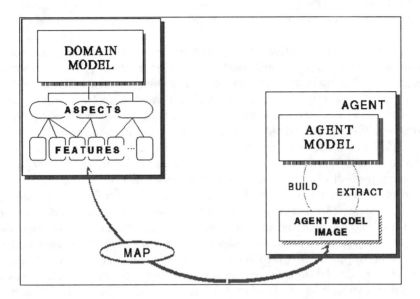

Figure 13.7 Agent Model Image and feature-mapping (Park et al., 1994)
(Reprinted with the permission of Cambridge University Press).

13.4.2.3 Inter-Agent Interaction / Communication

This subsection summarizes how the agents communicate using Agent Model Images
and a mechanism called the feature-mapping mechanism.

First-Link

The interaction mechanisms for agents in First-Link were developed to meet the
needs of human designers working on a collaborative project. For example, the
agents must allow designers to work either independently or in groups or a
combination of both. While working alone, a designer may want to test several what-
if scenarios using only a local model, without bothering other designers. This
temporary isolation postpones design propagation until the design meets local
constraints, thus reducing backtracking. Once the design looks promising, it is made
available to other designers for further input and analysis. Since in a distributed
design environment the experts may not be available for immediate feedback, the
agent architecture must address the asynchronous nature of the design process.

The interaction among agents is realized by the *feature-mapping*[3] mechanism derived from Next-Cut, in which all agents operated directly on a single Central Node.

Thus, an agent shares information with other agents via components of an Agent Model Image (AMI), which are mapped to Domain Features (DF's) (as shown in Figure 13.7). The translation between the Feature-Images and the Domain-Feature-Images provides a layer of isolation and makes it easier to scale First-Link to larger problems and to distribute it over multiple workstations. To insure autonomy, no rigid format is defined for the individual Agent Models and Feature-Images.

In First-Link, the Central Node is a key element. It ensures proper communications, records the structure of the system, and maintains its integrity, by managing the Adjacency Matrices, recording information about the artifact being designed. Actually, the Central Node performs three main functions: (1) maintenance of the Adjacency Matrices (U and G); (2) provision of a directory service for inter-agent notifications, mail and requests; and (3) version management of the various designs. A separate Central Node is used because it was not desired in First-Link that each agent keep track of which agents are on-line, and what each agent needs or produces. The Central Node role as a mail-handler and directory service also makes it the logical candidate for keeping track of design versions and of the status of the design process.

Adding and removing agents

Functions are available to add or remove agents. One of the effects of such functions is to send a message to the Central Node about the agent's input needs and output capabilities (expressed in terms of the Domain Model) so that the Central Node can modify the Adjacency Matrices (U and G).

---

[3] Feature-Mapping and Adjacency Matrices:

- For each Agent Model Image AMI_i (i.e. belonging to Agent A_i), there are n different feature-images, $\{FI_{i1}, FI_{i2}, ..., FI_{in}\}_v$ for a given design version, v. Thus, feature-image FI_{ij} represents the j-th component feature of an agent model A_i. A bi-directional mapping is used between the feature-images FI_{ij} and the features of the Domain Model, F_k. Two three-dimensional adjacency matrices U and G, capture the relationships among Agents (A_i), Feature-Images FI's (FI_{ij}), and Domain-Features (F_k).

- In the U matrix, u_{ijk} is an operator that translates the value of k-th domain-feature-image F_k into the j-th feature-image that is used by the agent A_i. A non-zero value for u_{ijk} for some i, j, and k, indicates that FI_{ij} is an input feature for local operations of A_i. Similarly, in the G matrix, g_{kij} represents an operator that generates a domain-feature F_k from FI_{ij}. A non-zero value for g_{kij} identifies FI_{ij} as an output feature of agent A_i.

- In most cases, the elements of the three dimensional arrays U and G take on the value of 1 or 0, but they can also be operators, T, that map a domain-feature into the corresponding feature-image, and vice versa.

Updating information

When an agent wants to publish a list of features (its AMI), a notification is sent to the Central Node. The Central Node uses its Adjacency Matrices to determine which agents use each of the features about to be published. For each of these agents, a list of the names of the altered features is sent via mail under the "Notify" header. Each affected agent can then act on the mail at its leisure.

Obtaining information

An agent can obtain new versions of features either indirectly, in response to a notification message from the central node, or directly, by invoking the central node's retrieval mechanism.

In either case, the retrieval mechanism is called with the design feature's name, the requesting agent's name, the cable design ID, and the version ID. An optional publishing-agent parameter is defined to specify where the data is located. If a publishing agent is not defined, the retrieval mechanism uses the directory of the central node to determine which agent(s) can supply the needed feature. Once the creator (publishing) agent is determined, the retrieval mechanism obtains the requested information from that agent. If the information is not immediately available, the requesting agent is prompted to decide if a request should be sent. This "Request" message contains the feature name, cable design ID, version ID, and requesting agent's name. It will provoke new computations on the part of the supplier agent. As the supplier agent publishes the requested information, the request messages are removed from its queue.

Next-Link, Process-Link

Next-Link and Process-Link use similar mechanisms for inter-agent interaction and communication. Several new interaction mechanisms were added to implement the additional functionalities of the *Redux' Server, Constraint Manager,* and *Project Manager,* to be described in the following section. In addition, the KQML communication protocol was introduced in the two projects in combination with the *Redux'* protocol.

13.4.2.4 Dynamics / Global Behavior

First-Link

In the First-Link system, domain-specific agents are capable of performing various tasks. Such tasks have a set of inputs and outputs, called *design features.* Given all the values of the input features, an agent can produce values for the outputs.

Agents interact by sending "publish/request" messages. All communications are channeled through the Central Node. The asynchronous nature of the interactions and the provision of version control, make it possible for agents to work on different versions of cable subsystems concurrently.

Next-Link

Based on the First-Link system, several new mechanisms were introduced in the Next-Link system:

- *Redux' services.* In Next-Link, *Redux'* keeps track of decisions, rejection of decisions, and of their rationale. By notifying agents of their position changes, and causes, *Redux'* performs a coordination service, allowing the agents to adjust to each other's changes.
- *Subtasking.* When tasks are further split into subtasks, the agent executing a particular task needs to know when all subtasks are completed. It can keep track of the situation locally (as in DIDE, see Section 13.5), or use an external service like a facilitator[4] (McGuire et al., 1993). In the Next-Link project, the *Redux'* agent is used as a facilitator to propagate the completion of tasks up the subtask tree, which is important when the tasks are performed by different agents. The *Redux'* agent also keeps track of propagating conditions that can affect subtasks like a change resulting from new facts, or the necessity to abort and clean up orphan subtasks.
- *Decision revision.* A design decision is a decision to accomplish a task in a particular way (perhaps involving subtasks). If a decision is revised, its subtasks become superfluous, and so do all the ones below them. For concurrent engineering, one agent stating a fact such as the unavailability of a part may invalidate a second agent's decision. The latter must be notified, and the effects of the decision revision propagated to other agents. It is also important that *Redux'* notify the designer agent of the need to reconsider the decision, but unlike the case of contingencies, not automatically retract the decision, which could possibly undo much design work for little or no gain.
- *Constraint violations.* In the design of cable harnesses by Next-Link, many problems are modeled as constraint violations, e.g., when a structural change invalidates free space for the cables; a module or package changes size; or a new heat or other electromagnetic radiation source is added for which current shielding is insufficient. Any such changes may conflict with the current design, necessitating modifications, such as a rerouting of a cable segment or shielding modifications. In the formal model, such conflicts are constraint modeled as violations and the remedies are resolutions of those violations. These are different from contingencies and loss of decision optimality, described in the previous paragraph. Constraint violation occurs

---

[4] Facilitator: proposed by the SHADE project as a program which coordinates the communication among agents. Facilitators provide a reliable network communication layer, route messages among agents based on the message content, and coordinate the control of multi-agent activities.

when feature value assignments conflict. A constraint violation may involve all the agents that made the decisions which led to the conflicting design features. Each agent needs to be notified of the problem. *Redux'* was also used for the resolution of these types of conflicts (constraint violations).

Process-Link

Process-Link has all the features of the Next-Link plus a new feature: the *Project Management* mechanism. The Central Node found in First-Link and Next-Link was replaced by the Agent Manager.

13.4.3 Implementation Details

The following paragraphs give some implementation details. Some of these are interesting for historical reasons, and can explain choices made in designing the systems.

First-Link

First-Link was tested as both a single-user, single-host system and a multi-user, multi-host system. In either case, agents communicate with one another via pointers to mutually accessible directories on a LAN, or by transferring flat files. While the former is preferable, this may not be possible or efficient if the design effort is geographically distributed. In First-Link, mail is used as the medium of information exchange, since it is easily scaled and explicitly asynchronous. Mail is used to notify agents of new information, to request design components and their status from agents, and to send unstructured messages among people concerning unmodeled constraints, design reviews, etc. A list of mail messages also forms the basis of a work-list.

The directory and dispatch mechanism for the mail was revised using DIS (Distributed Information Service) (Glicksman et al., 1991) to formalize the mail service functions with DIS tools, and to extend the architecture so that agents can easily reside on multiple workstations.

When testing First-Link on a single host, a 12 MIPS workstation with 36 megabytes of RAM and a CLOS-based object environment was used. In a typical design session involving one section of the cabling environment and a six-connector cable harness with associated components and paths, four agents generated approximately 630 objects. The total image size for the system, including the programming environment, and the user-interface was about 51 megabytes.

Whether implemented on a single workstation or on multiple workstations, the size of the Adjacency Matrices (**U** and **G**) for maintaining the dependency information grows as a product of the number of agents and the number of domain features. For very large design problems (e.g., aircraft or ship design), the approach taken in First-

Link is clearly inadequate. For smaller problems involving dozens of agents (the foreseeable upper bound for cable design problems, including links to structural and electronic package design organizations), the amount of time needed to search and operate on the matrices will be on the order of seconds and is negligible compared to computation times on the order of minutes needed for path planning.

The communications complexity does not grow as rapidly as the number of agents squared because not all agents interact with one another (especially when there are many agents) and because the number of domain features grows as the union rather than the sum of the features associated with all agents.

First-Link system contains the following components: a Coordination Agent; an Environment Editor; a Free Space manager; a Cable Editor; a Part Selector; and the Central Node.

-   The *Coordination Agent* initializes and evaluates the designs as they progress and specifies the raw input files for the Environment Editor and Cable Editor. Once the design is in progress, it can gather cost, part, and version information from the other agents. The Coordination Agent is considered as just another agent. Its role is to keep the design data organized rather than to control the process.

-   The *Environment Editor* is responsible for constructing and maintaining an internal geometric model. The Free Space Manager uses this agent's output to perform the routing.

-   The *Free Space Manager* is able to route cable harnesses in complex 3D environments. It uses the geometry information of the routing environment defined by the Environment Editor and the topology of the harness defined by the Cable Editor as inputs. It outputs both the path of the harness and the length of each of its bundles.

-   The *Cable Editor* is used to visualize and manipulate the topology of the cable harness using several high level operations. It uses a wiring list specification defined by the Manager Agent to determine the properties of each bundle as the harness topology changes. Once the Free Space Manager provides the bundle lengths, the harness representation updates to display the bundles in proper scale.

-   The *Part Selector* is responsible for selecting the components of a harness automatically and also for generating a complete part list for the cable. It can also compute the exact material cost and weight of the cable harness. It uses both the topology information from the Cable Editor and bundle lengths from the Free Space Manager as inputs. Its output is used directly by the production system.

-   The *Central Node* is not an agent but is a mechanism responsible for forwarding messages between agents, notifying agents of newly published design data (features), and filling feature requests from agents. It also maintains a directory of the inputs and outputs of all the agents.

13.4.4 Most Significant Results

First-Link

First-Link differs from a blackboard system in two ways. First, its Central Node is not a repository of shared information, but merely a directory service. The information resides in the Agent Models, and features are extracted from such models for transmission to other agents. The agents do not broadcast unsolicited information to the Central Node for use in cooperative problem solving.

The First-Link approach is less flexible than an undirected broadcast system, but, according to the authors, is potentially more efficient, because task planning and optimization can be done ahead of time based on the known dependencies among design tasks and variables. In fact, cable harness design can proceed under a closed-world assumption since the domain model, the potential users, and the suppliers of information are all known.

First-Link was tested as both a single-user, single-host system and a multi-user, multi-host system. As a result of the experiments, First-Link was found to be inadequate for large design problems, but efficient for small and middle-sized design problems.

According to the experience of the First-Link project, one benefit that agent-based implementation provides is that the designers can act on high-level features and operations without the danger of ignoring the details. The agents ensure that proper operations are applied to the elements of the internal model. The communication among agents and human users involves high-level features.

Next-Link

It is important to note that *Redux'* functions as a Central Node with the other Next-Link agents but with no understanding of the cable harness design domain. It can nevertheless coordinate domain-specific agents, because it works with the *Redux* model of the design process to which each Next-Link agent conforms. This feature was not designed into the agents.

Process-Link

The characteristics of the Process-Link project are the integration of planning, scheduling and plan execution; the support for incremental revision of the plan and schedule as well as the artifact design; and the notification system which keeps the concerned agents up to date on any changes.

13.4.5 Position with Respect to Other Projects or Future Work

First-Link, Next-Link and Process-Link are SHARE projects at the Center for Design Research of Stanford University in collaboration with several industrial partners and universities. They are also related to the SHADE project (McGuire et al., 1993). Several important results of the SHADE project, such as the KQML communication protocol (Finin et al., 1993), were used in the Next-Link and the Process-Link projects.

The predecessor of First-Link was a mechanical design system called Next-Cut in which the agents relied on a common data format and freely accessed a central model. Accordingly, the agents depended heavily on the system implementation and had little autonomy. At the other extreme, are architectures such as those associated with the PACT (Cutkosky et al., 1993) and Carnot (Collet et al., 1991) projects. Such systems promote total autonomy and are real open systems (see Figure 13.8).

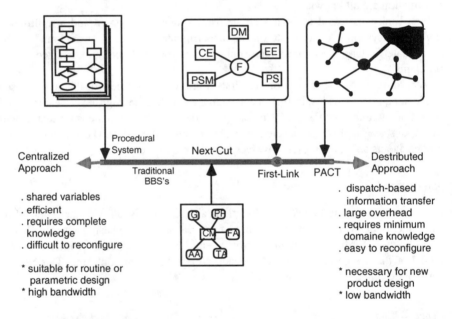

Figure 13.8 Categories of computer support systems (Park et al., 1994) (Reprinted with the permission of Cambridge University Press).

As mentioned previously, Next-Link was developed from First-Link by adding the *Redux'* services and other additional functionalities.

An important project related to the First-Link and Next-Link projects is the Agile Cable Production Service (ACaPS) project at Lockheed and EIT (Park et al., 1993). The goal of this project was to explore Agile Manufacturing with negotiation over the Internet.

Another related important project is CABLER (Collaborative, Agent-Based, Learning through Evolution, Router) (Conru and Cutkosky, 1993). The CABLER system is a collection of automated and manual agents that tackle routing problems synergistically. Each of these agents interact with other agents by supplying design hints in a shared workspace. A key feature of the system is an interactive, physically based routing representation which enables designers to move, clamp, and reroute a design in real-time.

As a part of Process-Link project, the Procura model (Goldmann, 1996) is based on the *Redux* model of revisable design that can be categorized as a model for conflict management. It was developed as a framework for distributed applications of generic engineering services. Future work on Procura includes automating the scheduling with the help of constraint management. Another future extension is the implementation of hierarchical control and authority, by introducing a notion of organizational structure among the agents involved in the planning and execution process, and limiting the authority of individual agents to plan and/or do specific tasks. The possibility to decentralize the system by splitting up the bookkeeping responsibilities among several *Redux'* agents is also envisioned.

At the end of 1996, Process-Link replaced the Next-Link project as part of the DSC (Design Space Colonization) project (see Section 13.3) at the CDR of Stanford University. Figure 13.9 shows the new Process-Link System Infra-Structure. At the stage of preparation of this book, the new Process-Link prototype system is still under implementation.

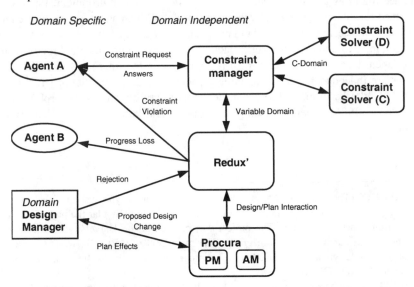

Figure 13.9 Process-Link system infrastructure [http://cdr.stanford.edu/ProcessLink/].

13.5 DIDE

13.5.1 Origin, Domain of Application, Research Objectives

Started in 1992 and based on previous experience in the ARCHIX[5] and EXPORT[6] projects, DIDE (Distributed Intelligent Design Environment) was a multi-agent experimental environment developed for large engineering projects, at the University of Technology of Compiègne. DIDI originally was developed for engineering design, but its general architecture can also be used to develop advanced distributed manufacturing systems or distributed integrated CAD/CAM systems. The objective was to integrate existing engineering tools, like CAD/CAM tools, engineering databases, knowledge-based systems, simulation systems, or other special purpose computational tools, into a truly open system for engineering design and intelligent manufacturing (Shen and Barthès, 1996).

13.5.2 Technical Features

13.5.2.1 Overall Approach and System Architecture

DIDE is organized as a population of asynchronous cognitive agents for integrating engineering tools and human specialists, within an open environment. Each tool (or interface for human specialists) can be encapsulated as an agent and connected to the system. Engineering tools and human specialists are connected by a local network and communicate via this network. Each one can also communicate directly with any other agent located in the other local networks using the Internet (Figure 13.10). All agents are independent and autonomous. There is neither a facilitator structure as in PACT (Cutkosky et al., 1993) nor a mediator structure as in MetaMorph (Maturana and Norrie, 1996). Agents exchange design data and knowledge via a local network or Internet.

---

[5] ARCHIX (Thoraval and Mazas, 1990) was a blackboard type system, developed at the Régie Renault to assist the design phase of vehicle design, in particular the preliminary design of the front wheel drive train. The project was pursued at the University of Technology of Compiègne, in the ICAD research group. The main goal of ARCHIX was to implement a model of the design process, in addition to integrating an expert system environment with commercial CAD TOOLS

[6] EXPORT (Monceyron and Barthès, 1992) was developed at the University of Technology of Compiègne (UTC) in the ICAD research group. It was a joint project between UTC and the Ministry of the Sea, involving the preliminary design of harbors. The main goal of EXPORT was to examine coupling between symbolic and numeric computations, and to address legacy issues.

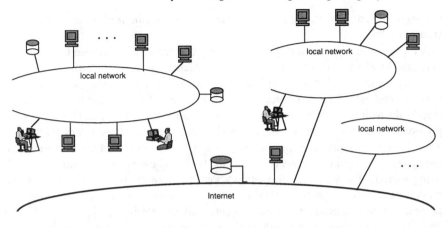

Figure 13.10 General architecture of DIDE (Shen and Barthès, 1996).

Several issues need to be clarified. Firstly, the system was not intended to run automatically. On the contrary, human beings are part of the system. The system cannot be organized independently of the company structure. Thus, a project manager will run the project, and each local group will in turn have a local project manager. Within the local group, human specialists will control some agents and other agents will provide services automatically. Note that in this type of system, human specialists have no direct control over the other agents or the whole system. Each human specialist works with an interface that is encapsulated as an agent connected to the system. For example, the project manager can start or stop the design process, may take some decisions for appropriately replying to the requests of the other agents, but has no control over the other agents or the whole working environment.

A second issue relates to the control of the design task. Apart from the project manager, there is no central control of the design task. Subtasks are created for answering requests as needed, once the work is initiated. There is no global planning (neither central nor distributed). From this point of view, the agents are purely reactive. However, some subtasks may have predefined sequences (e.g., scenarios, amounting to local plans). When this is the case, there is no reason why such sequences should not be used.

A third issue concerns the global consistency of the design. Consistency is kept at a minimum, considering that each agent has a local model of the designed device. However, in mechanical design it is sometimes possible to show a model of the designed product. In this case, each subgroup is allowed to modify the model independently, using its own design space. The results are kept in different versions of a model database. Then, at each review meeting, it is the responsibility of the product manager to merge all the proposed versions into a unique design. Of course, the environment has to support the merging (*reconciliation*) process efficiently (Barthès, 1993).

A fourth issue concerns the agent behavior. All agents are connected by means of a network, local network or Internet. Locally, each agent can reach any other active agent by means of a broadcast message. All agents receive messages. They may or may not understand such messages. When they do not understand a message, they simply do nothing. Whenever they understand a message, they start working on it, provided the priority of the message is higher than the current work they were doing. Thus, agents are multi-threaded. When a new agent is introduced and connected into the system, it informs other working agents, builds its own representation of the task being solved, and acquires information about other agents to establish its symbolic models of the other agents. Thereafter, it receives messages like any other agents. It becomes more interesting when some agents actively offering services know the final goal. Then the agents can not only reply to the requests of the other agents, but also can give some suggestions or advice relating to the current design project, according to their knowledge.

A fifth issue is related to the legacy problem. In this design environment, an agent offers some specific service, usually by encapsulating a particular engineering tool. Agent interaction relies on: shared concepts and terminology for communicating knowledge across disciplines; an *interlingua* (Gruber, 1993) for transferring knowledge among agents; and a communication and control language that enables agents to request information and services. This technology allows agents working on different aspects of a design to interact at the knowledge level, sharing and exchanging information about the design independently of the format in which the information is encoded internally.

13.5.2.2 Agent Structure

In the DIDE environment, agents are autonomous cognitive entities, with deductive, storage, and communication capabilities. Autonomous here means that an agent can function independently from any other agents.

A DIDE agent is composed of five parts (as shown in Figure 13.11): (i) a network interface (shaded portion in Figure 13.11); (ii) a communication interface; (iii) symbolic models of the other agents, and associated methods to use them; (iv) a model of its own expertise with its internal knowledge bases; (v) a model of the task to be performed, or of the current context (local knowledge).

The network interface simply couples the agent to the network.

The communication interface of an agent is composed of several methods or functions for treating all incoming and outgoing messages. A message box is used to temporarily store all received messages.

The symbolic models of the other agents are constructed by means of a knowledge base containing information about the other agents, obtained during interaction. The information includes an agent's name, address, and skills or competencies. This knowledge helps the current agent to select one or more agents as sub-contractors for processing tasks.

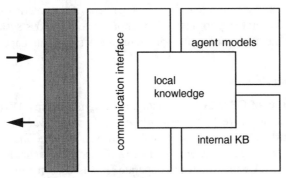

Figure 13.11 Internal structure of an agent in DIDE
(Shen and Barthès, 1996).

The model of "own expertise" is also a knowledge base, composed of self-information such as name, address and competencies or skills. The latter may be methods for activating tasks corresponding to the received requests.

The local knowledge is a knowledge base composed of the information about the working project and the subtasks. It contains also the historic information of the agent, implemented by a cache mechanism.

At first, when a new agent is connected to a group of active agents, only its communication interface and its own expertise contain information. The part recording facts about the work to be done, or the capabilities of the other agents, is empty. In the case of slave agents (e.g., a local database), this will not change, i.e., it will remain empty. In other cases, each agent must build its own image of both the work to be done and the capabilities of the other active agents, by extracting information from the various messages it receives.

13.5.2.3 Inter-Agent Communication

Communication Language

In general, communication among DIDE agents can be synchronous or asynchronous, and the communication mode can be point-to-point (between two agents), broadcast (one to all agents), or multicast (to a selected group of agents). DIDE uses the Inter-Agent Communication Language (IACL) developed under the OSACA project at the University of Technology of Compiègne (Scalabrin et al., 1996). IACL has two layers: a message layer and a content layer. The message layer encodes a set of communication features that describe the parameters at the lower level of the message, such as the identifiers of the sender and the receiver, a unique identifier and the used protocol. In addition, the message layer contains the minimal vocabulary necessary for a common language for inter-agent communication. Having this type of communication language, independent of the internal language used by the agents, is the first step to interaction among heterogeneous systems. The content layer is the actual content of a message, in the agent's own language representation.

Communication Primitives

As an application of OSACA for engineering design problems, DIDE allows for five simple actions (primitives): *request, inform, call-for-bid, bid* and *notice*. They are grouped into two categories: requests and assertions (*request, inform, notice*) and call for bids and offers (*announce, bid*). *Announce* is used by a sender to express intention to find a partner(s) with whom to cooperate. *Bid* is used for expressing an agent's willingness to cooperate with the sender of an *announce*. *Bid* can be used to update information on a given proposal. *Request* is used by a sender to ask a specific agent to accomplish a service. The receiver of this type of message has to reply to it. *Inform* is used for notifying the result of an operation, and in general, it is used as an answer to a *request*. *Notice* is used by a sender to notify an event. Table 13.1 gives further information about these types of message.

Table 13.1 Five types of message (Shen, 1996)

Primitives	Description	Wait for reply	Need to reply
request	Asking for executing a task	Yes	Yes
inform	Distributing information or results	No	No
notice	Announcing an event	No	No
announce	Sending an invitation to tender	Yes	Yes
bid	Replying to an announce	No	No

Communication Management

Each agent has a message-handling mechanism for incoming and outgoing messages, to retrieve the data from an incoming message and to convert outgoing information to a common format which will be described in the next sub-section. Each message is a MOSS object (Barthès, 1989). Note that agents exchange only the information attached to the message objects rather than the message objects. A mechanism was developed to remove the outgoing message object as soon as it is sent.

The messages are sent asynchronously without waiting for confirmation of reception except where this is necessary, which is possible because the agents are implemented in different computers and their activities are parallel. The communication protocol provides the possibility for sending a message to one only agent (point-to-point), or to a group of agents (multi-cast), or to all the agents in the systems (broadcast). It is the sender who decides the type of outgoing message.

Message Format

Figure 13.12 shows the hierarchy of the message classes. Other new message types may be created by adding new classes into this hierarchy.

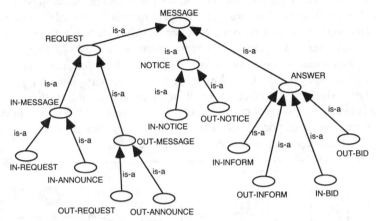

Figure 13.12 Hierarchy of the message classes (Shen, 1996).

In DIDE, the messages are formatted in an extended KQML format. The MESSAGE class is defined in MOSS environment as follows:

```
Entity-name        MESSAGE
Attribute          HAS-ID
                   HAS-PRIORITY
                   HAS-SENDER
                   HAS-RECEIVER
                   HAS-LANGUAGE
                   HAS-ONTOLOGY
                   HAS-COMMODE
                   HAS-CONTENT
```

Five types of message have the same structure, with common attributes and several private attributes. For example, only *request* and *announce* have a REPLY-WITH attribute and *inform* and *bid* have an IN-REPLY-TO attribute.

Message Processing by an Agent

Processing incoming messages requires two steps: (i) receiving, storing, and sorting messages; (ii) encoding message content for further processing by the agent in the context of a particular task. Processing an outgoing message similarly requires encoding of the information to be transmitted and actually mailing it in accordance with the exchange protocol.

The following describes how an agent handles different types of message (see Figure 13.13). When an agent receives a *request* message for some sub-task, the agent chooses a method according its internal KB (knowledge about itself), executes this method to complete the sub-task, and then sends the results by an *inform* message to the sender of the *request* message. During the execution of the method, this agent may have to ask another agent to complete part of the sub-task (sub-sub-task). It chooses an agent B according to its agent models (its knowledge about other agents) and sends a *request* message to agent B. After receiving the results by an *inform* message sent by agent B, the agent resumes its work. During this work, important information is saved in its local knowledge base for future use. A more complicated mechanism can be used to select an agent for carrying out a sub-task or sub-sub-task, which is described in detail in (Shen and Barthès, 1996). A *notice* message is similar to a *request* message, but when an agent sends a *notice* message, it does not wait for a response; and when an agent receives a *notice* message, it processes this message without sending any response.

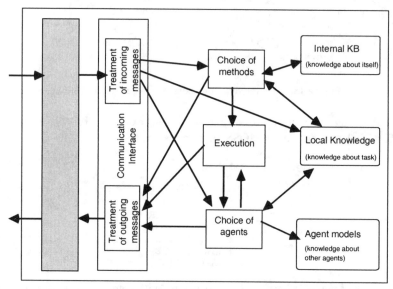

Figure 13.13 Treatment of messages by an agent (Shen and Barthès, 1996).

13.5.2.4 Dynamics / Global Behavior

DIDE is a true open system in the sense that one can freely add or remove agents without having to halt or reinitialize the work in progress. There is no global control mechanism. When a new agent connects to a group, it informs the other working agents, builds its own representation of the task being solved, and acquires information about other agents. It receives messages and processes them if it can.

If an agent leaves the working system normally, it should notify other agents in the system by sending a broadcast *notice* message so that these agents can update their knowledge bases. A simple surveillance mechanism was implemented to detect unexpected breakdown of the agents. An agent A (e.g. *Project Manager* in DIDE) sends a broadcast *request* message regularly to all other agents in the working system only asking if they are active. If it cannot receive a reply from an agent B, agent A considers that agent B is non-active, after repeating the *request* message several times. Then, agent A would update its knowledge base and also inform the other agents.

13.5.3 Implementation Details

An experimental prototype was developed for testing the basic feasibility of the DIDE approach, which could encapsulate traditional tools and allow agents to join or leave without halting the working system. The prototype was developed on a network of SUN and VAX workstations, using the OSACA platform developed in the DAI research group of University of Technology of Compiègne (Scalabrin et al., 1996). Although developed on a local network, DIDE has the capability of inter-group communication using the Internet. Some agents were implemented as MOSS objects (Barthès, 1989), a system of recursive frames capable of modeling objects with versions and well adapted to design activities, constructed on top of Common LISP. Other agents were developed in C/C++. Several graphic Interfaces were developed on CLIM and LispView for displaying system information, design results, and for user manipulations. AutoCAD™ was encapsulated as an agent for displaying design results. The web browser Netscape or HotJava was used to show information about the active agents and about the current situation of the design project. Two databases MATISSE™ (an object-oriented database, a commercial product of ADB) and EDBMS (an extended relational database developed in Chinese Academy of Sciences) were used for storing design data and knowledge.

In the prototype implementation, only three types of message (*request, inform, notice*) were used for inter-agent communication. *Request* messages can be sent by a method =request, and answers can be obtained by a method =get-result, e.g.,

```
(setq msg1-sent (-> *COM* '=request :receiver "agent3"
                    :ontology "nuts-and-bolts"
                    :content (list GETSTB TABB.112 14)))
```

Inform messages can be sent in the same fashion by a method =reply or by a method =inform with an option :in-reply-to. *Notice* messages can be sent in the same fashion by a method =inform but without the option :in-reply-to. Before being sent out, messages are also filled with default values such as :sender, :id, :language by the current agent.

As mentioned in Section 13.5.2.2, the model of an agent, including the symbolic models of other agents and the model of its own expertise, is very important for cognitive agents. In the experimental DIDE environment, the "model of own expertise" for an agent was implemented as a simple knowledge base composed of the information about itself such as its name, address and competencies or skills (which may be methods for activating actions corresponding to the received requests). This simple knowledge base is stored as a MOSS object, being an instance of the KNOWLEDGE class.

The symbolic models of other agents were implemented by means of a knowledge base containing information about other agents obtained during interaction with the other agents. This information includes the agent's name, address, and its skills or competencies. The same object structure as for the model of own expertise is used to store this information about other agents. This knowledge helps the current agent to select one or more agents as sub-contractors for processing actions. The message mechanism for obtaining this information is as follows: as soon as a new agent connects to the system, it broadcasts a message of *request* type and with context "REGISTER" to all other agents. Each of the other agents receives this message, extracts information of interest concerning the new agent such as its name, address and skills, and uses it to update its agent models, then sends an *inform* message to the new agent. The new agent receives the *inform* messages from all other agents, extracts important information and uses this to update its agent models.

The task model is created and updated by each agent during its interaction with other agents. When receiving and sending messages, it extracts important information such as the project name, sub-task names, situation of the each sub-task and the time estimated for a sub-task or the time used for a completed sub-task. It saves also some important traces such as the content of the sent messages and the results received from other agents for *request* messages. When an agent wishes to send a *request* message, it first verifies whether it sent a similar *request* previously and obtained a result. If so, it does not send the *request* message over again, but gets the results directly from its task model.

Using DIDE, a test case was carried out for a small design example – a simple gear box design. A detailed description of this test case can be found in (Shen and Barthès, 1996).

13.5.4 Most Significant Results

A typology for architectures of cooperative systems

During the DIDE project, a detailed study made of over 40 related systems developed by other laboratories allowed a distinction to be drawn between blackboard-type systems, in which agents are structured around a central database, and systems of independent agents in which each agent has its own representation independent from that of other agents. The first type of system is well suited to the solution of problems involving several specialists acting in a focused manner. It is less useful in complex systems in which many specialists must cooperate during a long time (e.g., several months or even several years). In the latter case, systems based upon independent agents are much more interesting and useful, in part because they allow more modular systems to be developed which can evolve dynamically.

Architecture for multi-agent systems

A two level architecture was proposed in the DIDE project for large engineering design projects. In the first level, agents are grouped locally and belong to the same LAN. The LAN structure allows an agent to see all messages. Thus the basic communication structure is of broadcast type. Each agent can look at messages, even if it is not the recipient of the message. It can look at any message and use the information to update its internal representation of the work in progress, or of the capabilities of the other agents. At a more global level, DIDE proposed inter-group communication, for the case when some needed service was not available locally. In this situation, the Internet has to be used. However, it is not necessary to channel all communications through a specialized agent like a facilitator, a mediator or a manager, which may cause a bottleneck. On the contrary, local agents are allowed to set up communication channels directly with other remote agents.

Advantages and drawbacks of the multi-agent systems for engineering design

DIDE demonstrated that multi-agent systems for engineering design have some advantages: (1) a multi-agent system can be developed into a real open system, which makes the design system dynamic; (2) autonomous and independent agents make the design system less complex than a blackboard system, and also allow an easy integration of existing engineering tools, resolving in part the legacy problem; (3) it is simple to integrate human elements, by developing "human-agent" user interfaces for human specialists; (4) it provides efficiency through a parallel design process; (5) it is flexible, because an application can be interfaced to existing ones by implementing it as a new agent. However, there are also some drawbacks: (1) it is not easy to test the overall behavior of the multi-agent system; (2) it takes some time to establish agent models and local knowledge models (task models), and inter-agent communication may be slow and therefore not efficient for small design projects; (3) it is not easy to integrate interactive tools; (4) it brings a number of difficulties such as the consistency of independent representations, reliability of communication, and sustainability of cooperation processes.

13.5.5 Position with Respect to Other Projects or Future Work

A related project called OADIS (**O**pen **A**rchitecture for **D**istributed **I**ntegrated **S**ystems) (Lin et al., 1996) derived from DIDE was developed in the CAD Laboratory of Institute of Computing Technology, Chinese Academy of Sciences, in Beijing, under an international cooperative project between the Institute of Computing Technology in Beijing and the University of Technology of Compiègne in France. DIDE was also involved with the multidisciplinary project CMCI

(Coopération et Mémorisation des Connaissances en Ingénierie), supported by the Ministry of Industry of France and the Peugeot Automobile Company. Several important features of the DIDE were also integrated in the MetaMorph II project (see Chapter 14, Section 14.2) for distributed intelligent design and manufacturing systems (Shen et al., 1997), at the University of Calgary.

13.6 SiFAs

13.6.1 Origin, Domain of Application, and Research Objectives

The concept of Single Function Agents (SiFAs) was developed in the AIDG research group (Artificial Intelligence in Design Group) of Worcester Polytechnic Institute, to investigate design problem-solving using multi-agent architectures. The premise is that Design Expert Systems can be built using many small, cooperating, limited function expert systems, called Single Function Agents (SiFAs) (Brown et al., 1995). The research goals (Dunskus et al., 1995) were:

- definition of a domain independent set of agents;
- investigation of agent negotiation for conflict resolution;
- analysis of communication patterns between agents;
- identification of appropriate agent knowledge and its representation;
- classification of conflicts.

The motivation was to explore elementary patterns of interaction, communication and conflict resolution. The work addressed such important issues as memory sharing, information passing between agents, conflict resolution via negotiation, and splitting a task into multiple subtasks and assembling the solution from sub-solutions.

Its application domain was engineering design.

13.6.2 Technical Features

13.6.2.1 Overall Approach and System Architecture

While investigating expert systems to support concurrent engineering tasks, the developers of SiFAs devised a model involving multiple agents that cooperatively produce a solution. The functionality of the entire design process is separated into many small functional types. One functional type is assigned to each individual agent. These agents are called Single Function Agents, SiFAs for short (Dunskus et al., 1995).

SINE is the platform developed for research into negotiating SiFAs (Brown et al., 1995). Every agent in SINE is an instance of some class. There is a specific class for every agent, where the class is based on features inherited from a specific type class, a target class and a point of view class.

The SINE system used a combination of rules and reactive objects to represent knowledge. Constraints and preferences are represented as objects. The knowledge referring to the artifact to be designed consists of the general rules, constraints and preferences that could be applied to the design problem, as well as the specific requirements of the current design task.

The SINE platform is most suitable for routine parametric design problems. These are characterized by the fact that the structure of the design, its features and the parameters influencing it are known in advance. The design task consists of making decisions about values for design attributes, which can be either parameters (lengths, widths, colors, materials, etc.) or general design decisions (e.g., selecting a structure type for the artifact or determining the class of material to use).

SINE allows a flat, totally connected structure. This means that each agent is able to communicate with any other, and no agent has a hierarchical priority over any other agent. The flat agent topology can be extended to support agent hierarchy in future research.

13.6.2.2 Agent Structure

A SiFA is an agent that performs a single function on a single target from a single point of view (Figure 13.14). Therefore, it is defined by specifying three parameters: Function, Target and Point-of-View. The *Function* of the agent defines what kind of work it performs. Possible functions are providing *Advice, Analysis, Criticism, Estimation, Evaluation, Planning, Selection* or *Suggestion*. The *target* defines the parameter or object on which the agent has an immediate effect. A target could be an artifact's material, or the process used to produce the object. The *Point-of-View* specifies the perspective that the agent takes, as it performs its function on the target. An agent will usually try to optimize the performance of the artifact with respect to its point of view. For example, an agent role could be selector (function) for a material (target) from a strength (point of view). Another role could be criticism of material from the point of view of cost. A third agent could select material from the perspective of thermal-conductivity, which would probably lead to different preferences than the first selector, and therefore to conflicts.

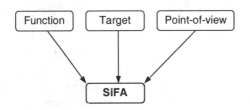

Figure 13.14 The components making up a SiFA (Dunskus et al., 1995)
(Reprinted with the permission of Cambridge University Press).

Since each agent has three aspects, agents can be viewed as points on a three dimensional space, where the axes are labeled with these three aspects. Figure 13.15 shows examples of the wide range of choices than can be made when specifying agent roles.

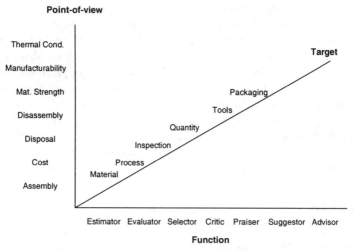

Figure 13.15 SiFA dimensions (Dunskus et al., 1995) (Reprinted with the permission of Cambridge University Press).

In a particular design problem, there can be at most one agent for any point in the space. More than one agent would mean that there are two identical agents in the system. Agents corresponding to some of the possible combinations of function, target, and point of view will not exist, either because they are not meaningful or because the expertise to build them does not exist.

Agents need various kinds of problem-solving knowledge to be able to carry out their tasks: design knowledge, conflict resolution knowledge, and communication knowledge, and so on.

The design knowledge contains all the functionality that the agent needs to perform its design task. This knowledge can be encoded in rules or in a programming language. The method of encoding is not the main issue for this work. In addition to this knowledge, there is also knowledge devoted to redesigning, to be used after design decisions are retracted.

In order to resolve conflicts, the agents need conflict resolution knowledge. This comprises all the methods that are used to detect and classify conflicts, as well as the strategies for conflict resolution. One part of this knowledge is used to analyze which agents have generated the conflict. This "blame assignment" task in SiFA systems consists of tracing the origins of values, or in using knowledge about value dependencies, to find the "culprit". The resulting information is used to analyze the conflict type. In general, this could involve tracing conflict history and/or design history information and/or querying the other agent. In SINE, agents can also use their knowledge about agent interactions to gather information about the agent type with which they conflict.

The knowledge structures about conflicts also have a memory for storing historical information about previous CR interaction with other agents, such as their goals, previous proposals that failed, preferences of other agents etc. (see Figure 13.16).

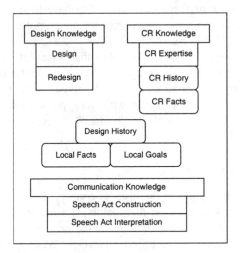

Figure 13.16 SINE agent structure (Dunskus et al., 1995) (Reprinted with the permission of Cambridge University Press).

The communication knowledge consists of two parts: speech act construction knowledge and interpretation knowledge. The first is used to produce and dispatch messages to other agents and the second analyzes messages that are addressed to the agent and provides immediate reaction to the message.

13.6.2.3 Inter-Agent Communication

The communication mechanism was implemented through messages. The message format was based on speech acts, i.e., it captures a complete utterance in one message. The utterances are similar to the sentences found in human discussion. The format is similar to the KQML message structure (see Chapter 16), with slots for storing the primitive, sender, receiver, time stamp, language and ontology information, as well as references to previous and future messages.

The SINE communication language defines how the primitive, subject, and parameters can be filled in to form meaningful sentences. Here are some sample speech acts (Dunskus et al., 1995):

- *Speech acts using alternatives:* All agents that produce values by selection, or by elimination, from a set of possible choices (e.g., Selector, Advisor, Suggestor) are able to communicate their alternatives, e.g., "Add the following alternative to your list of considered choices"; "Remove the following alternative from your list of considered choices."

- *Speech acts using constraints:* Selectors restrict their set of choices with constraints. Critics use them to trigger specific messages. Other agents can use them to check their preconditions. Constraints can be failed, satisfied or unknown, e.g., "The constraint is satisfied/unsatisfied"; "Add the constraint to your list of things to consider"; "Remove the constraint from your list".
- *Speech acts using preferences:* Agents that have to search a set of choices for the best option can use preferences for the purpose, e.g., "What are your preferences?"; "My preferences are...".
- *Speech acts using proposals:* A value that an agent asserts for a design parameter is considered a 'proposal', since it can be contested by other agents through negotiation, e.g., "I agree with your proposal for design-value x"; "I don't agree with your proposal for design-value x"; "Give me another/all possible proposals(s)"; "Here is another proposal / are all other proposals ...?".

13.6.2.4 Dynamics / Global Behavior

In a SiFA system, the execution of the tasks is controlled by an agenda. The SINE platform has a scheduling mechanism that maintains a list of agents to be run next. The agenda manager is the top-level controller of the system. An entry is proposed by an agent sending a message to the agenda manager. When no agent proposes an entry, the agenda manager assumes that the design is either finished or stuck unrecoverably, and it terminates, returning control to the user. In the case of a conflict, the agenda system is inactive. Instead, the communication mechanism that passes the information back and forth between the agents ensures that each agent receives control to be able to evaluate and react to the message.

The current state of the design resides on a blackboard structure which is accessible to all agents, i.e., all the information about the design parameters is located in one place. Agents also communicate each other directly, which differentiates a SiFA system from typical blackboard type systems. The results of direct agent interaction through negotiation are used to update the design status on the blackboard. Even though agents exchange messages directly, design data flows through the blackboard. This is shown in figure 13.17.

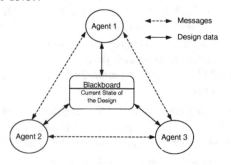

Figure 13.17 Communication and data flow.

13.6.2.5 Agent Types

According to this research, there are only a limited number of functions needed for design problems and a set of agents with these functions is sufficient for most design problem solving activities. These agents types include: Selector, Advisor, Estimator, Evaluator, Critic, Praiser and Suggestor.

Selector/Advisor

A Selector picks one item from a set of alternatives. In doing so, it can use preferences to rank alternatives, and constraints to restrict the alternatives to valid choices. It could pick a choice of a material from a list, or a value for some parameter of the design.

The Advisor is a more general form of the selector. It is not bound to use an enumerated list of possible choices, but it has an abstract description of what it can choose from. For example, the options could be described by a set of constraints on a numeric parameter, requiring the advisor to use an appropriate method to produce a specific value.

Estimator

An Estimator produces approximate values (usually numeric values with error margin and probability) derived from values of attributes of the design. These values are approximate, because not necessarily all the input parameters are available, or some of the input values are by themselves approximated or statistical. In addition to that, the agent can use an estimation function that is only close to, but is not exactly the function that applies in real life. This is often done to save time and effort in early design stages.

Evaluator

The Evaluator uses design goals and attributes of the artifact to evaluate the design from one point of view. The result is information about the quality of the design, i.e., how well the design meets the requirements or some explicit goal. The evaluation can be expressed numerically (e.g., in percent), or in a qualitative way (e.g., 'good', 'average' or 'poor'). The evaluation does not make any statements about whether the achieved quality is satisfactory or not.

Critic/Praiser

A Critic points out potential problems, sub-optimal decisions and poor choices that stand in conflict with the preferences expressed in the requirements. Positive comments are generated by Praiser agents.

Suggestor

The Suggestor takes a criticism, and its context, and suggests alternative solutions or recommended actions to achieve a solution.

13.6.2.6 Learning

A multi-agent design system includes many agents, with numerous potential interactions between them. Conflicts considerably increase the number of interactions needed during design. In SiFAs systems, such situations were improved through implementing learning mechanisms, by reducing the number of conflicts, and/or by reducing the number of messages during interactions. A more detailed description of learning in SiFAs systems can be found in (Grecu and Brown, 1995, 1996a, 1996b), and also in Chapter 5 of this book.

13.6.3 Implementation Details

SNEAKERS, WPI's first SiFA system, is a single-user demonstration system that teaches principles of concurrent engineering. The user's task is to design towers composed of pieces. The system assists by using various expert systems running in the background that offer criticisms and suggestions concerning design decisions (Douglas et al., 1993).

SNEAKERS has several agent types. Advisors make recommendations to the user for the next decision; Critics compare the design to standards, and offer criticism on the last user action; Suggestors take the criticism and offer suggestions for satisfying them; Analysts offer numerical analyses to derive attributes, such as strength, cost or size; Evaluators evaluate the whole design and determine how well it satisfies a particular perspective. Agents can have different points of view, such as Design, Manufacturing, Assembly, Cost, and Marketing. As this work was an initial version of SiFA, it only separated the agents by function and point of view. There was no distinction by target.

I3D, a design aid for powder ceramic components, interacts with a designer at a workstation (Victor et al., 1993). As the designer moves through the requirements specification, conceptual design, and detailed design of a part, the system graphically displays the state of the design. In both design phases, the system provides feedback about the material, the production process and the inspection points. There are different agent types, and Manufacturing, Cost, Reliability, and Durability points of view. The agents are partially dependent on one another, so a rigid sequencing of agent execution was adopted. The system does not deal with conflicts. These were eliminated at development time, by not allowing more than one agent to select a value for the same design parameter.

In I3D+ an attempt was made to remove I3D's restrictions (Victor, 1993). The potential for conflicts was allowed and a mechanism was implemented to demonstrate how conflict resolution might be handled. The rigid agent sequencing was replaced by a flexible agenda mechanism, which schedules the agents based on tasks that they announce. Since the agents handle conflict resolution by negotiation, they communicate directly with each other using a language based on speech acts.

Another implementation used the platform to produce advice to a sailboat designer. This system required a larger number of agents, points-of-view and targets. The solutions generated by using the SINE platform were good, even though the design problem was of a less structured type. This provides a strong argument for using conflict resolution in such problems.

SINE was developed using the CLIPS (C Language Integrated Production System) language. The UNIX X Windows version of CLIPS was used. The system was developed on a DEC Alpha™ 3000-300, using GNU C.

13.6.4 Most Significant Results

SiFA is easy to use and facilitates system development. The small agent granularity allows increased inheritance of knowledge and function. Conflict resolution can usually be performed within a few question/answer cycles, which is due to the small agent size. The SiFA approach to negotiation in multi-agent design systems introduces new ways of discovering and testing the components of negotiation, patterns of communication, functional primitives of design, and types of knowledge needed. By restricting the ability of agents, their behavior can be studied and categorized, which is more difficult with larger, general-purpose agents.

The investigation of learning issues in agent-based design systems, during this project, has suggested how the performance of multi-agent design systems can be improved, by avoiding conflicts and reducing the interaction needed to find a design solution.

13.6.5 Position with Respect to Other Projects or Future Work

This project using small simple single-function agents is somewhat unique, and must be contrasted with the larger more complex agents projects.

13.7 OTHER PROJECTS

In addition to the projects described in the previous sections of this chapter, there are also other significant projects relating to agent-based collaborative engineering design. We selected the five projects in the previous sections for detailed description, because these projects provide insights into different approaches. The following projects are also interesting and significant, but space precludes a more detailed description.

13.7.1 RAPPID

RAPPID (Responsible Agents for Product-Process Integrated Design) (Parunak et al., 1997), developed at the Industrial Technology Institute through ARPA's MADE program, was a community of agents (which the research team defined as active software objects with varying degrees of intelligence) that help human designers manage product characteristics across different functions and stages in the product life cycle. Agents represent not only design tools and humans with a stake in the design (including designers, manufacturing engineers, and marketing and support staff), but also the components of the design itself, and the characteristics of each component. These agents trade with one another for design constraints, requirements, and manufacturing alternatives, and the resulting marketplace provides a self-organizing dynamic that is said to yield more rational designs faster than conventional techniques.

The main features of the RAPPID approach are:

- small agent granularity approach;
- characteristics agents representing the characteristics (e.g., weight, cost, etc.) of a component;
- use of computerized tools in partnership with humans;
- markets as a mechanism for coordinating distributed decision-masking;
- set-based design[7].

13.7.2 ANARCHY

ANARCHY was a working prototype of an asynchronous design environment at Carnegie Mellon University (Quadrel et al., 1993). Its 20 agents are capable of generating, evaluating and modifying designs of medium- and high-rise office buildings.

Asynchronous Teams (a generic term in the Anarchy project) are characterized by five attributes: autonomous agents; broadcast communications; network structure; dynamic planning; and concurrent/asynchronous execution.

Agents in Anarchy are activated in a design system by modules that control the execution of individual agents and perform the roles of self-activation and modification of the data environment in the system.

The most interesting feature of ANACHY is its use of simulated annealing (Kirkpatrick et al., 1983), a traditional method of optimization, in the multi-agent

---

[7] A product's characteristics can be thought of as the dimensions of a Cartesian space within which the product is defined. Traditionally, designers seek to build a design to fit a predefined subspace of characteristics, without knowing in advance whether any acceptable design fits. The set-based design process developed by Toyota begins with a design space much larger than necessary, and then shrinks it incrementally to discover where the desired design is.

arena, for contracting searches. Its use in architecture design problems appears well suited. However, its application to other design areas such as mechanical design still has to be verified.

13.7.3 ABCDE

The Agent Based Concurrent Design Environment (ABCDE) (Balasubramanian et al., 1996) was a multi-agent architecture for implementing concurrent engineering in manufacturing. It was developed in parallel with MetaMorph I (see Chapter 14 Section 14.2 for details). A proof-of-concept system based on this architecture has design, manufacturability analysis, process planning, routing and scheduling as concurrent interacting activities.

The system includes a feature-based design sub-system for prismatic components, implementing "intelligent features". Using a simulated environment of four production machines, the system was tested with prismatic components being simultaneously designed. Manufacturability evaluation and shop floor planning were carried out concurrently. Valid process plans, routing, and scheduling were generated. This system was also integrated with the MetaMorph I system through a Design Mediator.

13.7.4 ACDS

ACDS (Automated Configuration Design Service) (Darr and Birmingham, 1994) is a good example of agent technology applied to configuration design (see Chapter 2 Section 2.2.6).

ACDS used four species of agents: a system agent, a bid agent, catalog agent, and constraint agent. The system agent represents the specifications of the overall system. Each catalog agent represents a set of parts of a given category, such as motors, cables, etc. Catalog agents record the attributes for the different parts of each type. Constraint agents represent the constraints among components based on these attributes. A single bid agent is used to monitor the evolution of the feasible set of solutions as the constraints are evaluated, tightening underconstrained spaces and backtracking if the space becomes overconstrained.

ACDS agents communicate through point-to-point messages over a fixed network of agents when the problem is configured. Coordination among ACDS agents is achieved through constraint propagation.

13.7.5 Agent-Based Blackboards (ABB)

As mentioned in Chapter 7, Agent-Based Blackboards were proposed by Lander et al (1996) to manage agent interactions using grouping. In this approach, each local group of agents shares a data repository that is provided specifically for the efficient storage and retrieval of active shared data. Within an agent group, a *control shell* (analogous to a facilitator or a mediator) notifies appropriate agents of relevant events. The *control shell* can be programmed to implement some coordination strategies.

This approach is very similar to the multiple blackboard approach proposed by Branki and Bridges (1993) in the CAAD project. It is in the middle portion of the spectrum (cf. Figure 13.8) that has at one end, pure blackboard concurrent design systems like ARCHIX (Thoraval and Mazas, 1990), CE Testbed (Londono et al., 1990), DICE (Sriram et al., 1992), and EXPORT (Monceyron and Barthès, 1992), and at the other end, pure agent-based concurrent design systems like DIDE (Shen and Barthès, 1996) and PACT (Cutkosky et al., 1993).

13.8 SUMMARY

In this chapter, we have described five significant projects in some detail, giving information about project name, research group, application area, communication protocol, representation and implementation technologies, and main characteristics. Several other projects were briefly described (see also Table 13.2).

In addition, a number of other significant projects related to cooperative engineering design, which use blackboard-like architectures (also called multi-expert systems) are listed in Table 13.3. Note that the collection of the projects presented in this chapter and summarized in Tables 13.2 and 13.3 is by no means complete. Since this is a rapidly developing field, the recent literature should also be consulted.

REFERENCES

Balasubramanian, S., Maturana, F. and Norrie, D.H. (1996) Multi-Agent Planning and Coordination for Distributed Concurrent Engineering, *International Journal of Cooperative Information Systems*, 5(2-3):153-179

Barthès, J.P. (1989) MOSS: a multi-user object environment, In *Proceeding of 2nd Symposium on Artificial Intelligence*, Monterrey, Mexico.

Barthès, J.P. (1993) La problèmatique de reconciliation en ingénierie simultanée, In *Actes 01 DESIGN'93*, Tunis.

Baudin, C., Underwood, J.G. and Baya, V. (1993) Using device Models to Facilitate the Retrieval of Multimedia Design Information, In *Proceeding of the 13th IJCAI*, Chambery, France, pp 1237-1243.

Branki, N.E. and Bridges, A. (1993) An Architecture for Cooperative Agents and Applications in Design, In M.R. Beheshi and K. Zreik, eds., *Advanced Technologies*, pp. 221-230.

Brown, D.C., Dunskus, B., Grecu, D.L. and Berker, I. (1995) SINE: Support For Single Function Agents, In *Proceedings of Applications of AI in Engineering*, Udine, Italy.

Collet, C., Huhns, M.N. and Shen, W.M. (1991) Resource Integration Using a Large Knowledge Base in Carno, *IEEE Computer*, 24(12):55-63.

Conru, A.B. and Cutkosky, M.R. (1993) Computational Support for Interactive Cable Harness Routing and Design, In *Proceedings of 1993 ASME Design Automation Conference*, Albuquerque, NM.

Coyne, R., Finger, S., Konda, S., Prinz, F.B., Sievorek, D.P., Subrahmanian, E. and Tenenbaum, M.J. (1994) Creating an Advanced Collaborative Open Resource Network, In *Proceeding of the sixth International ASME Conference on Design Theory and Methodology*, Minneapolis, MN, pp. 375-380.

Cutkosky, M., Brown, D., and Tenenbaum, J. (1991) Next-Cut: A Second Generation Framework for Concurrent Engineering, In D. Sriram, R. Logcher and S. Fukuda, eds., *Comput-Aided Cooperative Product Development*, Springer, pp. 1-18.

Cutkosky, M.R., Engelmore, R.S., Fikes, R.E., Genesereth, M.R., Gruber, T.R., Mark, W.S., Tenenbaum, J.M. and Weber, J.C. (1993) PACT: An Experiment in Integrating Concurrent Engineering Systems, *IEEE Computer*, 26(1):28-37.

Cutkosky, M.R., Tenenbaum, J.M. and Glicksman J. (1996) Madefast: Collaborative Engineering over the Internet, *Communication of the ACM*, 39(9):78-87.

Darr, T.P. and Birmingham, W.P. (1994) Automated Design for Concurrent Engineering, *IEEE Expert*, 9(5):35-42.

David, B.T. and Sadou, S. (1990) Un modèle d'architecture pour les systèmes CAO intelligents, In *Actes du MICAD'90*.

Douglas, R.E. Brown D.C. and Zenger, D.C. (1993) A Concurrent Engineering Demonstration and Training System for Engineers and Managers, *International Journal of CADCAM and Computer Graphics*, special issue on AI and Computer Graphics, 8(3):263-301.

Dunskus, B., Grecu, D.L., Brown, D.C. and Berker, I. (1995) Using Single Function Agents to investigate Negotiation, *AI EDAM*, Special Issue: Conflict Management in Design, 9(4):299-312.

Fenves, S.J., Flemming, U., Hendrickson, C., Maher M.L. and Schimitt, G. (1990) An Integrated Software Environment for Building Design and Construction, *Computer-Aided Design*, 22(1):27-36.

Finger, S., Fox, M., Prinz, F.B. and Rinderle, J.R. (1992) Concurrent Design, *Applied Artificial Intelligence*, 6:257-263.

Finin, T., Fritzon, R., McKay, D. and McEntire, R. (1993) KQML – A Language and Protocol for Knowledge and Information Exchange, Tech. Report, University of Maryland, Baltimore, Maryland.

Fruchter, R., Clayton, M., Krawinkler, H. and Teicholz, P. (1993) Interdisciplinary Communication Medium for Collaborative Conceptual Building Design, Tech. Report, Dept. of Civil Engineering and Center for Integrated Facility Engineering, Stanford University.

Gauchel, J., Bhat, R.R., Van Wyk, S., Asada, T., Bounds, N., Duerig, P. and Swonger, R. (1993) A Communication Framework for Cooperative Building Design, In M.R. Beheshi and K. Zreik, eds., *Advanced Technologies*, pp. 203-210.

Genesereth, M., (1991) Designworld. In *Proceedings of IEEE Conference on Robotics and Automation*.

Genesereth, M. and Fikes, R. (1992) Knowledge Interchange Format, Version 3.0 Reference Manual, Tech Report Logic-92-1, Computer Science Department, Stanford University.

Glicksman, J., Hilton, B.L., Pan, J. and Tenenbaum J.M. (1991) MKS: A Conceptually Centralized Knowledge Service for Distributed CIM Environment, *Journal of Intelligent Manufacturing*, 2(1):27-42.

Goldmann, S. (1996) Procura: A Project Management Model of Concurrent Planning and Design, In *Proceeding of WET ICE '96*, Stanford, CA.

Grecu, D.L. and Brown, D.C. (1995) Design Agents That Learn, In *Proceedings of IJCAI-95 Workshop on Adaptation and Learning in Multiagent Systems*, Montreal, Canada.

Grecu, D. and Brown, D. (1996a) Learning to design together, In *Proceedings of the 1996 AAAI Spring Symposium on Adaptation, Co-evolution and Learning in Multiagent Systems*, Stanford, CA.

Grecu, D. and Brown, D. (1996b) Learning by Single Function Agents during Spring Design, In J. Gero and F. Sudweeks, eds., *Artificial Intelligence in Design '96*, Kluwer Academic Publishers, Netherlands.

Gruber, T. (1993) A Translation Approach to Portable Ontology Specification, *Knowledge Acquisition*, 5(2):199-220.

Gupta, L., Chionnglo, J.F. and Fox, M.S. (1996) A Constraint Based Model of Coordination in Concurrent Design Projects, In *Proceeding of WET ICE '96*, Stanfrod, CA.

Hong, J., Toye, G. and Leifer L. (1995) Personal Electronic Notebook with Sharing, In *Proceeding of 4th WET ICE, West Virginia*, IEEE Press, pp. 88-94.

Iffenecker, C. (1994) Un système multi-agents pour le support des activités de conception de produits, Thèse de l'Université Paris VI, France.

Iwasaki, Y. and Low, C. (1991) Model Generation and Simulation of Device Behavior with Continuous and Discrete Changes, Technical Report KSL-91-69, Knowledge Systems Laboratory, Stanford University.

Kirkpatrick, S., Gelatt, C.D. and Vecchi, M.P. (1983) Optimization by Simulated Annealing, *Science*, 220:671-680.

Lander, S.E., Staley, S.M. and Corkill, D.D. (1996) Designing Integrated Engineering Environment: Blackboard-based Integration of Design and Analysis Tools, *Concurrent Engineering: Research and Applications*, Special Issue on the Application of Multiagent Systems to Concurrent Engineering, 4(1):59-72.

Lee, K.C., Mansfield, W.H.Jr. and Sheth, A.P. (1993) A Framework for Controlling Cooperative Agents, *IEEE Computer*, 26(7).

Lin, S., Lin, Z. Guo, Y. (1996) An Open Architecture for Distributed Integrated Systems, In *Proceedings of International Workshop on CSCW in Design*, Beijing, China.

Londono, F., Cleetus, K.J. and Reddy, Y.V. (1990) A Blackboard Scheme for Cooperative Problem-Solving by Human Experts, In D. Sriram, R. Logcher and S. Fukuda, eds., *Computer-Aided Cooperative Product Development*, Springer, pp. 26-50.

Maher, M.L. (1989) Building Design Using Multiple Expert Systems, In J. Gero, ed., *Expert Systems in Engineering, Architecture and Construction*, Sydney, pp. 149-164.

Mantyla, M. (1993) Towards Open Architecture Concurrent Engineering Framework, Otaniemi 1993/TKO-B96, Dept. of Computer Science, Helsinki University of Technology.

Maturana, F. and Norrie, D.H. (1996) Multi-Agent Mediator Architecture for Distributed Manufacturing, *Journal of Intelligent manufacturing*, 7:357-270.

Monceyron, E. and Barthès, J.P. (1992) Architecture for ICAD Systems: an Example from Harbor Design, *Revue Sciences et Techniques de la Conception*, 1(1):49-68.

McGuire, J., Huokka, D., Weber, J., Tenenbaum, J., Gruber, T. and Olsen, G. (1993) SHADE: Technology for Knowledge-Based Collaborative Engineering, *Journal of Concurrent Engineering: Applications and Research,* 1(3).

Morse, D.V. (1990) Communication in Automated Interactive Engineering Design, PhD thesis, Carnegie Mellon University.

Object Management Group (OMG), (1991) The Common Object Request Broker: Architecture and Specification, OMG Document #91.12.1, Framingham, MA.

Olsen, G., Cutkosky, M., Tenenbaum, J.M. and Gruber, T. (1994) Collaborative Engineering based on Knowledge Sharing Agreements, In *Proceeding of the 1994 ASME Database Symposium,* Minneapolis, USA.

Park, H., Tenenbaum, J.M. and Dove, R. (1993) Agile Infrastructure for Manufacturing Systems (AIMS): A Vision for Transforming the US Manufacturing Base, *Defense Manufacturing Conference '93.*

Park, H., Cutkosky, M., Conru, A. and Lee, S.H. (1994) An Agent-Based Approach to Concurrent Cable Harness Design, *AIEDAM,* 8(1):45-62.

Parunak, V., Ward, A., Fleischer, M. and Sauter, J. (1997) A Marketplace of Design Agents for Distributed Concurrent Set-Based Design, In *Proceedings of 4th ISPE International Conference on Concurrent Engineering,* Troy, Michigan.

Petrie, C. (1993) The *Redux'* Server, In *Proceeding of International Conference on Intelligent and Cooperative Information Systems (ICICIS),* Rotterdam.

Petrie, C., Cutkosky, M., and Park, H. (1994a) Design Space Navigation as a Collaborative Aid, In *Proceedings of the Third International Conference on Artificial Intelligence in Design,* Lausanne.

Petrie, C., Cutkosky, M., Webster, T., Conru, A. and Park, H. (1994b) Next-Link: An Experiment in Coordination of Distributed Agents, Position paper for the AID-94 Workshop on Conflict Resolution, Lausanne.

Quadrel, R., Woodbury, R., Fenves, S. and Talukdar, S. (1993) Controlling Asynchronous Team Design Environment by Simulated Annealing, *Research in Engineering Design,* 5(2):88-104.

Reddy, J.M., Chan, B. and Finger, S. (1995) Patterns in Design Discourse: A Case Study, In T. Tomiyama, M. Mantyla and S. Finger, eds., *Knowledge Intensive CAD,* Volume 1, Chapman & Hall, pp. 321-344.

Rehg, J., Elfes, S., Talukdar, S., Woodbury, R., Eisenberger, M. and Edahl, R. (1988) CASE: Computer-Aided Simultaneous Engineering, EDRC 18-05-88, Carnegie Mellon University.

Scalabrin, E., Vandenberghe, H., De Azevedo, H., and Barthès, J.P. (1996) A Generic Model of Cognitive Agent to Develop Open Systems, In *Proceeding of 18th Brazilian Symposium on Artificial Intelligence,* Curitiba, Brazil.

Shen W. (1996) *De l'utilisation d'agents cognitifs pour les systemes de CAO intelligents,* PhD Thesis, Université de Technologie de Compiègne, France.

Shen, W. and Barthès, J.P. (1996) An experimental environment for exchanging design knowledge by cognitive agents, In M, Mantyla, S. Finger and T. Tomiyama, eds., *Knowledge Intensive CAD,* Volume 2, Chapman & Hall, pp. 19-38.

Shen, W., Maturana, F., Norrie D., and Barthes, J.P. (1997) Agent-based approach for advanced CAD/CAM systems, In *Proceedings of Fifth International Conference on CAD/CG,* Shenzhen, China.

Sriram, D., Logcher, R., Groleau, N. and Cherneff, J. (1992) DICE: An Object Oriented Programming Environment for Cooperative Engineering Design, In C. Tong and D. Sriram, eds., *AI in Engineering Design*, Vo.3, Academic Press.

Stephanopoulos, G., Johnston, J., Kriticos, T., Lakshmanan, R., Mavrovouniotis, M. and Siletti, C. (1987) DESIGN-KIT: An Object-Oriented Environment for Process Engineering, *Computer in Chemical Engineering*, 11(6):655-674.

Thoraval, P. and Mazas, Ph. (1990) ARCHIX, an Experiment with Intelligent Design in the Automobile Industry, In *Proceeding of 4th Eurographics Workshop on Intelligent CAD Systems*, Mortefontaine, France.

Tomiyama, T., Umeda, Y., Ishii, M., Yoshioka, M. and Kiriyama, T. (1995) Knowledge Systematization for a Knowledge Intensive Engineering Framework, In T. Tomiyama, M. Mantyla and S. Finger, eds., *Knowledge Intensive CAD*, Volume 1, Chapman & Hall, pp. 55-80.

Tou, I., Berson, S., Estrin, G., Eterovic, Y. and Wu, E. (1994) Prototyping Synchronous Group Applicqtions, *IEEE Computer*, 27(5):48-56.

Toye, G., Cutkosky, M., Leifer, L., Tenenbaum, J. and Glicksman, J. (1993) SHARE: A Methodology and Environment for Collaborative Product Development, In *Proceeding of 2nd Workshop on Enabling Technologies: Infrastructure for Collaborative Enterprises*, IEEE Computer Society Press, pp. 33-47.

Trousse, B. (1993) Towards a Multi-Agent Approach for Cooperative Distributed Design Assistants, In M.R. Beheshi and K. Zreik, eds., *Advanced Technologies,* pp. 451-460.

Victor, S.K., Brown, D.C., Bausch, J.J., Zenger, D.C., Ludwig R. and Sisson, R.D. (1993) Using Multiple Expert Systems With Distinct Roles in a Concurrent Engineering System for Powder Ceramic Components, In G. Rzevski, J. Pastor and R.A. Adey, eds., *Applications of AI in Engineering VIII. Vol. 1: Design, Methods and Techniques*, Elsevier, pp. 83-96.

Victor, S.K. (1993) Negotiation between Distributed Agents in a Concurrent Engineering System, Master Thesis, Worcester Polytechnic Institute.

Weber, J. C., Livezey, B. K., McGuire, J. G. and Pelavin, R. N. (1992) Spreadsheet-Like Design through Knowledge-based Tool Integration, *International Journal of Expert Systems: Research and Applications*, 5(1).

Table 13.2 Agent-based concurrent design systems.

Project	Group	Application Area	Implementation Technologies	Main Characteristics or Remarks
PACT	Cutkosky et al, 1993 CDR, Stanford, USA	electro-mechanical design	CL, KIF, KQML, Ontolingua	facilitators for each group of agents
SHARE	Toye et al., 1993 CDR, Stanford, USA	mechanical design	CL, KIF, KQML, Ontolingua, Web	PENS, Web general architecture
First-Link	Park et al.,1994 CDR, Stanford, USA	cable harness assemblies	CL, KIF, KQML, Ontolingua	has a central node, but it is different from a shared database
Next-Link	Petrie et al., 1994b CDR, Stanford, USA	cable harness assemblies	CL, KIF, KQML, Ontolingua	First-Link + *Redux'*
Process-Link	Goldmann, 1996 Kaiserslautern University, Germany	mechanical design		multiple managers and *Redux'*
DIDE	Shen et al., 1996 UTC, France	mechanical design	CL, MOSS, CLOS, CLIM, C++, Web, e-mail	autonomous agents, open architecture
SINE /SiFAs	Brown et al., 1995 WPI, USA	mechanical design	GNU C, CLIPS	single function agents
RAPPID	Parunak et al., 1997 ITI, USA	mechanical design		characteristic agents, marketplace approach, set-based design
Anarchy	Quadrel et al., 1993 CMU, USA	architectural design	CL, EDESYN	broadcast communication, simulated annealing for A-team control
ABCDE	Balasubramanian et al, 1996, Univ. of Calgary, Canada	mechanical design & manufacturing		managers similar to facilitators in PACT
ACDS	Darr et al., 1994 CMU, USA	configuration design		coordination through constraint propagation
ABB	Lander et al., 1996 Blackboard Tech.	mechanical design		local data sharing through a blackboard
A4-Approach	Gauchel et al., 1993 CMU, USA	architectural design	Knowledge Craft	client/server architecture, shared objects
ACORN	Coyne et al., 1994 EDRC, CMU	electro-mechanical design	C++, AutoCAD AME and API	layered system architecture
CONDOR	Iffenecker, 1994 Paris VI, France	electro-mechanical design	Knowledge Craft EUCLID, Design View	individual agents and collective agents
OACEF	Mantyla, 1993 HUT, Filand	general architecture		Life-Cycle Assessment (LCA) agents
OADIS	Lin et al., 1996 Academia Sinica	mechanical & architectural design	C/C++, CL, MOSS	autonomous agents + groupware
Object World /CoSARA	Tou et al., 1994 UCLA, USA	general purpose architecture	CL, CLOS, C++	autonomous agents, open and dynamic environment

Table 13.3 Cooperative design sytems using blackboard-like architectures

Project	Group	Application Area	Implementation Technologies	Main Characteristics or Remarks
ANAXA-GORE	Trousse, 1993 INRIA, France	spacecraft design	SMECI, MAILY, XCAD	geometric constraint manager
ARCHIX	Thoraval et al., 1990 UTC, France	automobile design	SMECI, E-mail, EUCLID CAD	hierarchical task decomposition
CAAD	Branki et al., 1993 CCU/US, UK	architectural design		multi-blackboard architecture
CASE	Rehg et al., 1988 CMU, USA	mechanical design	Lucid CL, LOOPS, VEGA, FORTRAN	semantic networks and constraints networks for design representation
CE Testbed	Londono et al., 1990 CERC, WVU, USA	mechanical design	C++, ObjectiveC CLOS, ROSE, PPO, LASER...	an application of DARPA DICE
CIAO	David et al., 1990 EC Lyon, France	general environment		intellligent user interface
Design Fusion	Finger et al., 1992 CMU, USA	architectural design	Knowledge Craft	feature graph grammars, constraints networks
Design-Kit	Stephanopoulos et al., 1987, MIT, USA	chemical factory planning	CL, KEE	implemented on a symbolic machine
DICE	Sriram et al., 1992 MIT, USA	mechanical design	C++, OODMS, GEMSTONE	active objects and active database
EXPORT	Monceyron et al., 1992, UTC, France	harbor design	GKS standard, X Windows	supervisor control
GUIDE	Morse, 1990 CMU, USA	architectural design	Lucid CL, KEE	direct and dynamic access to global solution database
IBDE	Fenves et al., 1990 CMU, USA	architectural design	CL, C, Fortran, OPS 5	supervisor control
ICM	Fruchter et al., 1993 Dept. of Civil Eng., Stanford, USA	architectural design	CL, KEE	central, and shared graphic model
KAD	Gupta et al., 1996 UT, Canada	mechanical design	CLIPS, ECLiPSe, ...	knowledge network
KIEF	Tomiyama et al., 1995, Tokyo, Japan	mechanical design	Objectworks/ Smalltalk	metamodel mechanism multiple ontology functional knowledge
MATE	Maher, 1989 Sydney, Australia	architectural design	AutoCAD, video conferencing tool	two shared workspaces
Next-Cut	Cutkosky et al, 1991 CDR, Stanford, USA	electro-mechanical design	Lucid CL, OOP, Symbolic Env.	base of other projects PACT, First-Link, Next-Link, ...

14

Agent-Based Manufacturing Planning, Scheduling and Control

14.1 INTRODUCTION

> Planning is the process of selecting and sequencing activities such that they achieve one or more goals and satisfy a set of domain constraints. Scheduling is the process of selecting among alternative plans and assigning resources and times to the set of activities in the plan. These assignments must obey a set of rules or constraints that reflect the temporal relationships between activities and the capacity limitations of a set of shared resources. The assignments also affect the optimality of a schedule with respect to criteria such as cost, tardiness, or throughput. In summary, scheduling is an optimization process where limited resources are allocated over time among both parallel and sequential activities.
>
> Zweben and Fox, 1994

Following the above definition, we consider manufacturing planning as the process of selecting and sequencing production operations for a product (which can be an assembly, subassembly, or a part). The result is a process plan that can be represented in graph structure, with each node representing an operation. Manufacturing scheduling is the process of assigning shop floor resources (e.g., machines, tools, operators) and time slots to the process plans. The result is a schedule for a shop floor or a plant, usually in Gantt Chart format.

Manufacturing scheduling is a difficult problem, particularly when it takes place in an open, dynamic environment where things rarely go as expected. The system may be asked to do additional tasks that were not anticipated, and sometimes is allowed or requested to omit particular tasks. The resources available for performing tasks are subject to change. Some resources can become unavailable, and additional resources introduced. The beginning time and the processing time of a task are also subject to variation. A task can take more time than anticipated or less time than anticipated, and tasks can arrive early or late. Because of its highly combinatorial aspects (NP-complete), dynamic nature, and practical interest for manufacturing systems, the scheduling problem has been widely studied. The literature lists numerous scheduling methods: heuristics, constraint propagation techniques, constraint satisfaction formalisms, simulated annealing, Taboo search, genetic algorithms, neural networks, and so on. Recently, agent technology has also been used in attempts to resolve this problem and has shown considerable promise.

Manufacturing control relates to the strategies and algorithms used to activate a manufacturing plant so that it makes the desired products in accordance with production planning and scheduling. Manufacturing control can be divided into low-level and high-level control. At the low-level, individual manufacturing resources are controlled to deliver the unit-processes expected by the high-level control functions. High-level manufacturing control relates to the overall coordination of manufacturing resources, to make desired products. In agent-based manufacturing systems, agent technology is usually applied to high-level manufacturing control. Wang et al (1998), however, describe its application to low-level control.

In this chapter, we give a detailed introduction to representative agent-based manufacturing planning, scheduling, and control systems, using the same format as was used in Chapter 13 for agent-based concurrent design systems. A state-of-the-art survey of about 40 related projects, with an extensive annotated bibliography, can be found in (Shen and Norrie, 1999). Its extended version in HTML format is also available at: http://www.ucalgary.ca/~wshen/papers/survey-abm.htm.

14.2 METAMORPH

14.2.1 Origin, Domain of Application, and Research Objectives

MetaMorph (also referred to MetaMorph I) (Maturana and Norrie, 1996; Maturana, 1997; Maturana et al., 1999) was devised as an adaptive agent-based architecture to address system adaptation and extended-enterprise issues at four fundamental levels: virtual enterprise, distributed intelligent systems, concurrent engineering, and agent architecture. *Virtual enterprise* strategic partnership issues are associated with the unification of heterogeneous manufacturing subsystems into a large, dynamic, virtual coalition of cooperative subsystems. *Distributed intelligent systems* enable increased autonomy for each member of the manufacturing enterprise network. Each manufacturing partner (or manufacturing subsystem) can pursue individual goals while satisfying both local and external constraints. Applying *concurrent engineering* to product/process design and production management then becomes fundamental for managing manufacturing information and reducing time to market. To cope with increasing complexity and organizational issues, *intelligent agent architecture* is envisaged as necessary for these extended manufacturing enterprises.

The architecture was named MetaMorphic, since a primary characteristic is its changing form, structure, and activity as it dynamically adapts to emerging tasks and changing environment.

14.2.2 Technical Features

14.2.2.1 Overall Approach and System Architecture

MetaMorph uses an agent-based mediator-centric federation architecture (Gaines et al., 1995). In this particular type of federation organization, intelligent agents can link with mediator agents (also called mediators) to find other agents in the environment. Additionally, mediators assume the role of system coordinators by promoting cooperation among intelligent agents and learning from the agents' behavior. Mediators provide system associations without interfering with low-level decisions unless critical situations occur. Mediators are able to expand their coordination capabilities to include other mediation behaviors, including high-level policies to break decision deadlocks. Mediation actions are performance-directed behaviors.

Mediators can use brokering and recruiting communication mechanisms (Decker, 1995) to find appropriate agents for collaborative subsystems (also called coordination clusters or virtual clusters). The brokering mechanism comprises receiving a request message from an intelligent agent, understanding the request, finding suitable receptors for the message, and broadcasting the message to the selected group of agents. The recruiting mechanism is a superset of the brokering mechanism, since it uses the brokering mechanism to match agents. However, once appropriate agents have been found, these agents can be directly linked. The mediator then can step out of the scene to let the agents proceed with the communication themselves. Both mechanisms have been used in MetaMorph. To efficiently use these mechanisms, mediators need to have sufficient organizational knowledge to match agent requests with needed resources. Organizational knowledge at the mediator level is basically a list of agent-to-agent relationships that is dynamically enlarged.

The brokering and recruiting mechanisms generate two relevant types of collaboration subsystems. The first corresponds to an indirect collaboration subgroup, since the requester agent does not need to know about the existence of other agents that temporarily match the queries. The second type is a direct collaboration subgroup, since the requester agent is informed about the presence and physical location of matching agents so as to be able to continue with direct communication.

Common activities for mediators involved in either type of collaboration are interpreting messages, decomposing tasks, and providing processing times for every new subtask. These capabilities make mediators very important elements in achieving the integration of dissimilar intelligent agents. Federation multi-agent architectures require a substantial commitment to supporting intelligent agent interoperability through mediators.

14.2.2.2 Agent Structure

There are two main types of agent in MetaMorph: resource agents and mediator agents (also called mediators). Resource agents are used to represent manufacturing devices and operations, while mediator agents are used to coordinate the interactions among agents (resource agents and also mediator agents). In the MetaMorph architecture, the hybrid agent models used to build both resource and mediator agents can be classified as soft-hybrid agent models because none of the reactive-agent levels are strongly implemented in the agent structures.

Different levels of intelligence and behavior are associated with the two different types of agents. Resource agents are autonomous and cooperative. Mediator agents are autonomous, cooperative, and learning. Figure 14.1 shows the basic characteristics of resource and mediator agents.

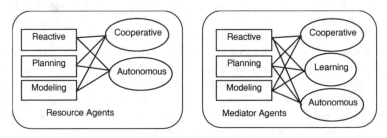

Figure 14.1 Basic characteristics of resource and mediator agents (Maturana, 1997).

14.2.2.3 Inter-Agent Communication

Communication among mediators in MetaMorph used the TCP/IP protocol. The KQML protocol (Finin et al., 1993 – see also Chapter 16) was used as the high-level agent communication language.

In the simulation implementation, resource agents were modeled as objects, and therefore the communication among resource agents was, in fact, message-passing among objects.

14.2.2.4 Dynamics / Global Behavior

Agent Coalition (Clustering)

In order to work cooperatively, agents form coalitions (clusters) that bond dissimilar agents into harmonious decision groups. Multistage negotiation and coordination protocols that can efficiently maintain the stability of these coalitions are required. Each agent has its individual representation of the external world, goals, and constraints, so diverse heterogeneous beliefs interact within a coalition through distributed cooperation models.

In MetaMorph, the core collaboration mechanism is based on task decomposition and dynamically-formed agent groups (clusters). High-level tasks are initially decomposed by mediators acting at the corresponding information level. Each subtask is subsequently distributed to determine the best solution plan. Mediators learn dynamically from the agent interactions and identify coalitions that can be used to establish distributed searches for the resolution of tasks.

Agent coalitions is incorporated in the main problem-solving mechanism in MetaMorph. Appropriate agents are dynamically and virtually incorporated within a problem-solving group (cluster). Where the agents in the problem-solving group (cluster) are only able to partially complete the task requirements, new subclusters of agents are formed to solve the remaining subtasks. This process is repeated, with further sub-clusters being formed as needed, in a dynamically interlinked structure. When the respective tasks and subtasks are all solved, the cluster structure is dissolved. This clustering process provides scaleability and aggregation properties to the system.

Agent Cloning

In MetaMorph, resource agents are cloned as needed for concurrent information processing. These clone agents are included in the virtual coordination clusters and subclusters. Within the clusters, the agents negotiate with each other to find the best solution for a production task or subtask. The clustering and cloning mechanisms and the coordination of agents by mediators will further be discussed in the following section.

When the system is running in simulation mode, the resource agents are active objects with goals and associated motivations. They are, in general, located in the same computer. Clone agents are, in fact, clone objects. In the case of real on-line scheduling, the cloning mechanism can be used to "clone" resource agents from remote computers (for NC machines, manufacturing cells and so on) to the local computer where resource mediators reside, so as to reduce communication time and consequently reduce the scheduling/rescheduling time.

Different Types of Mediators

There are two different coordination tasks in MetaMorph, which requires two different types of mediator. The first specializes in coordination of virtual groups (inter-coordination) and can be called a high-level mediator. The second is specialized for coordination of the intelligent agents within a virtual group (intra-coordination) and can be called a low-level mediator. However, these mediators are not restricted to these operations alone.

Every mediator encapsulates functionality to allow local coordination and interaction with other dissimilar mediators. Mediators responsible for mapping system entities are classified as static mediators. They are characterized as high-level mediators directly connected to physical objects. In addition, there are dynamic mediators for coordinating dynamic interaction among agents. Dynamic mediators are dynamically created through agent-replication action.

Each manufacturing enterprise needs at least one high-level mediator to act as the system's integrator (called an enterprise mediator). This enterprise mediator is able to recognize all sub-level mediators, platforms, and resources in the enterprise. The enterprise mediator supplies a global view of a system during integration of plans.

Coordination in MetaMorph

Coordination initially involves two main stages of activity: task decomposition and creation of virtual coordination clusters. These activities are supported by the static mediators, the Data-Agent Managers (DAM), and the Active Mediators (AM), each of which coordinates a specific level in the overall coordination task.

At the resource community level, shop floors are provided with high-level static mediators, which constitute the enterprise model of the system. Static mediators use their classification mechanisms to learn from the system's activity and to update the organizational knowledge of the system. A high-level task initially passes through the static mediator for recognition and decomposition. According to the shop-floor capability, the subtasks are each assigned to separate coordination clusters. Each coordination clusters will also incorporate clone agents obtained from active manufacturing agents. These entity interrelationships form a concurrent coordination framework.

Active Mediators

The Active Mediators (AMs) carry out control and mediation actions on clone agents. In the first stage of coordination, the AM broadcasts the unsolved task (or subtask) to the clone agents under its supervision. Each clone agent analyzes the task and prepares a bid (plan) or else it rejects the task. The clone agents, one by one, reply with their bids back to the AM. The AM analyzes the messages and incorporates agents with acceptable bids within the decision group and dismisses those other agents that are unsuitable , as shown in Figure 14.2.

In the second stage, the AM coordinates the respective activities of the remaining (clone) agents within the coordination cluster, to complete the task or subtask, which may involve further bidding among these agents.

The communication focus during both coordination stages is on selective broadcasting. For example, the AM needs to recognize both the senders of bids and the receptor of bids. Because the coordination actions form a very complex branching pattern, the instantiation of decentralized mediators has been chosen as a basic mechanism for distributed coordination (Figure 14.2).

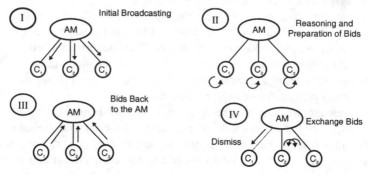

Figure 14.2 Active Mediator Coordination (Maturana and Norrie, 1996).

If a coordination cluster has many clone agents, the following guidelines enhance intra-agent coordination efficiency:

-   The AM regulates the number of messages to be exchanged according to variable priorities. When there are many agents in a coordination cluster, the AM is very strict in the selection of messages. In cases of high message traffic, exchanges are limited to two messages at a time.
-   In regulating message exchange, cost tables are implemented to prioritize agent communication.
-   When the number of agents is decreased (the clone agents progressively dismiss themselves from the negotiation group), the AM relaxes the selection of messages.

Data Agent Managers

To find and coordinate additional resources in a way that does not overload the AM, inter-cluster and other "external" coordination activities are provided by Data Agent Managers (DAMs). Thus the DAM for the cluster receives any request for additional resources and searches for appropriate agents to fulfill it. The DAM communicates with the static mediator for this resource community, and the overall system is searched for additional resources. The static mediator locates the resources and notifies the DAM with the respective address (of an individual agent or coordination cluster). Subsequently, the clone agent is informed about the new resources and direct communication between it and the new resource may be established if additional transactions are needed. AM and DAM each have separate computing processes with individual threads of execution.

Clone Agents

The resource agent connected to a manufacturing resource maintains real-time information about the resource's physical activity. This information includes the operational status of the resource, committed plans, schedules, and temporal interactions with other resources. This resource agent also uses cloning meta-level mechanisms to replicate itself in different coordination clusters as needed, as shown in Figure 14.3. Through this mechanism, clone agents can be involved in distributed and concurrent planning.

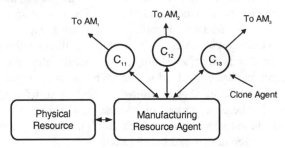

Figure 14.3 Clone Agent Coordination (Maturana, 1997).

Each clone agent initially maintains the current state of its resource when it is created. Clone agents are subsequently affected by dynamic changes in the task priorities. In this manner, variations in the state of the resource are dynamically introduced into the planning activity.

The clone agents can develop only promissory plans for tasks but cannot *commit* the resource to these (only the manufacturing resource agent can commit to a task since it manages and executes the resource's schedule). The complexity of the tasks affects both the solution overheads and the number of clone agents that arrive at a promissory plan stage at any specific time. The manufacturing resource agent assesses these promissory plans in its clone agents at discrete periods.

14.2.2.5 Learning in MetaMorph

Two fundamental learning mechanisms have been implemented in MetaMorph to enhance the system's performance and responsiveness, with mediators playing an essential role in both mechanisms. First, a mechanism that allows mediators to learn from history is developed at the resource mediator level to capture significant multi-agent interactions and behaviors. Second, a mechanism for propagating the system's behavior into the future is implemented to help mediators 'to learn from the future.'

The context of manufacturing has static and dynamic aspects. The static context encompasses the physical configurations of machines and products. The dynamic context changes with time, system loads, system metrics, costs, customer desires, etc. During manufacturing, a wide spectrum of emergent behaviors will be observable, which can be separated into specific behavioral patterns. A 'learning from history' mechanism based on distributed case-based learning was developed during MetaMorph to capture such behavioral patterns at the resource mediator level and storing these in its knowledge base. This knowledge is then subsequently reused by an extended case-based reasoning mechanism, as described in the following paragraph.

A manufacturability or manufacturing request sent to a resource community is first filtered by the respective resource mediator to decide whether the request can be recognized as associated with a previously considered product or is unknown. If it is recognized, the resource mediator retrieves the learned patterns to send to the select group of agents identified in the patterns. For an unknown request, the resource community's mediator uses its standard matchmaking actions to specify the primary set of resource agents to be contacted regarding their capability to satisfy this request. The solution to this unknown request is then obtained through the propagation of coordination clusters and decision strategies previously described. During this process, the resource mediator involved learns from partial emergent interactions at the coordination cluster level. This learning process is distributed over several coordination clusters. The plan aggregation process then enables the classification of various feature-machine-tool patterns, which are encoded and provided to the community's mediator for storage and future reuse.

The main purpose of 'learning from the future' is to modify promissory schedules at the resource agent level for otherwise unforeseen perturbations or changes in production priorities on the shop floor. At a selected time, the shop floor situation (including promissory schedules) is captured and "fast-forwarded" into the future through a separately-run simulation. The shop floor situation at the end of this future period is then analyzed and used to adjust the current promissory schedules, to avoid future conflicts and obtain more optimal use of resources. Experimental work in the MetaMorph project showed this procedure to be quite effective for adjusting and enhancing the system's performance.

A detailed description of these MetaMorph learning mechanisms can be found in (Shen et al., 1998b).

14.2.3 Implementation Details

The MetaMorph architecture and coordination protocols were used in implementing a distributed concurrent design and manufacturing system in simulated form. This virtual system incorporates heterogeneous manufacturing agents in different shop floors that are dynamically interconnected, and carries out concurrent manufacturability evaluation, production planning and scheduling. The system comprises the following multi-agent modules: Enterprise Mediator, Design System, Shop Floors, and Execution Control & Forecasting, as shown in Figure 14.4. Each multi-agent module uses common enterprise integration protocols to allow agent interoperability.

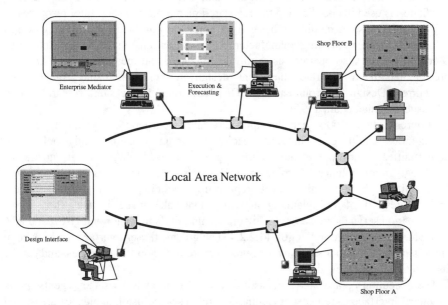

Figure 14.4 Prototype Implementation of MetaMorph Architecture (Maturana et al., 1999).

The multi-agent modules are implemented within a distributed computing platform consisting of four HP Apollo 715/50 workstations, each running an HP-UX 9.0 operating system. The workstations communicate with each other through a Local Area Network (LAN) and TCP/IP protocol. Graphical interfaces for each multi-agent module were created in the VisualWorks 2.5 (Smalltalk) programming language, which was also used for programming the modules. The KQML protocol (Finin et al., 1993) is used as high-level agent communication language. The whole system is coordinated by high-level mediators, which provide the integration mechanisms for realizing the extended enterprise (Maturana and Norrie, 1996).

The Enterprise Mediator acts as the coordinator for the enterprise, and all of the manufacturing shop floors and other modules are registered with it. Registration processes are carried out through macro-level registration communications. Each multi-agent manufacturing module offers its services to the enterprise through the Enterprise Mediator.

A graphical interface has been created for the Enterprise Mediator. Both human users and agents interact with the Enterprise Mediator and the registered manufacturing modules via KQML messages. Decision rules and enterprise policies can be dynamically modified by object-call protocols through input field windows by the user. Action buttons support quick access to any of the registered manufacturing modules, shown as icon-agents, as well as to the Enterprise Mediator's source code.

The Enterprise Mediator offers three main services: integration, communication, and mediation. Integration permits the registration and interconnection of manufacturing components, thereby creating agent-to-agent links. Communication is allowed in any direction among agents and between human users and agents. Mediation is used in coordinating the registered mediators and shop floor resources.

The design system module is basically a graphical interface for retrieving design information and requesting manufacturability evaluations through the Enterprise Mediator (which also operates as shop-floor manager and message router). Designs are created in a separate intelligent design system named the Agent-Based Concurrent Design Environment (ABCDE), developed in the same research group (Balasubramanian et al., 1996).

Different shop floors can be modeled and incorporated in the system as autonomous multi-agent modules each containing machine and tool agent communities. Shop-floor resources are registered in each shop floor using macro-level registration policies. Machine and tool agents are incorporated into the resource communities through micro-level registration policies. A shop-floor module incorporates all of the planning activities that take place in that shop floor. Coordination services are supported by the community's mediator and DAM and AM mediators as described in Section 14.2.2.4. The graphical user interface for each shop floor is provided with a set of icon-agents to represent shop-floor devices, and can be used to obtain and view device states.

The execution control and forecasting simulation module uses icon-agents on its graphical interface to represent machines, warehouses, collision avoidance areas, and AGV agents. When a shop floor is "captured", its resources are introduced into the module, thereby instantiating icon-agents in the graphical interface and specifying data files for each resource. This module includes Standard Operation times for loading, processing, unloading, and transportation that can be scaled as desired for each resource. Each resource can enforce a specific dispatching rule (i.e., weighted shortest processing time, earliest due date, shortest processing time, FIFO, LIFO, etc). Parts are modeled as part agents that are implemented as background processes. A local execution mediator is embedded in the module to integrate and coordinate the shop-floor resources. This local execution mediator communicates with the resource mediator to get promissory plans and to broadcast forecasting results. The system can be run in different time modes: real-time and forecasting. In the real-time mode, the speed of the shop-floor simulation is proportional to the execution speed of the real-time system. In the forecasting mode, the simulation speed is 40–60 times faster than the real-time execution.

14.2.4 Most Significant Results

MetaMorph uses a mediator-centric federation architecture. This type of architecture is suited for complex distributed integrated manufacturing systems with a large number of agents. It facilitates multi-agent coordination by minimizing communication and processing overheads. Adaptation in MetaMorph is through organizational structural change (to suit both external and internal (evolving) task requirements) and through two learning mechanisms: learning from past experiences and learning future agent interactions by simulating future dynamic, emergent behaviors.

14.2.5 Position with Respect to Other Projects or Future Work

The Agent-Based Concurrent Design Environment (ABCDE) was developed in the same research group (Balasubramanian et al., 1996) and was integrated with the MetaMorph system via a Design Mediator. The ABCDE architecture includes an environment manager, feature agents, part agents, and CAD physical layers. ABCDE agents interact with the shop-floor mediator to develop manufacturability assessments during the product design process. Human production managers may also request manufacturability evaluations using either the CAD system or the design system interface.

MetaMorph II was the successor to the MetaMorph project (also referred to as MetaMorph I). Its objective was to integrate the manufacturing enterprise's activities of design, planning, scheduling, simulation, execution, and product distribution, with those of its suppliers, customers and partners within a distributed intelligent open environment (Shen et al., 1998a).

14.3 AARIA

14.3.1 Origin, Domain of Application, and Research Objectives

AARIA stands for Autonomous Agents for Rock Island Arsenal (Parunak et al., 1998; Baker et al., 1999). The project, started in 1994, was part of a DARPA contract "Large Scale System Simulation and Resource Scheduling Based on Autonomous Agents," administered by the Air Force ManTech Directorate and awarded to Intelligent Automation, Inc. of Rockville, MD. The AARIA project members were Intelligent Automation, Inc., the Industrial Technology Institute, the University of Cincinnati, Flavors Technology, Inc., and Agent Technologies, Inc.

The project was to investigate large-scale resource allocation and system simulation using autonomous agents, in the context of factory scheduling. The purpose was to develop an autonomous agent-based factory scheduler for the Rock Island Arsenal. The project envisaged agentified manufacturing entities (e.g. people, machines, parts) interacting over a network as a manufacturing enterprise, whose performance and functionality could supersede that of existing centrally controlled manufacturing systems.

AARIA provides fundamental integrated ERP and MES functionality in a heterarchical multi-agent architecture. The MES functionality includes basic "what-if" simulation, finite capacity scheduling, and intelligent shop floor interfaces. The ERP functionality includes basic planning, order entry, purchasing, bill-of-materials management, inventory management, resource management, personnel management, integrated financials, and reporting. Each agent has some portion of this general functionality, and when placed together, these agents self-configure to emergently provide a fully integrated ERP and MES system.

AARIA was designed to dialog with customers and suppliers and allocate resources to new jobs as they enter the system, to optimize schedules across resources, to recover from faults in the factory, to dispatch work against the schedule, and to report results. It was designed to exhibit this functionality either when running an actual factory or when in a simulation or "what-if" mode.

14.3.2 Technical Features

14.3.2.1 Overall Approach and System Architecture

In AARIA, in discrete manufacturing, Parts move through a network of Unit Processes (UP's) and Buffers. Each UP acquires one or more input Parts from Buffers of the needed types, and engages certain Resources to produce one or more output Parts to be placed into appropriate Buffers. In AARIA, linguistic case theory (Parunak, 95) was used over a set of declarative sentences describing the domain to identify candidate agents. This population of agents was then refined using the requirements for these. Figure 14.5 shows the resulting community, with its flows of Parts and Engagements for Resources along orthogonal axes that intersect at UP's. In

AARIA, agents represent parts, resources and unit processes with substantially equal intelligence and responsibility in each type of agent.

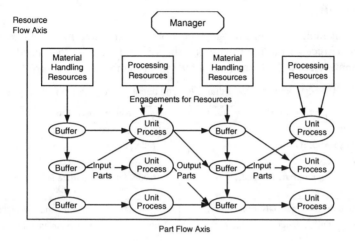

Figure 14.5 AARIA Architecture (Baker et al., 1999) (©1999 IEEE).

AARIA brings some important enhancements to agent-based manufacturing. The system incorporates new features for schedule optimization and fault recovery (see Section 14.3.2.6). The agent infrastructure allows true agent behavior by supporting true broadcast and multicast communication, subject-based mail handling, multithreaded agent activity, multi-platform instantiation, agent migration, and the implementation of multiple scheduled activities within an agent. Also, the system was implemented with dual functionality so it can either run a real factory or run in simulation mode.

14.3.2.2 Design Principles

AARIA was designed according to the following principles [http://www.aaria.uc.edu/]:

Emergent ERP

The aggregate behavior of the agent community subsumes functionality currently provided by manufacturing Enterprise Requirements Planning (ERP) systems. Table 14.1 shows the main differences between traditional ERP systems and agent-based emergent ERP systems.

Uniformity

An operation at the boundary of AARIA interacts with external suppliers or customers in the same way that it does with internal ones.

The traditional approach in supply chain management is to integrate factories within a supply chain. In AARIA, the factory is modeled as a supply chain. The walls between factories become transparent. Each agent seamlessly inter-operates with agents inside and outside its own factory.

Empowerment

Human stakeholders (including operators, manufacturing engineers, and management) receive the information (both as-is and what-if) they need to do their jobs, with interfaces to let them control the system rather than be controlled by it.

Table 14.1 Difference between traditional ERP and agent-based emergent ERP systems.

Traditional ERP	Agent-Based Emergent ERP
Client/Server: - Limited scaleability - Costly servers - Technology appropriate for medium and large companies	Fully Distributed: - full scalable - inexpensive computers - technology appropriate for large companies and proprietorships alike
Separate modules for financials, order entry, scheduling, planning, material management, purchasing...	Integrated financials, order entry, scheduling, planning, material management, and purchasing at each agent.
Supply-chain integration requires additional software and hardware	Agents act as a supply chain, requiring no additional software or hardware for supply chain integration
Commitments to customers not based on actual shop floor status	Commitments to customers made by shop floor
Poor integration with finite-capacity scheduling	Agents perform sophisticated finite-capacity scheduling algorithms
Heuristic management of company resources	Cost-based management of company resources
System provides no assistance when negotiating with customer over price, delivery, or quantity	Natural negotiation with customer over price, delivery, and quantity
System removed from decision implementors	System agents as decision implementors
"What's the Internet?"	Internet ready
Brute-force solution	Elegant solution

Dynamic Self-configuration

The system configures itself dynamically in response to internal and external changes.

According to the principle of increasing entropy from the second law of thermodynamics, any closed system will always become more disorganized as it operates. In order to obtain self-organization at a macro-level, a dissipative field was introduced at the micro-level. In particular, the dissipation of money was used to provide self-organization in AARIA. Agreements between agents are denominated in monetary units using mechanisms similar to those proposed by Baker (1991).

Least Commitment

The customer's statement of demand develops interactively, rather than being specified in detail at the outset. The system plans what it will do only as needed to meet its commitments to customers.

A customer's needs are often fuzzy. By letting the customer and supplier dialog about the customer's needs and the supplier's capabilities, the system provides more opportunities for customer satisfaction and greater flexibility to the supplier.

Modality Emergence

Whether a manufacturing capability is advance scheduled, dispatched, or pulled[1], depends dynamically on its capability, and its connections to the rest of the factory.

Advance scheduling, dispatching, and pulling can be thought of as being in a two-dimensional space of how constrained the system is, by demand from the customer and by commitments to the customer (see Figure 14.6). Advance scheduling is highly constrained by commitments that have been made to the customer but is not constrained by the demand placed by the customer. Pull is constrained both by commitments to deliver at a given rate and by demand. Dispatching is neither constrained by demand nor by commitment, since it arranges for production only as the resource becomes available.

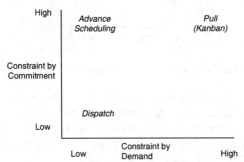

Figure 14.6 Advance scheduling, dispatching, and pulling
(Parunak, 1998) (©1998 Springer-Verlag).

In traditional manufacturing systems, the whole system is set-up to advance schedule, dispatch, or pull. In AARIA, each agent is allowed to decide which modality it will follow. The decision can be based on a number of different circumstances in which the agent may find itself: high utilization may imply scheduling is needed; an unreliable process may imply dispatching is needed; repetitive production may imply pull is the most cost effective modality. The

---

[1] An *Advance Scheduling* algorithm determines in advance when jobs will use factory resources. A *Dispatching* algorithm decides how to use a factory resource according to the availability of the resource. A *Pulling* algorithm uses the depletion of inventory to signal manufacturing production.

modality choice is left up to each agent, may change with time for each agent, and is allowed to rest anywhere in the two dimensional continuum between the three basic modalities. Agents with mismatched modalities are interfaced to each other using techniques which depend on the specific mismatch.

Metamorphosis

The system maintains continuity between different entities that represent different stages in a common life cycle (for example, an order for a part, the part itself, or its production history).

14.3.2.3 Agent Structure

AARIA agents have a body-head structure (Figure 14.7). That is, the physical entity being agentified is considered the body and the agent software is considered the head.

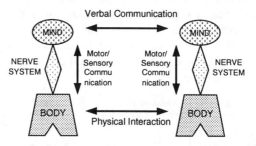

Figure 14.7 Internal structure of an agent (Baker et al., 1997).

A standard interface connects bodies and heads in this architecture. Figure 14.8 shows the structure of this system. An advantage of this structure is that any agent's body can be easily replaced with a simulation of that body which is implemented by another software agent (see Figure 14.9). Thus, the control agent is oblivious to whether it is controlling a real factory or a simulated one.

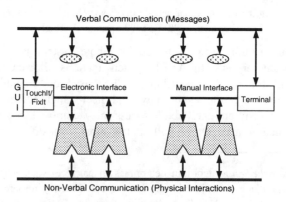

Figure 14.8 Body-Heads connections (Baker et al., 1997).

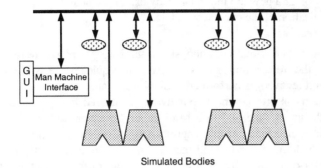

Simulated Bodies

Figure 14.9 Simulation of physical bodies (Baker et al., 1997).

One can run the same AARIA code that controls the factory against simulated bodies to decide how the factory could be reconfigured to improve performance, without the expense and potential inconsistency of developing a separate simulation model of the control algorithms.

14.3.2.4 Inter-Agent Communication

Inter-agent communication in the Cybele infrastructure for AARIA employs subject-based addressing (see Section 14.3.5 for details).

14.3.2.5 Dynamics / Global Behavior

At the finest grain, each agent represents a component competing with other like components for the rights to carry out some part of the manufacturing process. For instance, an agent may represent one of several similar milling machines each of which is trying to "maximize its own profit" by bidding on and trying to win new jobs. Bids and transactions are based on a currency model. An agent's (i.e., machine's) current schedule, the "cost" of the job up for bid and other related factors help determine a bid.

Each new piece of equipment (e.g., a milling machine, or a drilling machine, etc.) that is added to the factory is tied to a software agent that knows the equipment's abilities. These agents, and the equipment's controllers, are plugged into the factory's network. They then receive requests-for-bids, bid when appropriate and schedule jobs when bids are accepted. Similarly, if a machine breaks down, its agent is unavailable for bidding on, and hence scheduling, any new jobs. This results in the scheduler working around machine downtime as well as instantly bringing any new equipment into the factory fold.

14.3.2.6 Advance Scheduling in AARIA

Advance scheduling in AARIA is performed in three forms:

1) insertion of a new job into an already existing schedule,
2) optimization of the current schedule,
3) fault recovery.

Optimization is performed continuously in the background. New job insertion is performed in the foreground so as to have the quickest possible response to customers. Fault recovery is performed as significant faults occur.

This scheduling is on-line and interactive as compared to most other scheduling systems which run off-line and in a batch mode. Insertion, optimization, and fault recovery are performed using net-change (dynamic/network) rescheduling techniques. Scheduling is performed recognizing the finite capacities of people and machines. Costs of holding inventories and of using different routings at different times are directly integrated in the scheduling decisions.

AARIA uses a mixture of heuristic scheduling techniques: forward/backward scheduling, simulation scheduling, and intelligent scheduling. The forward/backward scheduling is more precisely forward/backward continuum scheduling, where not only the forward and backward schedules are considered but also the continuum of schedules in-between these schedules. The interaction with the customer determines which schedule in the continuum is chosen. Scheduling is performed by job (during insertion, optimization, and recovery), by resource (during optimization, and recovery), and by operation (during optimization and recovery). Scheduling decisions are made to minimize costs over time and production quantities.

14.3.3 Implementation Details

A demonstration was developed by the project team, which:

- was based on the operations at Rock Island Army Arsenal [http://www.ria.army.mil/]
- used about 35 machines, 15 operators, 275 parts
- had about 500 agents evenly distributed over 8 Pentium PCs

The AARIA system was composed of three basic building blocks:

- agents (control agents)
- simulator (plant agents) / plant and user interfaces
- infrastructure for agent creation, communication, and migration (see Section 14.3.5)

14.3.4 Most Significant Results

AARIA was one of earliest projects in distributed manufacturing and electronic commerce to use agent technology both in theoretical studies and practical application development. The most significant results and contributions include:

- the Cybele agent infrastructure with its interesting concepts of 'generic agent superclass' for agent creation, 'household' agent communities for resource agent organization, 'duplication' and 'termination' mechanisms for agent migration, and subject-based addressing for inter-agent location-independent communication.
- The advance scheduling and schedule optimization mechanism (see Baker et al., 1997 for details).

14.3.5 Position with Respect to Other Projects or Future Work

The Cybele Agent Infrastructure

Implementing multi-agent software systems such as AARIA requires infrastructure services including dynamic agent creation, migration, termination, inter-agent communication mechanisms, and so on. To meet such requirements, an infrastructure called Cybele was developed by the AARIA project team members (Baker et al., 1997) (also see Web page: http://www.i-a-i.com/projects/cybele/).

Cybele was designed with emphasis on high-performance, scaleability, flexibility and ease of use. It provides location-independent communication among agents, with efficient, flexible, dynamic multicasting. It provides services for creating, migrating, and terminating agents at run time. It supports threaded and non-threaded agents and both proactive and reactive behavior. It also provides an activity model where an agent can work on multiple tasks concurrently.

The prototype of Cybele was implemented on the NextStep platform in Objective C, with some support for C and C++. It can run heterogeneously on Sun and Intel hardware. It was also being tested to support other programming languages and platforms, such as C++ and JAVA on UNIX and Windows NT.

Cybele provides a generic Agent superclass that implements the generic data and behavior common to all agents. Custom agents are implemented as subclasses of the Agent superclass. This approach facilitates code reuse and rapid development of agent applications. A typical agent needs to deal with multiple activities at the same time, such as incoming inquiries for bids, monitoring of current jobs, as well as background optimization. The activity model provided by Cybele allows an agent to concurrently deal with multiple activities in a simple, flexible way. It also provides a prioritization mechanism to let an agent decide how it should split its attention among its current activities.

In the agent framework of Cybele, each agent lives on an independent thread within an agent community. Each community is implemented as a process on the operating system. Figure 14.10 shows the basic agent household. Each household is managed by a mother agent that represents the availability of computational resources to her child agents. A mother agent is responsible for creation, termination, and migration of agents in her community. An agent is migrated by creating a duplicate on another mother agent, and then terminating the original agent. A mother agent can load the code for an agent class dynamically at run time.

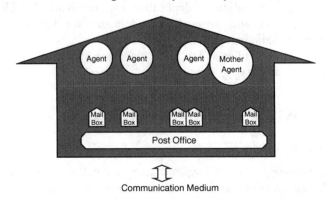

Communication Medium

Figure 14.10 Cybele agent communities (Baker et al., 1997).

Each computer on the network can have one or more agent households. The agents in these can then work together as a single agent society (see Figure 14.11). Each mother agent maintains what agents are in her household. However, child agents view the complete distributed system as a single virtual household of agents with which they can interact. The inter-agent communication mechanisms are location-independent, thus the dialogs among agents do not break as agents move from one computer to another.

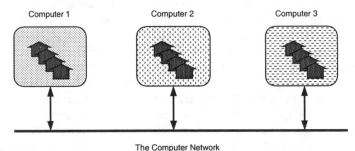

Figure 14.11 Cybele agent deployment (Baker et al., 1997).

Inter-agent communication in Cybele employs subject-based addressing, which provides flexible, location-independent communication, via efficient, dynamic multicasting. In this paradigm of subscribe/publish, an agent notifies his local mailman of the subjects in which he is interested. Each time someone publishes a message to one of these subjects, the local mailman delivers a copy of the message to the agent's mailbox, and notifies him. The agent can examine his mailboxes anytime. Optionally, he can register a callback with each subscription, specifying how to deal with incoming messages to that subscription. The callback is triggered automatically by the infrastructure. Subject-based addressing is independent of the location of the sender and receiver agents, in fact, the sender does not need to know how many receivers there are or where they are located on the network. A subject is an arbitrary string of characters with few syntactic restrictions. By utilizing hierarchies of subjects and wild cards, an agent can flexibly address a group of agents.

Physical message delivery was implemented as a true broadcast, where the message traverses the network only once, instead of a separate message going to each computer. The preliminary evaluations showed that the multicasting mechanism performed comparably to point-to-point communication mechanisms. The broadcast capability also has an inherent scaleability as messages do not have to be resolved through a centralized name server.

14.4 ADDYMS

14.4.1 Origin, Domain of Application, Research Objectives

ADDYMS stands for Architecture for Distributed Dynamic Manufacturing Scheduling (Butler and Ohtsubo, 1992). It was one of the earliest applications of agent technology to distributed dynamic manufacturing scheduling.

14.4.2 Technical features

14.4.2.1 Overall Approach and System Architecture

Scheduling in ADDYMS is decomposed into two levels: the first level assigns manufacturing workcells to tasks, and the second shares local resources, such as workers, portable equipment and tools, among workcells. Negotiation between agents is used for carrying out the first level of manufacturing scheduling in a distributed fashion. For the second level, a dynamic procedure is utilized for allocation of shared local resources.

Corresponding to above two levels, there are two kinds of agents: site agents and resource agents.

- A site agent is responsible for the scheduling of work at a particular manufacturing workcell. This workcell may be a robot, or collection of robots in the case of an FMS, or may simply be a large assembly area in the case of a ship construction or aircraft construction facility. There may also be sub-workcells derived from parent work cells. For each workcell or sub-workcell, there is a site agent responsible for scheduling.
- A resource agent (also called scheduling agent) is used to represent each local resource (a machine, a worker or a tool) that could be allocated to a workcell.

Figure 14.12 shows the architecture of ADDYMS.

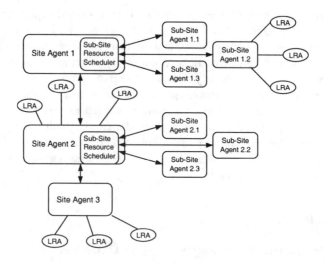

Figure 14.12 Architecture of ADDYMS (Butler and Ohtsubo, 1992) (©1992 AAAI).

14.4.2.2 Agent Structure

A site agent consists of the components shown in Figure 14.13. Associated with each site agent is a database of abilities representing the type of manufacturing production that can occur at the workcell. This database is referenced during scheduling when a site agent desires to know if it is capable of a job it has been notified of.

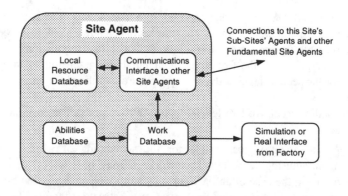

Figure 14.13 Composition of a site agent (Butler and Ohtsubo, 1992) (©1992 AAAI).

A site agent has knowledge both of the operations being executed at its location and being queued for execution. A site agent holds the addresses of the site agents corresponding to the previous and next sites if they exist, as well as of the head of the list of sub-sites of which it is composed.

14.4.2.3 Inter-Agent Communication

In its prototype implementation where site agents were represented by objects, inter-agent communication is realized through message passing among objects.

14.4.2.4 Dynamics / Global Behavior

With the proposed framework, scheduling may be performed incrementally in real time, or a predictive schedule may be generated through simulation. As input to the schedule, a hierarchical description of the work to be performed is provided as a tree of subtasks, thus there is no need for task decomposition capabilities within the scheduler. This input would be generated from the process planning stage of design.

Tasks are distributed to appropriate sites through recursive allocation. With recursive allocation, a site agent receiving a task for execution may further sub-divide the task and through communication with other site agents establish sites for executing parts of the task. If the site agent's site is composed of sub-sites, there will be a site agent corresponding to each sub-site.

The mechanism employed to coordinate the actions of the site agents and their sub-site agents is negotiation and the protocol of the negotiation is similar to that of the Contract Net (Smith, 1980).

When a sub-site agent receives a message from its parent agent announcing the availability of a task for execution, it first determines from its list of capabilities whether or not it is able to perform the task. If it is capable, it then determines whether or not it desires to perform the task. It next responds to the sender of the message and in the case of an affirmative answer includes requested information such as the time at which it can commence execution. All responses are evaluated by the sender agent, according to specified criteria, so that the most appropriate site is chosen.

Local resource allocation occurs dynamically. When an operation is completed, all local resource agents that had resources participating in the operation's execution are notified. Each local resource agent then examines the list of *waiting* and current operations of the site at which its local resource currently resides. If the local resource is shared among a number of sites, it considers the operations at all sites that have access to it. An operation is chosen from this list based on a heuristic, such as minimizing work in progress, and the operation is moved into the *current* list of the site. If the operation needs to employ more local resources, a copy of the operation is placed in the *current* list, with a copy remaining in the *waiting* list.

14.4.3 Implementation Details

ADDYMS was implemented in C++ and its graphical user interface was constructed using InterViews – a GUI toolkit developed at Stanford.

A prototype of ADDYMS was implemented by representing site agents as objects in an object-oriented framework. The site agents could be distributed across a network of workstations connected through a local network.

14.4.4 Most Significant Results

ADDYMS was used to investigate the ability of distributed schedulers to deal with new requirements, such as flexibility, imposed upon manufacturing scheduling by the rapid spread of FMS across the manufacturing industry. The most significant results of ADDYMS include:

- an agent-based prototype architecture for distributed dynamic manufacturing scheduling;
- a dynamic local resource allocation mechanism for dynamic scheduling.

14.5 OTHER RELEVANT PROJECTS

Shaw may have been the first to propose using agents in manufacturing scheduling and factory control. He proposed that a manufacturing cell could subcontract work to other cells through a bidding mechanism (Shaw and Whinston, 1983; Shaw, 1988). YAMS (Yet Another Manufacturing System) (Parunak, 1987) was another early agent-based manufacturing system, in which each factory and factory component was represented as an agent. Each agent had a collection of plans representing its capabilities. The Contract Net was used for inter-agent negotiation.

Shen and Norrie (1999) carried out an extensive survey of relevant projects in this area. Table 14.2 summarizes these projects and lists their main characteristics.

Table 14.2 Projects on agent-based manufacturing planning, scheduling and control.

Project	Group & Reference	Main characteristics
AARIA	Parunak et al., 1998 ITI	Uses autonomous agents to represent physical entities, processes and operations
ABACUS	McEleney et al., 1998 UCB	Uses functional agents; BDI approach for agent design
ADDYMS	Butler & Ohtsubo, 1992	Agents represent physical resources; dynamic local resource scheduling
AMACOIA	Sprumont & Muller 1996 U. of Neuchatel	Uses simulated annealing to search problem space
AMC	Goldsmith & Interrrante, 1998, Sandia Lab	Uses physical agents: part agents and machine agents
CAMPS	Miyashita, 1998	Repair-based methodology together with constraint-based mechanism
CORTES	Sadeh & Fox, 1989, Sycara et al., 1991, CMU	Micro-opportunistic techniques for solving scheduling problems
DAS	Burke & Prosser, 1994 U. of Strathclyde	Hierarchical architecture with agents representing resources, resource groups, and a whole scheduling process
I-Control	Brennan et al., 1997, Wang et al., 1998, U of Calgary	Partial Dynamic Control Hierarchy (PDCH); Uses agents to model IEC-1499 Functional Blocks (IEC 1997)
IFCF	Lin and Solberg, 1992 Purdue	Resource agents representing physical resources; market-like control model

Table 14.2 (continued)

LMS	Fordyce & Sullivan, 1994	Uses functional agents; voting protocol for communication
MAPP	Hayes, 1998 U. of Minnesota	Combination of sequential and blackboard architectures
MASCADA	Bruckner et al., 1998 Daimler-Benz AG, KULeuven	Emergent Behavior in Manufacturing Control; Proactive Disturbance Handling; Hot Plugable Agents
MASCOT	Parunak, 1993 ITI	A shared ontology & a base set of realistic modules
MetaMorph	Maturana & Norrie, 1996 U of Calgary	Mediator-centric architecture; dynamic clustering & cloning; learning
Reagere	Berry & Kumura, 1998 Penn State U.	Based on blackboard architecture
Sensible Agents	Barber et al., 1998 U of Texas at Austin	Implemented as CORBA objects communicating through ILU object environment
SFA	Parunak, 1996 NCMS, 1998	Real manufacturing applications
YAMS	Parunak, 1987 ITI	One of earliest applications in the domain
?[2]	Baker, 1991 U. of Cincinnati	Market-Driven Contract Net; forward & backward scheduling
?	Choi and Park, 1997	An economical method for developing intelligent agent systems
?	Duffie & Piper, 1986 Wisconsin	Agents represent physical resources, parts, and humans; part-oriented scheduling
?	Fischer, 1994 DFKI	Hierarchical layered architecture
?	Hasegawa et al., 1994 Toshiba	Uses HMS approach
?	Interrante & Goldsmith, 1998, Sandia Lab	Type-A, Type-B and Type-C agents
?	Saad et al., 1995 Vanderbilt	Production Reservation approach; machine-centered & part-centered negotiation
?	Kouiss et al., 1997	Each agent represents a work center
?	Liu & Sycara 1994, 1995 CMU	CP&CR (Constraint Partition and Coordinated Reaction) for constraint satisfaction; Anchor&Ascend for distributed constraint optimization
?	Murthy et al., 1997	A-team architecture
?	Ouelhadj et al., 1998 U. of Toulouse	Agents represent physical resources
?	Patriti et al., 1997, Schaefer et al., 1996, CRAN GGP	Layered architecture; uses different mechanisms at different levels
?	Sousa & Ramos, 1997 ISEP/IPP	Uses HMS approach; dynamic scheduling
?	Tseng et al., 1997 HKUST	Market-like model for manufacturing control with agents representing resources

---

[2] ? No project name found through available publications.

REFERENCES

Baker, A.D. (1991) Manufacturing Control with a Market-Driven Contract Net, PhD Thesis, Rensselaer Polytechnic Institute.

Baker, A.D., Parunak, H.V.D. and Erol, K. (1997) Manufacturing over the Internet and into Your Living Room: Perspectives from the AARIA Project, Tech. Report TR 208-08-97, Department of Electrical & Computer Engineering and Computer Science, University of Cincinnati.

Baker, A.D., Parunak, H.V.D. and Erol, K. (1999) Agents and the Internet: Infrastructure for Mass Customization, *IEEE Internet Computing*, 3(5):62-69.

Barber, S., White, E., Goel, A., Han, D., Kim, J., Li, H., Liu, T.H., Martin, C.E. and McKay, R. (1998) Sensible Agent Problem-Solving Simulation for Manufacturing Environments, In *Proc. of AI & Manufacturing Research Planning Workshop*, Albuquerque, NM, The AAAI Press, 1998, pp. 1-9.

Berry, N.M. and Kumura S. (1998) Evaluating the Design and Development of Reagere, In *Working Notes of the Agent-Based Manufacturing Workshop*, Minneapolis, MN, pp. 5-13.

Balasubramanian, S., Maturana, F. and Norrie, D. (1996) Multi-agent planning and coordination for distributed concurrent engineering, *International Journal of Cooperative Information Systems*, 5(2-3):153-179.

Brennan, R.W., Balasubramanian, S. and Norrie, D.H. (1997) Dynamic Control Architecture for Advanced Manufacturing Systems, In *Proceedings of International Conference on Intelligent Systems for Advanced Manufacturing*, Pittsburgh, PA.

Brückner, S., Wyns, J., Peeters, P. and Kollingbaum, M. (1998) Designing Agents for the Manufacturing Process Control, In *Proc. of Artificial Intelligence and Manufacturing Research Planning Workshop – State of the Art & State of the Practice*, AAAI Press, Albuquerque, NM, pp. 40-46.

Burke, P. and Prosser, P. (1994) The Distributed Asynchronous Scheduler, In M. Zweben and M.S. Fox, eds., *Intelligent Scheduling*, Morgan Kaufman Publishers, San Francisco, CA, pp. 309-339.

Butler, J. and Ohtsubo, H. (1992) ADDYMS: Architecture for Distributed Dynamic Manufacturing Scheduling, In A. Famili, A., D.S. Nau and S.H. Kim, eds., *Artificial Intelligence Applications in Manufacturing*, The AAAI Press, pp. 199-214.

Choi, H.S. and Park, K.H. (1997) Shop-Floor Scheduling at Shipbuilding Yards Using the Multiple Intelligent Agent System, *Journal of Intelligent Manufacturing*, 8(6):505-515.

Decker, K. (1995) Environment Centered Analysis and Design of Coordination Mechanisms, Ph.D. Thesis, Dept. of Computer Science, University of Massachusetts, Amherst.

Duffie, N.A. and Piper, R.S. (1986) Non-Hierarchical Control of Manufacturing Systems, *Journal of Manufacturing Systems*, 5(2):137-139.

Finin, T., Fritzon, R., Mckay, D. and Mcentire, R. (1993) KQML – A Language and Protocol for Knowledge and Information Exchange, Tech. Report, University of Maryland.

Fischer, K. (1994) The Design of an Intelligent Manufacturing System, In *Proceedings of the 2nd International Working Conference on Cooperating Knowledge-based Systems*, University of Keele, pp. 83-99.

Fordyce, K. and Sullivan, G.G. (1994) Logistics Management System (LMS): Integrating Decision Technologies for Dispatch Scheduling in Semiconductor Manufacturing, In M. Zweben and M.S. Fox, eds., *Intelligent Scheduling*, Morgan Kaufman Publishers, San Francisco, CA, pp. 473-516.

Gaines, B., Norrie, D., and Lapsley, A. (1995) Mediator: an Intelligent Information System Supporting the Virtual Manufacturing Enterprise, In *Proceedings of 1995 IEEE International Conference on Systems, Man and Cybernetics*, New York, pp. 964-969.

Goldsmith, S.Y. and Interrante, L.D. (1998) An Autonomous Manufacturing Collective for Job Shop Scheduling, In *Proc. of AI & Manufacturing Research Planning Workshop*, Albuquerque, NM, The AAAI Press, pp. 69-74.

Hasegawa, T., Gou, L., Tamura, S., Luh, P.B. and Oblak, J.M. (1994) Holonic Planning and Scheduling Architecture for Manufacturing, In *Proceedings of the 2nd International Working Conference on Cooperating Knowledge-based Systems*, University of Keele.

Hayes, C.C. (1998) MAPP: an Agent Organization for Process Planning, In *Working Notes of the Agent-Based Manufacturing Workshop*, Minneapolis, MN, pp. 32-40.

Interrante, L. and Glodsmith, S. (1998) Emergent Agent-Based Scheduling of Manufacturing Systems, In *Working Notes of the Agent-Based Manufacturing Workshop*, Minneapolis, MN, pp. 47-56.

Kouiss, K., Pierreval, H., and Mebarki, N. (1997) Using multi-agent architecture in FMS for dynamic scheduling, *Journal of Intelligent Manufacturing*, 8(1):41-47.

Lin, G.Y.-J. and Solberg, J.J. (1992) Integrated Shop Floor Control Using Autonomous Agents, *IIE Transactions: Design and Manufacturing*, 24(3):57-71.

Liu, J. and Sycara, K.P. (1994) Distributed Problem Solving through Coordination in a Society Of Agents, In *Proceedings of the 13th International Workshop on DAI*.

Liu, J. and Sycara, K.P. (1995) Exploiting Problem Structure for Distributed Constraint Optimization, In *Proceedings of ICMAS95*, San Francisco, CA, pp. 246-253.

Maturana, F. and Norrie, D.H. (1996) Multi-Agent Mediator Architecture for Distributed manufacturing, *Journal of Intelligent Manufacturing*, 7:257-270.

Maturana, F. (1997) MetaMorph: An Adaptive Multi-Agent Architecture for Advanced Manufacturing Systems, Ph D thesis, The University of Calgary.

Maturana, F., Shen, W. and Norrie, D.H. (1999) MetaMorph: An Adaptive Agent-Based Architecture for Intelligent Manufacturing, *International Journal of Production Research*, 37(10):2159-2174.

McEleney, B., O'Hare, G.M.P. and Sampson, J. (1998) An Agent Based System for Reducing Changeover Delays in a Job-Shop Factory Environment, In *Proc. of PAAM'98*, London, UK, pp. 591-613.

Miyashita, K. (1998) CAMPS: A Constraint-Based Architecture for Multi-Agent Planning and Scheduling. *Journal of Intelligent Manufacturing*, 9(2):147-154.

Murthy, S., Akkiraju, R., Rachlin, J. and WU, F. (1997) Agent-Based Cooperative Scheduling, In *Proceedings of AAAI Workshop on Constrains and Agents*, AAAI Press, pp. 112-117.

NCMS. (1998) Shop Floor Agents. NCMS, http://www.ncms.org/dr/REP/Rep_Project.idc

Ouelhadj, D. and Bouzouia, B. (1998) Multi-Agent System for Dynamic Scheduling and Control in Manufacturing Cells, In *Working Notes of the Agent-Based Manufacturing Workshop, Minneapolis*, MN, pp. 96-105.

Parunak, V.D. (1987) Manufacturing Experience with the Contract Net, In M.N. Huhns, ed., *Distributed Artificial Intelligence*, Pitman, pp. 285-310.

Parunak, V.D. (1993) MASCOT: A Virtual Factory for Research and Development in Manufacturing Scheduling and Control, Tech Memo 93-02, Industrial Technology Institute.

Parunak, H.V.D. (1995) Case Grammar: A Linguistic Tool for Engineering Agent-Based Systems, Industrial Technology Institute, [http://www.iti.org/~van/casegram.ps]

Parunak, V.D. (1996) Workshop Report: Implementing Manufacturing Agents, Sponsored by the SFA project of NCMS in conjunction with PAAM'96, NCMS.

Parunak, H.V.D., Baker, A.D. and Clark, S.J. (1998) The AARIA Agent Architecture: From Manufacturing Requirements to Agent-Based System Design, In *Working Notes of the Agent-Based Manufacturing Workshop*, Minneapolis, MN, pp. 136-145.

Parunak, H.V.D. (1998) What Can Agents Do in Industry, and Why? An Overview of Industrially-Oriented R&D at CEC, In M. Klusch and G. Weiss, eds., *Cooperative Information agents II: learning, mobility and electronic commerce for information discovery on the Internet,* Lecture Notes in Computer Science 1435, Springer-Verlag, pp. 1-18.

Patriti V., Schaefer K., Ramos M., Charpentier P., Martin P. and Veron M. (1997) Multi-agent and manufacturing: A multilevel point of view, In *Proceedings of CAPE'97*, Detroit.

Saad, A., Biswas, G., Kawamura, K., Johnson, M.E. and Salama, A. (1995) Evaluation of Contract Net-Based Heterarchical Scheduling for Flexible Manufacturing Systems, In *Proceedings of Intelligent Manufacturing Workshop at IJCAI'95*, Montreal, pp. 310-321.

Sadeh, N. and Fox, M.S. (1989) CORTES: An Exploration into Micro-Opportunistic Job-Shop Scheduling, In *Proceedings of Workshop on Manufacturing Production Scheduling at IJCAI-89*, Detroit.

Schaefer, K., Patriti, V., Charpentier, P. Martin, P. and Spath, D. (1996) The Multi-Agent Approach in Scheduling and Control of Manufacturing Systems, In *Proceedings of PAAM'96*, London, UK.

Shaw, M.J. and Whinston, A.B. (1983) Distributed Planning in Cellular Flexible Manufacturing Systems, Tech. Report, Management Information Research Center, Purdue University.

Shaw, M.J. (1988) Dynamic Scheduling in Cellular Manufacturing Systems: A Framework for Networked Decision Making, *Journal of Manufacturing Systems*, 7(2):83-94.

Shen, W., Xue, D. and Norrie, D.H. (1998a) An Agent-Based Manufacturing Enterprise Infrastructure for Distributed Integrated Intelligent Manufacturing Systems, In *Proc. of PAAM'98,* London, UK, pp. 533-548.

Shen, W., Maturana, F. and Norrie, D.H. (1998b) Learning in Agent-Based Manufacturing Systems, In *Proc. of AI & Manufacturing Research Planning Workshop*, Albuquerque, NM, The AAAI Press, pp. 177-183.

Shen W. and Norrie, D.H. (1999) Agent-Based Systems for Intelligent Manufacturing: A State-of-the-Art Survey, *Knowledge and Information Systems: an International Journal*, 1(2):129-156.

Smith, R.G. (1980) The Contract Net Protocol: High-Level Communication and Control in a Distributed Problem Solver, *IEEE Transactions on Computers*, C-29(12):1104-1113.

Sousa, P. and Ramos, C. (1997) A Dynamic Scheduling Holon for Manufacturing Orders, *Journal of Intelligent manufacturing*, 9(2):107-112.

Sprumont, F. and Muller, J.P. (1996) AMACOIA: A Multi-Agent System for Designing Flexible Assembly Lines, In *Proceedings of PAAM'96*, London, UK, pp. 573-585.

Sycara, K.P., Roth, S.F., Sadeh, N. and Fox, M.S. (1991) Resource Allocation in Distributed Factory Scheduling, *IEEE Expert*, 6(1):29-40.

Tseng, M., Lei, M., and Su, C. (1997) A collaborative control system for mass customisation manufacturing, *CIRP Annals*, 46(1):373-376.

Wang, L., Balasubramanian S. and Norrie D. (1998) Agent-based Intelligent Control System Design for Real-time Distributed Manufacturing Environments, In *Working Notes of the Agent-Based Manufacturing Workshop*, Minneapolis, MN, pp. 152-159.

Zweben, M. and Fox, M.S. eds., *Intelligent Scheduling,* Morgan Kaufman Publishers, San Francisco, CA.

15

Agent-Based Enterprise Integration and Supply Chain Management

15.1 INTRODUCTION

Enterprise Integration means that each unit of the organization will not only have access to information relevant to its tasks but will understand how its actions will impact other parts of the organization, thereby enabling it to choose alternatives that optimize the organization's goals.

The Supply Chain of a manufacturing enterprise is the worldwide network of suppliers, factories, warehouses, distribution centers, and retailers through which raw materials are acquired, transformed and delivered to customers (Fox et al., 1993b). Improving supply chain management is a key strategy for increasing the enterprise's competitive position and profitability. Consequently, enterprises are moving towards increasingly integrating their activities with those of their suppliers, customers and partners within wide supply chain networks.

New approaches to this integration have been proposed and developed in the past few years: Virtual Enterprises, Virtual Factories, and Electronic Commerce. These embody many concepts of the Supply Chain, and thus overlap with it and with each other. There are so far no widely accepted definitions of these terms and, in fact, a variety of definitions exist, from different points of view (Camarinha-Matos et al., 1997; Larsen, 1999). We will not discuss these definitions further in this book, but refer the reader to the literature in this area.

The approaches presented in this chapter for Supply Chains are also applicable to Virtual Enterprises[1], Virtual Factories or Electronic Commerce. In fact, similar approaches can be used to implement both enterprise integration and supply chain networks. The main difference is their implementation levels: enterprise integration is at intra-organization level, while supply chain is at inter-organization level. Agent-based technology provides a natural way to design and implement such environments at both levels.

In the following sections, we review selected projects in agent-based enterprise integration and supply chain management.

---

[1] A Virtual Enterprise is a flexible dynamic network of independent organizations that collaborate to exploit worldwide production or marketing opportunities. It is supported by electronic commerce technologies and can reconfigure quickly in response to changing conditions.

15.2 ISCM (INTEGRATED SUPPLY CHAIN MANAGEMENT)

15.2.1 Origin, Domain of Application, and Research Objectives

The Integrated Supply Chain Management Systems (ISCM) project (Fox et al., 1993b; Barbuceanu and Fox, 1995a) is an extensive research program at the University of Toronto which considers the manufacturing enterprise as a network of operational nodes, and which facilitates the decentralization of control using agent technology. A major focus is on the flow of material through the supply chain during the operation phase of the life cycle. A recent description of this project can be found in (Barbuceanu and Fox, 1997).

15.2.2 Technical Features

15.2.2.1 Overall Approach and System Architecture

ISCM addresses the coordination problems at both the tactical and operational levels. The supply chain is organized as a network of cooperating intelligent agents, each performing one or more supply chain functions, and each coordinating their actions with other agents. According to the different enterprise functions, ISCM distinguishes eight types of agents (Figure 15.1): seven functional agents (Order Acquisition, Logistics, Scheduling, Resource, Dispatching, Transportation and Plant Management) and one Information Agent. The Information Agent is a central communication resource that provides knowledge management, conflict resolution, and coordination support to the other agents.

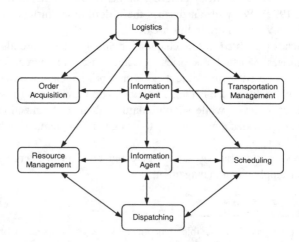

Figure 15.1 The ISCM agents (Fox et al., 1993b).

A supply chain has four agents at the enterprise level (Information Agent, Logistics, Order Acquisition and Transportation agents) and five at each plant (Information Agent, Plant Management, Resource Management, Dispatching and Scheduling agents), so a typical system can have 40–50 agents.

15.2.2.2 Agent Structure

The eight types of agents are distinct from one another, but make use of the common resources of the Agent Building Shell (Barbuceanu and Fox, 1996). A generic agent is composed of several layers of languages and services including Agent Communication Language, Information Distribution service, Organizational Model, Coordination Models and Language, Conflict Management service, Generic Interfaces to Applications, and the Knowledge Management System (Figure 15.2).

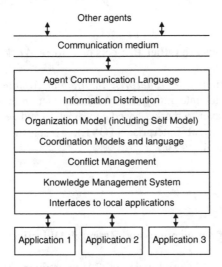

Figure 15.2 Architecture of a generic agent for ISCM
(Barbuceanu and Fox, 1994).

15.2.2.3 Inter-Agent Communication

KQML (Finin et al., 1993) and KIF (Genesereth and Fikes, 1992) are used as knowledge level communication protocols. TCP/IP is used as a low-level transport mechanism.

15.2.2.4 Dynamics / Global Behavior

ISCM agents model and reason explicitly with respect to one another and their environment, and use AI techniques to plan how to move from the current state of the world to a desired state. The relations among agents within a single plant, and between these agents and the enterprise level functional agents are defined by the architecture, but the course of a given conversation will vary dynamically depending on the state and capability of the individual agents. ISCM is coordinated by propagation and resolution of constraints among agents.

15.2.3 Most Significant Results

A generic Agent Building Shell (ABS) was developed to support agent construction by providing several layers of reusable services and languages (Barbuceanu and Fox, 1996). A domain independent COOrdination Language (COOL) was developed to provide services for defining coordination models and protocols (Barbuceanu and Fox, 1995b).

15.2.4 Position With Respect to Other Projects or Future Work

ISCM was based on the multi-layer *Enterprise Information Architecture* proposed by Roboam and Fox (1992) for integrating enterprise information processing, from the network communication layer to market based coordination and negotiation. ISCM is related to the Toronto Virtual Enterprise (TOVE), a sophisticated model of the business enterprise (Fox et al., 1993a).

15.3 CIIMPLEX

15.3.1 Origin, Domain of Application and Research Objectives

The Consortium for Intelligent Integrated Manufacturing Planning-Execution (CIIMPLEX) (Peng et al., 1998), consisting of several private companies and universities, was formed in 1996 to develop technologies for intelligent enterprise-wide integration of planning and execution for manufacturing. Recent information about this project can be found in (Peng et al., 1999) and through the Internet at: http://www.cs.umbc.edu/lait/research/ciimplex/.

15.3.2 Technical Features

15.3.2.1 Overall Approach and System Architecture

The system is composed of service agents (e.g., Agent Name Server, Broker Agent, Gateway Agent, etc.) and special agents each of which possesses specialized expertise for a particular application. Figure 15.3 shows the multi-agent system architecture for a CIIMPLEX enterprise.

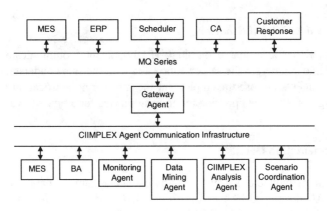

Figure 15.3 CIIMPLEX integration architecture (Peng et al., 1998).

15.3.2.2 Agent Types and Structure

There are two primary categories of agents: service agents and special agents (or domain specific agents). The service agents are to facilitate agent collaboration and are of three types: Agent Name Server; Broker Agent; and Gateway Agent. The Agent Name Server maintains an address table of all registered agents, accessible through the agents' symbolic names. The Broker Agent serves as a dynamic information hub or switchboard. The architectures of the Agent Name Server and the Broker Agent are quite simple, basically consisting of a communication interface for handling incoming and outgoing messages and a simple inference engine with a database/knowledge base. The Gateway Agent is used to provide an interface between the agent world and the application world. Its functions include making connections between the two transport mechanisms (TCP/IP and MQ Series) and converting messages between the two different formats (KQML and BOD). It is, basically, a specially developed program. However, it could be replaced by appropriate types of service agent (Peng et al., 1999).

Some of the special agents provide special services in a particular scenario. They are usually specially developed independent programs or programs for encapsulating existing tools.

15.3.2.3 Inter-Agent Communication

TCP/IP is used as a low-level transport mechanism, KQML as a common communication language, KIF as a common content format, BOD (Business Object Document by the Open Application Group) as the message format between applications, and MQ Series of IBM and VisualFlow of Envisionit for communication to and from applications.

15.3.2.4 Dynamics / Global Behavior

With KQML, KIF and a shared ontology (BOD), agents communicate with each other in the same manner, in the same syntax, and with the same understanding of the world. Broker agents are used to implement the monitoring/notification architecture to track the agents which can provide monitoring service for a type of event.

15.3.3 Implementation Details

The agent system architecture is supported by the agent communication infrastructure called JACKAL developed by the consortium. JACKAL stands for **J**ava to support **A**gent **C**ommunication using the **KQML A**gent communication **L**anguage. The main components of JACKAL are the Cache, Message Handler, Conversation Arena, Splitter, Distributor and Switchboard for handling incoming and outgoing messages.

15.3.4 Most Significant Results

This research project took an important step toward agent based enterprise integration for manufacturing planning and execution, through experimentation with a real manufacturing scenario involving a real legacy MES and scheduler.

15.4 AIMS

15.4.1 Origin, Domain of Application and Research Objectives

The AIMS (Agile Infrastructure for Manufacturing Systems) (Park et al., 1993) project was established to develop processes that would assist suppliers to successfully thrive in a customer-dominated marketplace characterized by continuous and unanticipated change. More specifically, this project was to facilitate agile manufacturing through negotiation over the Internet. Its objective was to develop an appropriate information infrastructure and to provide services to facilitate Agile Manufacturing. The AIMS project team, led by Lockheed Martin, included Texas Instruments, Rockwell, several small businesses, academia, and a trade association. AIMS was the first major pilot project in the DoD's Agile Manufacturing Program.

15.4.2 Technical Features

15.4.2.1 Overall Approach and System Architecture

AIMS is a network of certified manufacturing services, spanning numerous companies, linked via the Internet. A certified supplier is one that employs standard business processes, product data formats, and network interfaces. Customers interact with these suppliers using structured messages, requesting information about costs, capabilities, and availability, soliciting bids, placing orders, and so forth. They can communicate with a provider directly or via third-party directory and brokering services.

15.4.2.2 Agent Structure

AIMS agents encapsulate production resources at any level, transforming them into suppliers capable of responding to service requests.

15.4.2.3 Inter-Agent Communication

Structured messages via E-mail and TCP/IP protocol.

15.4.2.4 Dynamics / Global Behavior

A production service is itself a virtual enterprise, relying on other network services to do its work. AIMS agents can use brokers to locate appropriate providers, or to bid on jobs for themselves. They can also call on providers of manufacturing support services, such as process planning and scheduling. In AIMS, services can be added or taken out at any time, with incremental impact on capacity, simply by informing the appropriate directories and brokers. Resources can be allocated on-demand and loads continually balanced through an open bidding process.

15.4.3 Implementation Details

An experimental software architecture was developed to support the negotiation process over the Internet. This architecture used the World Wide Web as the medium of exchange, and the Web browser (initially Mosaic) as the user interface.

15.4.4 Most Significant Results

AIMS was the first major pilot project in the DoD's Agile Manufacturing Program.

15.4.5 Position With Respect to Other Projects or Future Work

AIMS is closely related to other research projects under DARPA MADE program.

15.5 MetaMorph II

15.5.1 Origin, Domain of Application, and Research Objectives

The MetaMorph II project commenced in early 1997. Its objective was to integrate the manufacturing enterprise's activities of design, planning, scheduling, simulation, execution, and so on, with those of its suppliers, customers and partners within a distributed intelligent open environment (Shen et al., 1998). It was based on previous research in MetaMorph I (also referred as MetaMorph, see Chapter 14, Section 14.2), DIDE (see Chapter 13, Section 13.5), and ABCDE (Balasubramanian et al., 1996).

15.5.2 Technical Features

15.5.2.1 Overall Approach and System Architecture

MetaMorph II uses a hybrid agent-based architecture. In this architecture, the manufacturing system is organized at the highest level through 'subsystem' mediators (Figure 15.4). Each subsystem is integrated within the system through a mediator. Each subsystem can be an agent-based system (such as an agent-based manufacturing scheduling system), or any other type of system (such as a feature-based design system or a knowledge-based material management system). Agents in a subsystem may also be autonomous agents at the subsystem level. Some of these agents may also be able to communicate directly with other subsystems or agents in other subsystems.

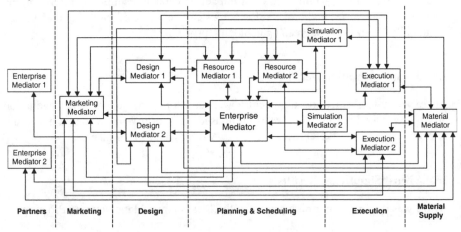

Figure 15.4 Functional architecture of MetaMorph II (Shen et al., 1998).

15.5.2.2 Agent Structure

As in MetaMorph I, mediators are also agents, so can also be called mediator agents. The basic functionality of a mediator has been discussed in Chapter 7 (Section 7.4.2.4) and also in Chapter 14 (Section 14.2). To have the desired functionality (basic functionality and additional functionality described in this section), a mediator needs the following components (Figure 15.5):

- a network interface for connection to the system
- a communication interface for handling incoming and outgoing messages
- knowledge about itself
- detailed knowledge about the agents which it coordinates
- knowledge about other mediators
- knowledge about its environment
- knowledge about the working projects or products to be manufactured
- reasoning mechanisms which use its knowledge
- learning mechanisms for updating its knowledge
- control mechanisms for controlling its actions and events

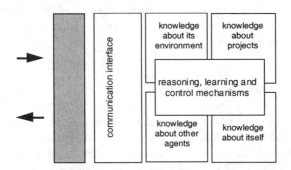

Figure 15.5 Internal structure of a mediator.

A mediator is an agent. A resource agent is also an agent. Some basic aspects of their architectures will therefore be similar. A resource agent, however, does not need all of the detailed knowledge about other agents and mediators, about its environment and the whole project or parts to be manufactured, that a mediator may need.

15.5.2.3 Inter-Agent Communication

Mediator-mediator and mediator-agent communications are asynchronous, and the communication mode can be point-to-point, broadcast, or multicast. The prototype implementation used the multicast mode which can be extended to point-to-point and broadcast. As in the DIDE project, five types of messages are used: *request, inform, announce, bid* and *notice*. The messages are formatted in an extended KQML format (Finin et al., 1993).

In order to realize asynchronous communication, each mediator and resource agent has an input message box. All incoming messages are first queued in this input message box. These incoming messages are treated in accordance with their priority. The message with highest priority is handled first. For messages with same priority, the first arriving message is handled first.

15.5.2.4 Dynamics / Global Behavior

In addition to the features related to dynamics and global behavior provided by MetaMorph I (see Chapter 14, Section 14.2.2.4), the hybrid architecture of MetaMorph II provides several new features which are very useful for enterprise integration and for complex supply chain implementation.

- At the system level, any type of (agent-based or non-agent-based) subsystem can be integrated into the system through a mediator which serves as an interface (similar to a wrapper in some other agent-based concurrent design and manufacturing systems) between this subsystem and other subsystems.
- Agents in an agent-based subsystem may also be autonomous agents at the subsystem level, i.e., some of these agents may be able to communicate directly with other subsystems or agents in other subsystems, rather than always through mediators.

15.5.3 Implementation Details

The prototype implementation of the architecture has four mediators: Enterprise Mediator, Design Mediator, Resource Mediator and Marketing Mediator.

The Enterprise Mediator can be considered as the administration center of the manufacturing enterprise. Other mediators register with this mediator. The Design Mediator is used to integrate a feature-based intelligent design system. The functions of the design module include: (1) generating design candidates from the design functional requirements; (2) modeling design geometry; and (3) representing the design using manufacturing features. The Resource Mediator is used to coordinate an agent-based dynamic manufacturing scheduling subsystem (see Shen and Norrie 1998 for details). In the present prototype implementation, at the resource level, only machine agents and worker agents have so far been implemented. The Marketing Mediator is used to integrate related customer services (the marketing subsystem) into the system. The functions of the marketing subsystem include: (1) using its class libraries (product catalogue) to show different product models on a 3D graphical interface according to the customer's requests; (2) if the requested product cannot be found in its libraries (product catalogue), sending design (functional) requirements to the Design Mediator which then asks the design subsystem to design an appropriate new or modified product; (3) requesting product price for any order with special product specifications or due time.

In this implementation, all scheduling requests derive from a feature-based intelligent design system (Yadav and Xue, 1998) via a Design Mediator. Each product corresponding to a customer order has been decomposed down into manufacturing features. Each manufacturing feature can be realized by a manufacturing task. These manufacturing tasks are organized in a graph data structure representing the sequence to produce the corresponding manufacturing features. Each product is modeled as a part agent containing the information about this product including the feature data, due time, and its plan (which will be assigned with times and resources during the scheduling process).

At the time of writing, the prototype is still under implementation. It is being developed on a network of PCs. The primary development language is VisualWorks™ (Smalltalk). The graphical interface for customer services is being developed using 3D studio™ and Visual C++™. Communication among mediators is realized using the TCP/IP protocol, and inter-mediator messages are formatted in KQML (Finin et al., 1993). The prototype implementation is being developed as a simulation.

15.4.4 Most Significant Results

The proposed hybrid architecture can facilitate the integration of legacy systems at the system level, and may help to avoid communication bottleneck at the subsystem level, since a mediator may become a bottleneck in a pure mediator-centric architecture.

15.4.5 Position With Respect to Other Projects or Future Work

As mentioned above, MetaMorph II was based on MetaMorph I (see Chapter 14), DIDE (see Chapter 13), and ABCDE (Balasubramanian et al., 1996). As a project in the Intelligent Manufacturing Systems Group (IMSG) at the University of Calgary, it is strongly linked with other IMSG projects such as Intelligent Concurrent Design (Yadav and Xue, 1998), Intelligent Control (Brennan et al., 1997, Wang et al., 1998), and HMS related projects (see below).

15.6 HMS (Holonic Manufacturing Systems)

In holonic manufacturing systems, agents can be used to model holons, which are software and hardware entities (Deen, 1994; Christensen et al., 1994; Hasegawa et al., 1994; Biswas et al., 1995). In this section, we introduce the main concepts of the holonic manufacturing systems and discuss some important features. A good discussion on agent technology for holonic manufacturing systems can be found in (Bussmann, 1998).

15.6.1 Origin, Domain of Application, and Research Objectives

The Intelligent Manufacturing Systems (IMS) Research Program is a ten-year research collaboration among industrial organizations, universities and research centers in major industrialized countries (Hayashi, 1993). The HMS project was launched in 1994 by the HMS consortium, under the IMS program. "*Holon*" is a word coined by combining '*holos*' (the whole) and '*on*' (a particle), following Koestler (1967). A holon is defined by the HMS consortium as "an autonomous and cooperative building block of a manufacturing system for transforming, transporting, storing and/or validating information and physical objects". Another important concept "*holarchy*" is defined as "a system of holons which can cooperate to achieve a goal or objective". A Holonic Manufacturing System (HMS) is "a holarchy which integrates the entire range of manufacturing activities from order booking through design, production and marketing to realize the agile manufacturing enterprise". A more detailed description of the HMS consortium and the main HMS concepts can be found in (Van Leeuwen and Norrie, 1997).

15.6.2 Technical Features

15.6.2.1 Overall Approach and System Architecture

In an HMS, each holon's activities are determined through cooperation with other holons, as opposed to being determined by a centralized mechanism. An HMS is a holarchy, comprised of holons which are autonomous and cooperative enitities. In other words, an HMS is a manufacturing system where key elements, such as raw materials, machines, products, parts, AGVs, have autonomous and cooperative properties (Christensen, 1994).

15.6.2.2 Agent Structure

Each individual holon consists of a physical processing component and an information processing component. The physical processing component can be a physical entity such as a product or a machine (or subsystem or device or part) and possesses the processes, structures, and mechanisms necessary for its functioning. The information processing component processes information and has mechanisms for interacting with and managing the physical processing component. A holon may consist of only an information processing component in special circumstances. A holon can be part of another holon.

Agent-oriented techniques can be used to design and implement the information processing part of a holon. Figure 15.6 shows the agent-oriented architecture for a holon proposed by Bussmann (1998).

15.6.2.3 Inter-Agent Communication

Different communication protocols can be used for information exchange among holons. For example, Hasegawa et al (1994) used broadcast communication.

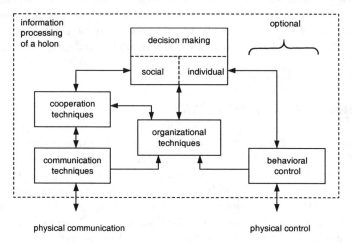

Figure 15.6 Agent-oriented architecture for a holon (Bussmann, 1998).

15.6.2.4 Dynamics / Global Behavior

Holarchy is an important concept for HMS. A holarchy is a system of holons that cooperate to achieve goals or objectives. A holarchy defines the basic rules for cooperation of the holons and thereby limits their autonomy. It is interesting to note that holons within a holarchy may dynamically create or change their relationships, thus, holarchies can be created or changed dynamically.

15.6.3 Implementation Details

Using the same basic holonic concepts, partners in the HMS consortium have developed several different testbeds, using various software and hardware. Most of the research results on HMS are only reported internally in the HMS consortium. However, some work has been reported publicly for materials handling (Christensen et al., 1994), manufacturing planning and scheduling (Hasegawa et al., 1994; Biswas et al., 1995; Sousa and Ramos, 1997), and intelligent manufacturing control (Brennan et al., 1997; Wang et al., 1998). The Proceedings of the Second International Workshop on Intelligent Manufacturing Systems IMS-99 (Van Brussel and Valckenaers, 1999) also contains papers on various aspects of holonic manufacturing systems and enterprise integration.

15.6.4 Most Significant Results

The HMS approach allows integration of appropriate elements of hierarchical and heterarchical systems, within an intelligent and open structure, and shows promise for new types of agile manufacturing system. As an architecture becomes more heterarchical, as is possible in an HMS, the complexity of centralized control is reduced, software modularity is increased, and flexibility and fault-tolerance can be enhanced.

15.7 OTHER RELEVANT PROJECTS

Early projects in this area include the Intelligent Agent (IA) framework developed by Pan and Tenenbaum (1991) for integrating people and computer systems in large, geographically dispersed manufacturing enterprises, and the Enterprise Management Network (EMN) developed by Roboam and Fox (1992) to support integration of enterprise manufacturing activities across the six levels of Network, Data, Information, Organization, Coordination and Market. The latter was subsequently integrated into the ISCM system (see Section 15.2). Later significant research work includes MADEFAST (Cutkosky et al., 1996), which was a DARPA DSO-sponsored project to demonstrate technologies for collaborative engineering developed under the MADE (Manufacturing Automation and Design Engineering) program.

A survey of relevant projects in this area was carried out by Shen and Norrie (1999). Table 15.1 provides a summary of some of these projects.

Table 15.1 Projects on agent-based enterprise integration and supply chain management.

Project	Reseaarch Group	Domain	Main Characteristics
ABMA	Budenske et al., 1998 Architecture Tech Co.	Enterprise Integration	middleware architecture
ADE	Mehra & Nissen, 1998 Gensym Co.	Supply Chain Management	delegation based event handling similar to JavaBeans
AIMS	Park et al., 1993 Lockheed Martin	Agile Manufacturing	uses the Internet
ATP	NIST, 1998 NIST	Agile Manufacturing	plug-and-play framework
CIIMPLEX	Peng et al., 1998 UMBC	Enterprise Integration	Service Agents (Name Server, Facilitator Agent, Gateway Agent)
CLAIM	Malkoun & Kendall, 1997, RMIT	Enterprise Integration	methodology for enterprise integration using agents
IA framework	Pan & Tenenbaum, 1991, Stanford	Enterprise Integration	large number of computerized assistants known as Intelligent Agents
IAO	Kwok & Norrie, 1994 U. of Calgary	Intelligent Manufacturing	a rule based object system for developing intelligent manufacturing software

Table 15.1 (continued)

iDCSS	Klein, 1995 CSS, MIT	Concurrent Engineering	an integrated model that combines existing coordination technologies
ISCM	Fox et al., 1993b U. of Toronto	Supply Chain Management	Agent Building Shell (ABS); Coordination Language (COOL); functional agents
KRAFT	Gray et al., 1998 KRAFT consortium	Transformation and Reuse of Knowledge	mediators as knowledge brokers
Madefast	Cutkosky et al., 1996 Stanford	Collaborative Engineering	uses the Internet
MADEsmart	Jha et al., 1998 Boeing	Collaborative Engineering	wrapper agents for legacy system encapsulation
MetaMorph I	Maturana & Norrie, 1996, U of Calgary	Intelligent Manufacturing	mediator-centric architecture; dynamic clustering & cloning; learning
MetaMorph II	Shen et al., 1998 U of Calgary	Intelligent Manufacturing, Supply Chain Management	hybrid architecture; mediators as subsystem coordinators and interfaces to the main system
?[2]	Brugali et al., 1998 Politecnico di Torino	Supply Chain Management	uses mobile agents for an industrial process
?	Fleury et al., 1996	Manufacturing System Optimization	'triple coupling' of multi-agent techniques, simulated annealing, and simulation
?	Pancerella et al., 1995 Sandia Lab	Agile Manufacturing	an agent is defined as an autonomous, encapsulated software component
?	Papaioannou & Edwards, 1998, Loughborough	Virtual Enterprise	uses mobile agents
?	Swaminathan et al., 1996, CMU	Supply Chain Management	supply chain library with structural elements (agents) and control elements for coordination
?	Wunderli et al., 1996 ETH Zentrum	CIM Systems	database agents for CIM systems
?	Yan et al., 1998 Leipzig	Project Management	uses mobile agents

REFERENCES

Balasubramanian, S., Maturana, F. and Norrie, D. (1996) Multi-Agent Planning and Coordination for Distributed Concurrent Engineering, *International Journal of Cooperative Information Systems*, 5(2-3):153-179.

Barbuceanu, M. and Fox, M.S. (1994) The Architecture of a Generic Agent for Collaborative Enterprises, EIL working paper, University of Toronto.

---

[2] ? No project name found through available publications.

Barbuceanu, M. and Fox, M.S. (1995a) The Architecture of an Agent Based Infrastructure for Agile Manufacturing, In *Proceedings of IJCAI'95 Workshop on Intelligent Manufacturing*, Montreal, Canada.

Barbuceanu, M. and Fox, M.S. (1995b) COOL: A Language for Describing Coordination in Multi Agent Systems, In *Proc. of ICMAS'95*, San Francisco, CA, The AAAI press/The MIT Press, pp. 17-24.

Barbuceanu, M. and Fox, M.S. (1996) The Architecture of an Agent Building Shell, In M. Wooldridge, J. P. Muller and M. Tambe (eds.), *Intelligent Agents II*, LNAI 1037, Springer, pp. 235-250.

Barbuceanu, M. and Fox, M.S. (1997) Integrating Communicative Action, Conversations and Decision Theory to Coordinate Agents, In *Proc. of Autonomous Agents'97*, ACM Press, Marina del Rey, CA, pp. 49-58.

Biswas, G., Bagchi, S. and Saad, A. (1995) Holonic Planning and Scheduling for Assembly Tasks, TR CIS-95-01, Center for Intelligent Systems, Vanderbilt University.

Brennan, R.W., Balasubramanian, S. and Norrie, D.H. (1997) Dynamic Control Architecture for Advanced Manufacturing Systems, In *Proceedings of International Conference on Intelligent Systems for Advanced Manufacturing*, Pittsburgh, PA.

Brugali, D. Menga, G. and Galarraga, S. (1998) Inter-Company Supply Chains Integration via Mobile Agents, In *Proceedings of PROLAMAT'98*, Kluwer Academic Publishers, Trento, Italy, pp. 43-54.

Budenske, J., Ahamad, A. and Chartier, E. (1998) Agent-Based Architecture for Exchanging Modeling Data between Applications, In *Working Notes of the Agent-Based Manufacturing Workshop, Minneapolis*, MN, 1998, pp. 18-27.

Bussmann S. (1998) An Agent-Oriented Architecture for Holonic Manufacturing Control, In *Proceedings of First International Workshop on IMS*, Lausanne, Switzerland, pp. 1-12.

Camarinha-Matos, L.M., Afsarmanesh, H., Garita, C. and Lima, C. (1997) Towards An Architecture for Virtual Enterprises, In *Proceedings of the 2nd World Congress on Intelligent Manufacturing Processes & Systems*, Budapest.

Christensen, J.H. (1994) Holonic Manufacturing Systems: Initial Architecture and Standards Directions, In *Proceedings of First European Conference on Holonic Manufacturing Systems*, Hanover, Germany.

Christensen, J.H., Struger, P.J., Norrie, D.H. and Schaeffer, C. (1994) Material Handling Requirements in Holonic Manufacturing Systems, In *Proceedings of International Symposium on Material Handling Research*, Grand Rapids, MI.

Cutkosky, M.R., Tenenbaum, J.M. and Glicksman J. (1996) Madefast: Collaborative Engineering over the Internet, *Communication of the ACM*, 39(9):78-87.

Deen, S.M. (1994) A Cooperation Framework for Holonic Interactions in Manufacturing, In *Proc. of the Second International Working Conference on Cooperating Knowledge Based Systems (CKBS'94)*, DAKE Centre, Keele University, 1994.

Finin, T., Fritzon, R., McKay, D. and McEntire, R. (1993) KQML – A Language and Protocol for Knowledge and Information Exchange, Tech. Report, University of Maryland, Baltimore.

Fleury, G., Goujon, J-Y., Gourgand, M. and Lacomme, P. (1996) Multi-Agent Approach for Manufacturing Systems Optimization, In *Proceedings of PAAM'96*, London, pp. 225-244.

Fox, M., Chionglo, J.F., and Fadel, F.G., (1993a) A Common Sense Model of the Enterprise, In *Proceedings of the 2nd Industrial Engineering Research Conference*, pp. 425-429.

Fox, M.S., Chionglo, J.F. and Barbuceanu, M., (1993b) The Integrated Supply Chain Management System, Internal Report, Dept. of Industrial Engineering, Univ. of Toronto.

Genesereth, M.R. and Fikes, R.E. (1992) Knowledge Interchange Format, Version 3.0, Reference Manual, Computer Science Department, Stanford University, Technical Report Logic-92-1.

Gray, P.M.D., Embury, S.M., Hui, K. and Preece, A. (1998) An Agent-Based System for Handling Distributed Design Constraints, In *Working Notes of the Agent-Based Manufacturing Workshop*, Minneapolis, MN, pp. 28-31.

Hasegawa, T., Gou, L., Tamura, S., Luh, P.B. and Oblak, J.M. (1994) Holonic Planning and Scheduling Architecture for Manufacturing, In *Proceedings of the 2nd International Working Conference on Cooperating Knowledge-based Systems*, University of Keele.

Hayashi, H. (1993) The IMS International Collaborative Program, In *Proceedings of 24th ISIR*, Japan Industrial Robot Association.

Jha, K.N., Morris, A. Mytych, E. and Spering, J. (1998) MADEsmart: Agents for Design, Analysis, and Manufacturability, In *Working Notes of the Agent-Based Manufacturing Workshop*, Minneapolis, MN, pp. 57-63.

Klein, M. (1995) iDCSS: Integrating Workflow, Conflict and Rationale-based Concurrent Engineering Coordination Technologies, *Concurrent Engineering: Research and Application*, 3(1):21-27.

Koestler, A. (1967) *The Ghost in the Machine*, Arkana Books, London.

Kwok, A.D. and Norrie, D.H. (1994) A Development System for Intelligent Agent Manufacturing Software, *Integrated Manufacturing Systems*, 5(4-5):64-76.

Larsen, K.R.T. (1999) Virtual Organization as an Inter-organizational Concept: Ties to Previous Research, *VoNet: The Newsletter*, 3(1):19-35, available also at: http://www.virtual-organization.net/

Malkoun, M.T. and Kendall, E.A. (1997) CLAIMS: Cooperative Layered Agents for Integrating Manufacturing Systems, In *Proceedings of PAAM'97*, London, UK.

Maturana, F. and Norrie, D. (1996) Multi-Agent Mediator Architecture for Distributed manufacturing, *Journal of Intelligent Manufacturing*, 7:257-270.

Mehra, A., and Nissen, M. (1998) Intelligent Supply Chain Agents using ADE, In *Proceedings of AI & Manufacturing Research Planning Workshop*, Albuquerque, NM, The AAAI Press, pp. 112-119.

NIST (1998) Advanced Technology Program, http://www.atp.nist.gov/.

Pan, J.Y.C. and Tenenbaum, M.J. (1991) An intelligent agent framework for enterprise integration, *IEEE Transactions on Systems, Man, and Cybernetics*, 21(6):1391-1408.

Pancerella, C., Hazelton, A. and Frost, R. (1995) An Autonomous Agent for On-Machine Acceptance of Machined Components, In *Proceedings of SPIE International Symposium on Intelligent Systems and Advanced Manufacturing*, Philadelphia, PA.

Papaioannou, T. and Edwards, J. (1998) Mobile Agent technology Enabling the Virtual Enterprise: a Pattern for Database Query, In *Working Notes of the Agent-Based Manufacturing Workshop*, Minneapolis, MN, pp. 106-116.

Park, H., Tenenbaum, J. and Dove, R. (1993) Agile Infrastructure for Manufacturing Systems: A Vision for Transforming the US Manufacturing Base, Defense Manufacturing Conference.

Peng, Y., Finin, T., Labrou, Y., Chu, B., Long, J., Tolone, W.J. and Boughannam, A. (1998) A Multi-Agent System for Enterprise Integration, In *Proceedings of PAAM'98*, London, UK, pp. 533-548.

Peng, Y., Finin, T., Chen, H., Wang, L., Labrou, Y., Cost, R.S., Chu, B., Cross, V., Russell, M., Tolone, B., Boughannam, A. and McCobb, J. (1999) An Agent System for Application Initialization in an Integrated Manufacturing Environment, In *Proceedings of SCI'99/ISAS'99*, Volume 7, Orlando, FL, pp. 415-421.

Roboam, M. and Fox, M.S. (1992) Enterprise Management Network Architecture, In A. Famili, D.S. Nau and S.H. Kim, eds., *Artificial Intelligence Applications in Manufacturing*, The AAAI Press, pp. 401-432.

Shen, W., Xue, D. and Norrie, D.H. (1998) An Agent-Based Manufacturing Enterprise Infrastructure for Distributed Integrated Intelligent Manufacturing Systems, In *Proceedings of PAAM'98*, London, UK, pp. 533-548.

Shen, W. and Norrie, D.H. (1999) Agent-Based Systems for Intelligent Manufacturing: A State-of-the-Art Survey, *Knowledge and Information Systems, an International Journal*, 1(2), 129-156.

Sousa, P. and Ramos, C. (1997) A Dynamic Scheduling Holon for Manufacturing Orders, *Journal of Intelligent manufacturing*, 9(2):107-112.

Swaminathan, J.M., Smith, S.F. and Sadeh, N.M. (1996) A Multi-Agent Framework for Supply Chain Dynamics, In *Proceedings of NSF Research Planning Workshop on AI & Manufacturing*, Albuquerque, NM.

Van Brussel, H. and Valckenaers, P. (1999) Proceedings of the Second International Workshop on Intelligent Manufacturing Systems IMS-99 (Eds), Leuven, Belgium, 22-24 September, 1999, 903pp.

Van Leeuwen, E.H. and Norrie, D.H. (1997) Intelligent manufacturing: holons and holarchies. *Manufacturing Engineer*, 76(2), 86-88.

Wang L., Balasubramanian S. and Norrie D. (1998) Agent-based Intelligent Control System Design for Real-time Distributed Manufacturing Environments, In *Working Notes of the Agent-Based Manufacturing Workshop*, Minneapolis, MN, pp. 152-159.

Wunderli, M., Norrie, M.C. and Schaad, W. (1996) Multi-database Agents for CIM Systems, *International Journal Computer Integrated manufacturing*, 9(4):293-298.

Yadav, S. and Xue, D. (1998) Feature-Based Concurrent Design for Improving Manufacturability, In *Proceedings of CSME Forum 1998*, Vol. 3, Toronto, pp. 7-14.

Yan, Y., Kuphal, T. and Bode, J. (1998) Application of Multi-Agent Systems in Project Management, In *Working Notes of the Agent-Based Manufacturing Workshop*, Minneapolis, MN, pp. 160-171.

Part IV: Guideline for System Implementation

16

Tools, Frameworks, Languages, Standards and Methodologies

16.1 INTRODUCTION

Most agent-based concurrent design and manufacturing systems have been implemented using traditional programming languages, such as C++, Java, Lisp, Smalltalk, and Prolog. For wider use of agent technology in design and manufacturing, powerful agent development tools and supporting standards and methodologies are much needed. Several such tools, languages, frameworks, and methodologies have been recently reported, some of which are already commercially available.

In this chapter, we briefly introduce the more widely known of these tools, frameworks, languages, standards, and methodologies, giving for each: (1) its name; (2) company, institution or research group name; (3) URL where available; (4) short description. Interested readers should consult the relevant web site and references in this chapter for more detailed descriptions. Note that this list is not complete since new products are being introduced quite quickly.

16.2 TOOLS AND FRAMEWORKS

This section introduces more well-known agent development tools and frameworks in alphabetic order, in two categories: commercial products and academic and research projects.

16.2.1 Commercial Products

The following are agent-based system development tools and frameworks that are widely available as commercial products.

Name: AgentBuilder
Company: Reticular Systems, Inc.
URL: http://www.agentbuilder.com/
Description: AgentBuilder is an integrated tool suite for constructing intelligent software agents. AgentBuilder consists of two major components – the Toolkit

and the Run-Time System. The Toolkit includes tools for managing the agent-based software development process, analyzing the domain of agent operations, designing and developing networks of communicating agents, defining behaviors of individual agents, and debugging and testing agent software. The Run-Time System includes an agent engine that provides an environment for execution of agent software.

Agents constructed using AgentBuilder communicate using KQML. In addition, AgentBuilder allows the developer to define new inter-agent communication commands that suit to particular needs.

All components of both the AgentBuilder Toolkit and the Run-Time System are implemented in Java.

Name: Agent Building Environment (ABE)

Company: IBM

URL: http://www.networking.ibm.com/iag/iagsoft.htm

Description: Agent Building Environment (ABE) is an extensible C++ class library for enhancing applications, especially network-centric applications, with embedded intelligent agents. ABE has powerful mechanisms for agents to perform reasoning and for agents to be embedded closely and flexibly with a variety of applications and software environments.

One of the major differences between ABE and most of the other Java-based agent environments is that it is focused on the "intelligent" part of intelligent agents. While IBM has Aglets for mobile Java agents, their ABE environment focuses on providing a powerful rule-based reasoning engine (called RAISE) to control agent behavior. This has a flexible approach to configuring sensors and effectors for use by the agent as it reasons about the world.

ABE is an architecture and toolkit for developing agents. It consists of several major components, including Engines, Knowledge, Library, Views, and Adapters. The engine in ABE is the RAISE rule-based engine, but the architecture supports the concept of pluggable engines. Knowledge is represented in KIF, and a Java-based GUI editor is provided for rule authoring. ABE also provides a flexible Library component for storing named KIF rule sets and metadata. The View component defines how editors and dialogs can be attached through ABE Adapters to provide clean model-view separation between intelligent agent processing and agent interactions with users. Adapters are used to interface ABE agents and the RAISE inference engine to applications. For example, ABE comes with adapters to the World Wide Web (HTTP), Internet news servers (NNTP), a Time adapter that triggers events based on time, and a File adapter that triggers events when files are changed. Sample adapters to interface with e-mail systems and to watch stock prices at a Web site are also included. These adapters can be written in Java and they can fully interoperate with other Java or C++ adapters through the ABE.

Versions are available for OS/2, Windows, AIX and OS/390.

Name: Agent Development Environment (ADE)

Company: Gensym Corporation

URL: http://www.gensym.com/successstories/adepaper.htm

Description: ADE provides a predefined class hierarchy of agents and agent parts, an agent communications middleware, and a graphical language for designing and developing agent behavior based on the Grafcet[1] standard. In addition, ADE provides a distributed simulation environment to test agent-based applications, and a centre to deploy agents on the network.

ADE provides the foundation classes for a multi-agent application as well as a powerful user interface that facilitates development of multi-agent applications. The foundation classes are Agent, Host, Message, Environment and Activity. Agents are located on Hosts and communicate through Messages. Hosts are responsible for the reliable delivery of Messages. The Activities performed by Agents can be defined using Grafcet Charts. Agents can first be simulated and, once the desired behavior is achieved, deployed in G2[2] or as JavaBeans in a Java virtual machine. The G2/Java Beans agents once deployed, can interact with agents written in other languages using CORBA or Java RMI (Remote Method Invocation).

Name: Aglets

Company: IBM Japan

URL: http://www.trl.ibm.co.jp/aglets/

Description: Aglets are Java-based autonomous agents developed by IBM Japan. They are provided with the basic capabilities required for mobility. Each aglet has a globally unique name. A travel itinerary is used to specify the destinations to which the agent must travel and what actions it must take at each location. In order for an aglet to run on a particular system, that system must be running an aglet host application. This provides a platform-neutral execution environment for the aglet. The aglet workbench includes a configurable Java security manager, which can restrict the activity of an aglet on the system in the same way that the default security manager is used to restrict the activities of an applet.

Aglets can communicate using a whiteboard that allows agents to collaborate and share information asynchronously. Synchronous and asynchronous message passing is also supported for aglet communication. Aglets are streamed using standard Java serialization or externalization. A network agent class loader is supplied which allows an aglet's byte-code stream and state to travel across a network.

---

[1] Grafcet is a graphical language that has been accepted as an industrial standard (IEC 848 and IEC 1131-3) for local PLC-level sequential logic control (David 1995).

[2] G2 is a core product of Gensym Corporation.

Name: Concordia

Company: Mitsubishi Electric Information Technology Center America (ITA)

URL: http://www.meitca.com/HSL/Projects/Concordia/

Description: Concordia is a full-featured framework for development and management of network-efficient mobile agent applications for accessing information anytime, anywhere and on any device supporting Java. With Concordia, applications:

- process data at the data source;
- process data even if the user is disconnected from the network;
- access and deliver information across multiple networks (LANs, Intranets and Internet);
- use wire-line or wireless communication;
- support multiple client devices, such as Desktop Computers, Notebook Computers, Personal Digital Assistants, and Smart Phones.

Concordia delivers a rich set of features that support the implementation of agents for a wide range of industries and end users. Agents implemented with Concordia are mobile objects. They can travel to different locations and devices on a network and perform work at those locations. The mobility of Concordia Agents sets them apart from distributed objects and Java applets.

A Concordia Agent's network travels are defined by its Itinerary. The Itinerary specifies where the agent is to travel and what task it should perform when it arrives. Concordia Itineraries are specified at run-time. Agents may change their own Itineraries based upon information and events discovered as the agent travels.

Name: JACK Intelligent Agents

Company: Agent Oriented Software Pty. Ltd. (AOS), Melbourne, Australia,

URL: http://www.agent-software.com/

Description: JACK Intelligent Agents is a Java framework for multi-agent system development. This framework supplies a high performance, light-weight implementation of the Belief-Desire-Intention (BDI) architecture, and can be easily extended to support different agent models or specific application requirements.

JACK agents can be used as components of larger environments and will be visible simply as objects by the non-agent software. JACK agents are not bound to any specific agent communications language and high-level protocols such as KQML may be used. However, JACK has been geared towards industrial object-oriented middleware (such as CORBA) and message passing infrastructures (e.g. PVM or DIS in simulation environments). JACK provides a native lightweight communication infrastructure for situations where high performance is required.

From an engineering perspective, JACK consists of architecture-independent facilities, plus a set of plug-in components that address the requirements of specific agent architectures. An example of such a plug-in is the default BDI reasoning model supplied with JACK. From a programming perspective, JACK consists of three main extensions to Java.

Name: Odyssey
Company: General Magic
URL: http://www.genmagic.com/technology/odyssey.html
Description: Odyssey is a set of Java class libraries. When incorporated in a network application, the libraries enable the creation of agent-accessible places[3] in the network, the stationing of agents in those places, and facilitating agents to travel between places whenever they need to.

Agents are created by subclassing either the Agent class or the Worker class. Each Agent executes in its own Java thread, and can travel from place to place. At each new place, however, the thread must be restarted. An instance of the Place class defines the execution environment for an agent on each host system. A Worker is a specialization of the Agent class. A Worker contains an itinerary that is a set of tasks and destinations. The Worker executes one task at each destination host. The Odyssey agent system also provides support classes such as Ticket, which defines how and where an agent may travel, and Petition, which identities other Agents with whom the Agent wishes to communicate.

Odyssey is transport-independent. Whenever an agent travels from one place to another, Odyssey calls upon a transport API designed by General Magic for that purpose. Odyssey comes complete with implementations of the API for Sun's Remote Method Invocation (RMI), the OMG Internet Inter-ORB Protocol (IIOP), and Microsoft's Distributed Component Object Model (DCOM). The RMI implementation is the default.

Odyssey is written entirely in Java and requires JDK. Odyssey does not include its own Security Manager, and relies only on the standard Java security mechanisms.

Name: Voyager
Company: ObjectSpace Inc.
URL: http://www.objectspace.com/
Description: Voyager is an agent-enhanced Object Request Broker (ORB) written entirely in Java. An ORB provides the capability to create objects on a remote system and invoke methods on those objects.

Voyager augments the traditional ORB with agent capabilities. It combines the power of mobile autonomous agents and remote method invocation with complete CORBA support and comes complete with distributed services including directory, publish, subscribe, and multicast. Voyager allows Java programmers to quickly and easily create network applications using both traditional and agent-enhanced distributed programming techniques.

---

[3] A *place* is similar to the term *location* used in the descriptions of other tools and frameworks.

Voyager agents have mobility and autonomy, both provided in the base class, Agent. An Agent can move itself from one location to another and can leave behind a forwarding address with a "secretary" so that future messages can be forwarded to its new location. Specialized agents, called Messengers, are used to deliver messages. Messages can be synchronous, one-way (similar to asynchronous), or future, which are asynchronous but return a placeholder that can be used to retrieve a return value at a later time.

A Voyager agent's itinerary instructs the agent on what operations it needs to perform at each location. Voyager uses serialization to stream the agent's state as it moves from location to location. Voyager also includes a security manager that can be used to restrict the operations an agent can perform. Voyager is being enhanced with a rule-based security model that supplements the standard Java security mechanisms.

Voyager agents themselves are not inherently intelligent, but they can be augmented with any artificial intelligence techniques.

16.2.2 Academic and Research Projects

This subsection introduces academic and research projects related to the development of agent tools and frameworks, some of which are recently becoming commercialized.

Name: Agent Building Shell (ABS)
Company/Institution: Enterprise Integration Laboratory, University of Toronto
URL: http://www.ie.utoronto.ca/EIL/ABS-page/ABS-overview.html
Description: Agent Building Shell provides several reusable layers of languages and services for building agent systems:
- Agent Communication Language is the language the agents use to communicate. KQML/KIF is used for this purpose.
- Information Distribution service is a generic service whose major purpose is distributing information of interest to other agents, in a manner that relies on the content of the information.
- Organizational Model tells the agent what other agents exist, which roles they have, what goals agents pursue, what sort of communication can take place amongst them, and so on.
- Coordination Models (as shared conventions about exchanged messages during cooperative actions) are described in a special purpose coordination language (COOL – see Section 16.3).
- Conflict Management service enables the agent to make decisions when confronted with contradictory information derived or received from other agents.

- Generic Interface to Applications provides an interface allowing data, parameters and control specifications to be transmitted to or from applications.
- Knowledge Management service provides a powerful description logic representation of knowledge, extended with temporal reasoning and several processing mechanisms.

The approach is being used to develop multi-agent applications in the area of manufacturing enterprise supply chain integration.

Name: Cybele

Company/Institution: Intelligent Automation, Inc.

URL: http://www.i-a-i.com/projects/cybele/index.html

Description: Cybele is a software infrastructure for building agent-based applications. This infrastructure supports a number of services for agent-based applications, some of which are unique. These services include:

- agent creation and deployment over a network of varied platforms;
- a message addressing scheme for agent communication which is independent of the location of a sending or receiving agent;
- multicasting, broadcasting, and peer-to-peer messaging;
- the accumulation of messages intended for a busy recipient agent;
- the proper conversion of message data across platforms;
- the migration of agents across processors for performance optimization and/or fault tolerance.

Cybele has been used in the AARIA project (see Chapter 14) for developing agent-based manufacturing systems.

Name: dMARS

Company/Institution: Australian Artificial Intelligence Institute

URL: http://www.aaii.oz.au/proj/dMARS-prod-brief.html

Description: dMARS is an agent-oriented development and implementation environment for building complex, distributed, time-critical systems.

Based on the Procedural Reasoning System (PRS) developed by SRI International (California), dMARS takes advantage of the recent research into multi-agent, real-time reasoning.

dMARS is suited for the development of any application that requires both proactive goal directed behavior and reliable time-critical response to change. It is particularly suited to applications where a large number of complex but well-defined procedures or tactics exist for accomplishing particular tasks in a variety of situations.

dMARS provides a sophisticated suite of graphical tools for development and debugging. These tools not only provide an intuitive interface, but also address specific issues involved with large-scale development.

Name: FIPA-OS
Company: Nortel Networks
URL: http://www.nortelnetworks.com/products/announcements/fipa/info.html
  Description: FIPA-OS is an open source implementation of mandatory elements in the FIPA 97 specification for agent interoperability. FIPA-OS is an experimental agent framework, originating from research at Nortel Networks' Harlow Laboratories in the UK. The primary aim of FIPA-OS is to reduce the current barriers in the adoption of FIPA technology by supplementing the technical specification documents (available at Fipa.org) with managed open source code.

  FIPA-OS supports the development of agents systems to FIPA specifications. It utilizes ObjectSpace's Voyager (see Section 16.2.1) and Java for the agent system implementation. It has been used in European collaborative research projects and is being utilized in numerous other research systems.

Name: InfoSleuth
Company/Institution: MCC
URL: http://www.mcc.com/projects/infosleuth/
Description: InfoSleuth is a consortium project carried out by MCC on behalf of several participants. It uses agents to find, retrieve, and integrate information from heterogeneous data sources. Its architecture is based on cooperating autonomous agents which advertise their services and process requests for those services. It is based on KQML, KIF, HTTP, and Java technologies.

  KQML is used as the standard message format and protocol between agents. The InfoSleuth project includes a Java implementation of the KQML language and protocol. KIF is used for exchanging knowledge between agents. The agents convert information from their internal formats into KIF and wrap it in a KQML performative before sending the message to another agent.

  In the InfoSleuth architecture, rule-based agents are written in a variety of languages including Lisp, LDL++[4], CLIPS, and Java. On top of the cooperative agents is an ontology layer that includes agents that having a common vocabulary and semantic model for each particular problem domain. Applications are written on top of the ontology layer of the architecture.

  A number of different types of agent are used in the different layers of the architecture. User agents, written in Java, handle requests from users, route the requests to server agents, and return the results to the user. They are persistent and autonomous, which allows them to run tasks for users even when the user is no longer interacting with the agent. User agents also contain the intelligence to model the user's behavior. To do this, they interact with other Java agents called monitors. A monitor tracks a user's Web access and reports it to the user agent for use in inferencing and pattern detection.

  The broker agent is another Java agent that is similar to the Facilitator in

---

[4] LDL++ is a declarative database language that includes rule-based inferencing.

JATLite (see below). It matches requests from agents with ontology server agents that advertise the ontologies for which they are responsible. The ontology agents interact with execution agents responsible for executing ontology-based queries. The execution agents work with resource agents which are intelligent front ends to the underlying datastore. The data returned from a query may also be passed to data analysis agents that perform intelligent analysis, knowledge mining, and pattern recognition tasks on the data.

InfoSleuth provides some interesting features[5] such as diversity, content-addressability, user perspective, user power, monitoring capabilities, distributed processing, collaborative processing, dynamics, and scaleability.

Name: Intelligent Agent Library
Company/Institution: Bits & Pixels
URL: http://www.bitpix.com/business/main/bitpix.htm
Description: Intelligent Agent Library provides components for building intelligent agents, implemented entirely in Java. The library is a collection of hundreds of Java classes covering various aspects of building intelligent agents. In addition to agent communication and data handling facilities, the library contains rule-based and neural net-based processing modules. There is also an extensive modeling facility for prototyping agent interactions and graphical classes for creating static and animated interactive displays.

Name: Java Agent Template Lite (JATLite)
Company/Institution: Stanford University
URL: http://java.stanford.edu/java_agent/html/
Description: Java Agent Template Lite (JATLite) is a package of programs written in Java that allow users to quickly create new software agents that communicate robustly over the Internet. JATLite provides a basic infrastructure in which agents register with an Agent Message Router *facilitator* (using a name and password), connect/disconnect from the Internet, send and receive messages, transfer files, and invoke other programs or actions on the various computers where they are running.

JATLite especially facilitates the construction of agents that send and receive messages using KQML. The communications are built on open Internet standards, TCP/IP, SMTP, and FTP. However, developers may build agent systems with other agent languages when using JATLite.

Name: JAFMAS
Company/Institution: University of Cincinnati
URL: http://www.ececs.uc.edu/~abaker/JAFMAS/
Description: JAFMAS stands for a Java-based Agent Framework for Multi-Agent Systems. It provides a generic methodology for developing speech-act based

---

[5] See http://www.mcc.com/projects/infosleuth/introduction/special.html for details.

multi-agent systems, an agent architecture, and a set of classes for implementing the agents in Java. The methodology follows five stages: (i) identifying the agents; (ii) identifying the conversations; (iii) identifying the conversations rules; (iv) analyzing the conversation model; and (v) MAS implementation. JAFMAS provides communication, linguistic and coordination support through sixteen Java classes. Communication support is provided for both directed communication and subject-based broadcast communication. Linguistic support is provided through a speech-act (e.g., KQML) based communication language. Coordination support follows from Searle's approach (Searle, 1969). JAFMAS classes support each agent with multiple threads, one for the agent, one for each conversation in which the agent engages, and one for each subject to which the agent subscribes. Though JAFMAS provides sixteen Java classes, the user needs to extend just four of these classes to develop a multi-agent application. Multi-agent application development using JAFMAS has been demonstrated through example applications of the N-Queens problem and supply chain integration. These two examples use only 567 and 1276 lines of additional code respectively for implementation.

Name: Open Agent Architecture (OOA)
Company/Institution: SRI International
URL: http://www.ai.sri.com/~oaa/
Description: Open Agent Architecture (OOA) is a framework for integrating a community of heterogeneous software agents in a distributed environment. OOA has the following characteristics:
-   open: agents can be created in multiple programming languages and interface with existing legacy systems;
-   extensible: agents can be added or replaced individually at runtime;
-   distributed: agents can be spread across any network-enabled computers;
-   parallelism: agents can cooperate or compete on tasks in parallel;
-   mobility: lightweight user interfaces can run on handheld PDA's and most applications can be run through a telephone-only interface;
-   multi-modal: communication with agents, handwriting, speech, pen gestures and direct manipulation (GUIs) can be combined in a natural way.

OAA agent libraries exist for different operating systems including SunOS, Solaris, SGI IRIX, Windows, and different programming languages including ANSI C, Common Lisp, Quintus Prolog, Java, Borland Delphi, and Visual Basic.

Name: Swarm
Company/Institution: Santa Fe Institute
URL: http://www.santafe.edu/projects/swarm/
Description: Swarm is a multi-agent software platform for the simulation of complex adaptive systems. In the Swarm system the basic unit of simulation is the swarm, a collection of agents executing a schedule of actions. Swarm

supports hierarchical modeling approaches whereby agents can be composed of swarms of other agents in nested structures. Swarm provides object-oriented libraries of reusable components for building models and analyzing, displaying, and controlling experiments on those models. At the time of writing, Swarm is available as a beta version free, in full source code form. It requires the GNU C Compiler, Unix, and X Windows. The Swarm libraries are distributed under the GNU Library Public License.

Name: ZEUS
Company/Institution: British Telecom Labs
URL: http://www.labs.bt.com/projects/agents/research/collaborative.htm
Description: ZEUS is a collaborative agent tool-kit. It consists of four components: an Agent Component Library, Agent Building Tools, a Development Methodology, and Visualization Tools. The Agent Component library is a collection of classes written in Java, which provide the means for a developer to create an agent application whilst writing as little program code as possible. The Agent Building Tools include a Visual Agent Generator, a Code Generator, and a Legacy system API. The Development Methodology has six stages: (1) Domain Study and Agent Identification; (2) Agent Definition and Task Identification; (3) Task Definition; (4) Agent Organization; (5) Agent Coordination; (6) Agent Implementation. Its Visualization Tools include a society tool, a report tool, an agent viewer, a control tool, and a statistics tool.

Each ZEUS agent consists of a definition layer, an organizational layer and a coordination layer. The definition layer comprises the agent's reasoning (and learning) abilities, its goals, resources, skills, beliefs, preferences, etc. The organization layer describes the agent's relationships with other agents; what agencies it belongs to, what abilities it knows other agents possess, and so on. In the coordination layer, the agent is modeled as a social entity, in terms of the coordination and negotiation techniques it possesses. Built on top of the coordination layer are the communication protocols that implement inter-agent communication and beneath the definition layer is the API that links the agent to the physical realizations of its resources and skills.

16.3 AGENT LANGUAGES

By an agent language, we mean a programming language which can be used to develop agents and agent-based systems. Such a language is usually called an agent-oriented programming language (AOP) (Shoham, 1990). The key idea of an AOP is to be able to directly program agents in terms of mentalistic and intentional notions. Shoham proposed that a fully developed AOP should have the following three components:

-   a logical system for defining the mental state of agents;
-   an interpreted programming language for programming agents;

- an 'agentification' process for compiling agent programs into low-level executable systems.

Here, we introduce some of the more widely known agent-oriented programming languages. Although agent communication languages are not included in this chapter, we note here that the most common communication language, KQML, will be briefly considered in Section 16.4.

Name: Agent0
Research Group: Shoham et al., Stanford University
URL: n/a
Description: Agent0 was the Shoham's first attempt at an AOP language. In Agent0, an agent is specified in terms of a set of capabilities (things the agent can do), a set of initial beliefs and commitments, and a set of commitment rules. The key component, which determines how the agent acts, is the commitment rule set. Each commitment rule contains a message condition, a mental condition, and an action. In order to determine whether such a rule fires, the message condition is matched against the messages the agent has received; the mental condition is matched against the beliefs of the agent. If the rule fires, the agent becomes committed to the action. Actions may be private, correspond to an internally executed subroutine, or communicative (i.e., for sending messages). More information can be found in (Shoham, 1990).

Name: Agent-K
Research Group: Edwards et al., University of Aberdeen
URL: http://www.csd.abdn.ac.uk/~pedwards/publs/agentk.html
Description: Agent-K is an integration of Agent0 and KQML. It provides for interoperable (or open) software agents that communicate via KQML and are programmed using the AOP approach. More information can be found in (Davies and Edwards, 1994).

Name: APRIL and MAIL
Research Group: ESPRIT IMAGINE[6] project group
URL: n/a
Description: The Agent PRocess Interaction Language (APRIL) (McCabe and Clark, 1995) and MAIL (Steiner, 1996) are two languages for multi-agent applications, which were developed as part of the ESPRIT IMAGINE project. The two languages were intended to fulfill quite different roles. APRIL was designed to provide the core features required to implement most of the desired agent architectures and systems. Thus APRIL provides facilities for multi-tasking, communication, and pattern matching and symbolic processing capabilities. The generality of APRIL comes at the expense of abstraction; an APRIL system builder must implement an agent or system architecture *ab initio*

---

[6] IMAGINE: Integrated Multi-Agent Interaction Environment.

using APRIL's primitives. In contrast, the MAIL language provides a rich collection of pre-defined abstractions, including plans and multi-agent plans. APRIL was originally envisaged as the implementation language for MAIL.

Name: CONGOLOG

Research Group: Lesperance et al., York University and University of Toronto

URL: n/a

Description: CONGOLOG is an agent-oriented extension of the GOLOG logic programming language which was implemented in PROLOG. In CONGOLOG, agents communicate via message passing; each agent has associated with it a message queue, which it can access via message-read actions. The messages that agents send look very much like KQML performatives, with a logical semantics defined within the language. More information can be found in (Shapiro et al., 1998).

Name: COOL

Research Group: Barbuceanu and Fox, University of Toronto

URL: http://www.ie.utoronto.ca/EIL/ABS-page/ABS-intro.html

Description: COOL is a COOrdination Language developed for specifying coordination activities in multi-agent systems. In COOL, coordination is conceptualized as a structured conversation among agents. As a basic component of the ABS (Agent Building Shell – see Section 16.2.2), COOL allows programmers to build multi-agent applications by defining:
- distributed execution environments that manage the execution of agents;
- the agents, together with their behavior and interaction patterns;
- the communication language used by agents (a form of KQML);
- specifications of agent interaction by means of conversation plans and conversation rules.

In COOL, conversation plans are specifications of how agents act, react and interact in specific situations. A conversation plan consists of states and transitions. Transitions are expressed by several types of conversation rules. A conversation rule specifies patterns of received messages that may trigger it as well as patterns of messages that will be transmitted in response by the agent. Conversation rules also specify the local decision-making and actions taken by agents. In particular, this includes the use of legacy software the agent has access to. A detailed description of COOL can be found in (Barbuceanu and Fox, 1995).

Name: LALO

Research Group: Centre de recherche informatique de Montréal

URL: http://www.CRIM.CA/sbc/english/lalo/

Description: LALO is an Agent-Oriented Programming (AOP) language and a framework for developing intelligent multi-agents systems. The architecture is extensible and supports the creation of multi-agent systems, including reactive agents and deliberative agents. A program written in LALO is translated into

C++ source code, and then may be compiled using a regular C++ compiler. The inter-agent communication language used is KQML. The framework is available on Windows and UNIX platforms.

Name: PLACA
Research Group: Thomas et al., University of Aberdeen
URL: n/a
Description: PLACA is an extension of the original AOP concept. It expanded the underlying logic of mental states to include *intentions* (a commitment to achieve a state of the world). In parallel with this, actions may be planned. This means that multiple actions (private actions, message sending, etc.) are composed into plans, which are based on the agent's beliefs and capabilities. Once the agent determines a suitable plan, it will make a commitment to execute it. This contrasts with Agent0, where a single commitment is generated for each successful rule match. More details can be found in (Thomas, 1993).

Name: TKQML
Research Group: Finin et al., University of Maryland Baltimore
URL: http://www.csee.umbc.edu/kqml/tkqml/
Description: TKQML is an integration of KQML into Tcl/Tk. It can be used to build KQML-speaking agents that run within a TKQML shell. It can also be used to bind together diverse applications into a distributed framework, using KQML as a communication language. Tcl's embeddable nature allows one to easily add agent communication facilities to existing code. Thus, TKQML can be used to enhance the functionality of new or existing systems built using a Tcl framework, by allowing easy integration with agent-based systems.

16.4 STANDARDS

Wide use of agent technology in industry depends on the availability of development tools and platforms that save developers having to implement the same basic functionality for each system. These tools and platforms should desirably be based on standards that reflect the agreement of developers on what that basic functionality should be and how it should be available. Much effort has been devoted to standards for agent-based systems, but no approved international standards exist at the time of writing. However, in this chapter, we introduce several proposed standards that have already become widely accepted for developing agent based concurrent design and manufacturing systems.

Note that COM/DCOM and CORBA are not standards for developing agent-based systems, but are widely used in developing agent-based applications in the absence of accepted standards for agent-based system development. For object-oriented implementations, they will continue as intermediate standards until agent-oriented standards (e.g., the FIPA standards) are available.

Name: COM/DCOM
Company/Research Group: Microsoft
URL: http://www.microsoft.com/com/default.asp
Description: The Component Object Model (COM) is a software architecture that allows applications to be built from binary software components. COM is the underlying architecture that forms the foundation for higher-level software services, like those provided by OLE. OLE services span various aspects of commonly needed system functionality, including compound documents, custom controls, inter-application scripting, data transfer, and other software interactions.

The Distributed Component Object Model (DCOM) is a protocol that enables software components to communicate directly over a network in a reliable, secure, and efficient manner. Previously called "Network OLE," DCOM was designed for use across multiple network transports, including Internet protocols such as HTTP. DCOM is based on the Open Software Foundation's DCE-RPC specification and works with both Java applets and ActiveX™ components through its use of the Component Object Model (COM).

More information about COM/DCOM is available through Mircosoft's COM/DCOM web site. An excellent architectural comparison of DCOM and CORBA can be found in (Chung et al., 1998).

Name: CORBA
Company/Research Group: OMG
URL: http://www.omg.org/corba/
Description: The Common Object Request Broker Architecture (CORBA) has been introduced in Chapter 3 (see Section 3.4.2). It is the Object Management Group's solution to the need for interoperability among the rapidly proliferating number of hardware and software products available today. CORBA allows applications to communicate with one another no matter where they are located or who has designed them. CORBA 1.1, introduced in 1991 by the Object Management Group (OMG), defined the Interface Definition Language (IDL) and the Application Programming Interfaces (API) that enable client/server object interaction within a specific implementation of an Object Request Broker (ORB). CORBA 2.0, adopted in December of 1994, defines true interoperability by specifying how ORBs from different vendors can interoperate.

In fielding typical client/server applications, developers may use their own design or a recognized standard to define the protocol used between the devices. Protocol definition depends on the implementation language, network transport and numerous other factors. ORBs simplify this process. With an ORB, the protocol is defined through the application interfaces via a single implementation language-independent specification, the IDL. ORBs also provide flexibility. They let programmers choose the most appropriate operating system, execution environment, and even programming language to use for each component of a system under construction. More importantly, they allow the integration of existing components. In an ORB-based solution, developers simply model the legacy component using the same IDL they use for creating new objects, then write "wrapper" code that translates between the standardized bus and the legacy interfaces.

Name: FIPA
Company/Research Group: FIPA Consortium
URL: http://www.fipa.org
Description: The Foundation for Intelligent Physical Agents (FIPA) is a non-profit organization established in August 1996 and registered in Geneva, Switzerland. Its purpose is to promote the development of specifications of generic agent technologies that maximize interoperability within and across agent-based applications.

Within FIPA, a number of standards (specifications) are being proposed and evaluated, including Benchmarking, Communication, Input/Output, Information Representation, Privacy/Security, Agent Societies, Tools and Architectures, Personal information and personalization, Execution environment for mobile agents, and Agent Management. It is interesting to note that Part 9 of the FIPA '98 Application Specification is the Specification of Product Design and Manufacturing Agents.

The FIPA specifications are becoming increasingly used and accepted by industry. For example, it is FIPA standards that are being used in the Nortel Agent Development Environment, known as FIPA-OS (http://www.nortelnetworks.com/fipa-os).

At the time of writing, the development and evaluation of specifications and standards is still in progress. Updated information on FIPA standards can be found at the FIPA's web site.

Name: KIF
Company/Research Group: ARPA-sponsored Knowledge Sharing Effort
URL: http://www.cs.umbc.edu/kse/kif/
Description: Knowledge Interchange Format (KIF) is a computer-oriented language for the interchange of knowledge among disparate programs. Its main features are:
- declarative semantics (i.e. the meaning of expressions in the representation can be understood without appeal to an interpreter for manipulating those expressions);
- logically comprehensive (it provides for the expression of arbitrary sentences in first-order predicate calculus);
- provides for the representation of knowledge about the representation of knowledge;
- provides for the representation of non-monotonic reasoning rules;
- provides for the definition of objects, functions, and relations.

Name: KQML
Company/Research Group: ARPA-sponsored Knowledge Sharing Effort
URL: http://www.cs.umbc.edu/kqml/
Description: KQML (Knowledge Query and Manipulation Language) is a language
and protocol for exchanging information and knowledge. KQML is both a
message format and a message-handling protocol to support run-time knowledge
sharing among agents. KQML can be used as a language for an application
program to interact with an intelligent system or for two or more intelligent
systems to share knowledge in support of cooperative problem solving.

KQML focuses on an extensible set of performatives which define the
permissible operations that agents may attempt on each other's knowledge and
goal stores. The performatives comprise a substrate on which to develop higher-
level models of inter-agent interaction such as contract nets and negotiation. In
addition, KQML provides a basic architecture for knowledge sharing through
agents called communication facilitators which coordinate the interactions of
other agents.

KQML is based on speech-act theory; a message is a performative indicating
what the receiver is expected to do with the message. For example, a simple
"tell" message indicates that the receiver is expected to believe the fact(s)
provided. An *"ask"* message expects an answer to a question.

KQML provides an extended list of performatives, which deal with belief
revision, querying, knowledge base maintenance, actions and services. The
syntax is that of a performative followed by an unordered list of keyword-value
pairs. For example:

```
(ask-one    :receiver weather-station :sender forecaster
            :content rain(today, X) :language prolog :reply-with day10 )
```

may elicit the response:

```
(reply      :receiver forecaster :sender weather-station
            :content rain(today, no) :language prolog :in-reply-to day10 )
```

Name: NIIIP
Company/Research Group: NIIIP Consortium
URL: http://www.niiip.org/
Description: The National Industrial Information Infrastructure Protocols (NIIIP)
Consortium is a team of organizations having a cooperative development
agreement with the U.S. Government to develop open industry software
protocols to enable manufacturers and their suppliers to effectively interoperate
as if they were part of the same enterprise.

These protocols will enable a new form of collaborative computing in support
of efficient and globally competitive Virtual Enterprises. The consortium expects
to accomplish this through a set of consensus protocols based on object-oriented
technology and programming adapters that makes the intricacies of the
communications network transparent to the user. This will make it easy for

geographically dispersed organizations of any size to share information and efficiently collaborate on projects.

The NIIIP protocols are considered by the Advanced Research Project Agency (ARPA) as being among fundamental building blocks of the National Information Infrastructure (NII). Therefore, the NIIIP protocols are potential standards for use in agent-based concurrent design and manufacturing systems, especially in North America.

16.5 METHODOLOGIES

Analysis and design methodologies for multi-agent systems are quite rare in the literature. Most agent-oriented analysis and design methodologies are extensions of those for object-oriented analysis and design (Kinny et al., 1996; Burmeister, 1996; Brückner et al., 1998; Bryson and McGonigle, 1998) or extensions of knowledge engineering methodologies (Ovalle and Garbay, 1992; Dieng, 1995; Glaser, 1996; Iglesias et al., 1998; Gustavsson, 1998). In this section, we survey methodologies developed recently for multi-agent system design.

Name: Agent-Oriented Design Methodology
Research Group: Kinny et al., AAII
URL: http://www.aaii.oz.au/
Description: This is an agent-oriented methodology for the design and specification of agent systems. It advocates system decomposition based on key roles. Identification of roles and their relationships guides the specification of the agent class hierarchy; agents are particular instance of these classes. Analysis of the responsibilities of each agent class leads to the identification of the services provided and used by an agent, and hence its external interactions. Consideration of issues such as the creation, duration, and interaction of roles determines the control relationships between agent classes.

Although this methodology is also an extension of object-oriented design methodologies, it is quite subtle in its interpretation and leads to a substantially different approach. Interested readers may refer to (Kinny et al., 1996; Kinny and Georgeff, 1997).

Name: CLAIMS
Research Group: Kendall et al., RMIT
URL: http://www.cse.rmit.edu.au/~rdsek/
Description: CLAIMS (Cooperative Layered Agents for Integrating Manufacturing Systems) is a methodology for developing agent-based systems for enterprise integration, based upon the IDEF (Integration DEfinition for Function modeling) approach for workflow modeling and analysis, the CIMOSA (Computer Integrated Manufacturing Open System Architecture) enterprise modeling

framework, and the use case driven approach to object-oriented software engineering.

With this methodology, agents can be identified, along with the plans, goals, beliefs, sensors, and effectors that allow them to deal with objects existing in their outside environment. The methodology for the analysis and design of agent collaboration is based on use cases and use case abstraction. More details about CLAIMS can be found in (Malkoun and Kendall, 1997).

Name: DESIRE

Research Group: Brazier, Treur et al., Vriji University Amsterdam

URL: http://www.cs.vu.nl/vakgroepen/ai/projects/desire/

Description: DESIRE (DEsign and Specification of Interacting REasoning components) is a modeling framework for developing agent-based systems, which includes:

(1) a design method for compositional multi-agent systems;

(2) a formal specification language for system design – supporting conceptual design through to detailed (but implementation-independent) design;

(3) software tools to support system design;

(4) an implementation generator to automatically translate specifications into code in a specific implementation environment;

(5) verification tools for static properties of components such as consistency, correctness, and completeness.

The structure of DESIRE specifications is based on the notion of compositional architecture: an architecture composed of components with hierarchical (sub-super) relationships between them. Each component has its own input and output interface specification and task control knowledge, thus specifying the dynamics of the whole system in a structured, decentralized manner. Components (composed or primitive) can be (re)-used as building blocks; their input and output interfaces are defined, but the internal structure is hidden from the rest of the system. Agents that perform a part of a complex task are modeled as composed components. The functionality of a composed component is specified by the way in which it is composed from its sub-components (by means of a temporal declarative specification of task control and information exchange between its sub-components). The functionality of a primitive component is specified in a declarative manner by a knowledge base or by reference to another type of specification.

A generic agent model based on the notion of weak agency[7] is supported, whose main components are:

- *Own Process Control*: relates to an agent's ability to reason about its own processes, goals and plans, and to control these processes;

- *Agent Interaction Management*: relates to the ability to communicate with

---

[7] The characteristics of weak agency include autonomous, social, reactive and pro-active behaviours (Wooldridge and Jennings, 1995).

other agents;

- *World Interface Management*: relates to the ability to interact with the external world;
- *Maintenance of World Information*: relates to the information an agent has about the world (its environment);
- *Maintenance of Agent Information*: relates to the information an agent has about other agents;
- *Agent Specific Task*: an empty generic component.

Name: MAS-CommonKADS
Research Group: Iglesias et al., Univ. de Valladolid, Univ. Politécnica de Madrid
URL: n/a
Description: MAS-CommonKAD extends the knowledge engineering methodology CommonKAD (Schreiber et al., 1994) with techniques from object-oriented and protocol engineering methodologies. The methodology involves developing seven models:

- *Agent Model*: describes the characteristics of each agent;
- *Task Model*: describes the tasks that the agents carry out;
- *Expertise Model*: describes the knowledge needed by the agents to achieve their goals;
- *Organizational Model*: describes the structural relationships between agents (software agents and/or human agents);
- *Coordination Model*: describes the dynamic relationships between software agents;
- *Communication Model*: describes the dynamic relationships between human agents and their respective personal assistant software agents;
- *Design Model*: refines the previous models and determines the most suitable agent architecture for each agent, and the requirements of the agent network.

More information can be found in (Iglesias et al., 1998).

Name: Multi-Agent Scenario-Based Method
Research Group: Moulin et al., University of Laval
URL: http://barber.ift.ulaval.ca/~moulin/
Description: The Multi-Agent Scenario-Based Method is a multi-agent system design approach based on the analysis and design of scenarios involving both human and artificial agents. This method is suited for model systems involving human/computer cooperative work. Various design techniques are proposed for scenarios and agent behaviors: behavior diagrams, data models, transition diagrams, object life cycles, and object behavior diagrams. This approach involves structured analysis and design following a philosophy familiar to those versed in software engineering, object-oriented design, or knowledge engineering. More information can be found in (Moulin and Cloutier, 1994) and (Moulin and Brassard, 1996).

16.6 EVALUATIONS AND SURVEYS

Kiniry and Zimmerman (1997) provide a comparison of three leading commercial products (IBM's Aglets, General Magic's Odyssey, and ObjectSpace's Voyager) in terms of ease of installation, feature set, documentation, and cost.

Levy et al. (1996) identify the primary infrastructure services required by agent-based applications, and evaluate current operating systems, programming languages and development tools to determine their suitability for implementing agent-based applications.

Iglesias et al. (1999) survey agent-oriented methodologies including some extensions to object-oriented methodologies and knowledge engineering methodologies.

AgentBuilder's web site (http://www.agentbuilder.com/AgentTools/) provides a comprehensive list of agent tools and frameworks within two categories: commercial products and academic and research projects.

Finally, note that the UMBC AgentWeb (http://www.cs.umbc.edu/agents/) maintained by Tim Finin provides a wealth of information on tools, frameworks, languages, standards and methodologies for developing agents and agent-based applications.

REFERENCES

Barbuceanu, M. and Fox, M. (1995) COOL: A Language for Describing Coordination in Multi-Agent Systems, In *Proceedings of the First International Conference on Multi-Agent Systems*, AAA Press/The MIT Press, pp. 17-25.

Brückner, S, Wyns, J, Peeters, P, and Kollingbaum, M. (1998) Designing Agents for the Manufacturing Process Control, In *Proceedings of Artificial Intelligence and Manufacturing Research Planning Workshop - State of the Art & State of the Practice*, AAAI Press, Albuquerque, New Mexico, pp. 40-46.

Bryson, J. and McGonigle, B. (1998) Agent Architecture as Object Oriented Design, In Singh, M.P., Rao, A. and Wooldridge, M.J., eds., *Intelligent Agents IV: Agent Theories, Architectures, and Languages*, Lecture Notes in Artificial Intelligence 1365, Springer, pp. 15-29.

Burmeister, B. (1996) Models and methodology for agent-oriented analysis and design, In Fischer, K., ed., *Working Notes of the KI'96 Workshop on Agent-Based Programming and Distributed Systems*, DFKI, Germany.

Chung, P.E., Huang, Y., Yajnik, S., Liang, D., Shih, J.C., Wang, C.-Y. and Wang, Y.M. (1998) DCOM and CORBA Side by Side, Step By Step, and Layer by Layer, *C++ Report*, 10(1):18-29,40.

David, R. (1995) Grafcet: A Powerful Toll for Specification of Logic Controllers, *IEEE Transactions on Control Systems Technology*, 3(3):253-268.

Davies, W. and Edwards, P. (1994) Agent-K: An Integration of AOP and KQML, In *Proceedings of the CIKM'94 Workshop on Intelligent Agents*, Gaithersburg, Maryland.

Dieng, R. (1995) Specifying a cooperative system through agent-based knowledge acquisition, In *Proceedings of the International Workshop on Cooperative Systems (COOP'95)*, Antibes-Juan-Les-Pins, France, pp. 141-160.

Glaser, N. (1996) Contributions to Knowledge Modeling in a Multi-Agent Framework (the CoMoMAS Approach), PhD Thsis, University of Nancy I, France.

Gustavsson, R.E. (1998) Multi-Agent Systems as Open Societies – A Design Framework, In Singh, M.P., Rao, A. and Wooldridge, M.J., eds., *Intelligent Agents IV: Agent Theories, Architectures, and Languages*, Lecture Notes in Artificial Intelligence 1365, Springer, pp. 329-337.

Iglesias, C.A., Garijo, M., Gonzalez, J.C. and Velasco, J.R. (1998) Analysis and Design of Multi-Agent Systems Using MAS-CommonKADS, In Singh, M.P., Rao, A. and Wooldridge, M.J., eds., *Intelligent Agents IV: Agent Theories, Architectures, and Languages*, Lecture Notes in Artificial Intelligence 1365, Springer, pp. 313-327.

Iglesias, C.A., Garijo, M. and Gonzalez, J.C. (1999) A Survey of Agent-Oriented Methodologies, In Muller, J.P., Singh, M.P. and Rao, A.S., eds., *Intelligent Agents V: Agent Theories, Architectures and languages*, Lecture Notes in Artificial Intelligence 1555, Springer, pp. 317-330.

Kiniry, J. and Zimmerman, D. (1997) A hands-on look at Java Mobile Agents, *IEEE Internet Computing*, 1(4).

Kinny, D., Georgeff, M. and Rao, A. (1996) A methodology and modeling technique for systems of BDI agents, In van der Velde, W. and Perram, J., eds., *Agents Breaking Away*, Lecture Notes in Artificial Intelligence 1038, Springer.

Kinny, D. and Georgeff, M. (1997) Modelling and Design of Multi-Agent Systems, In Muller, J.P., Wooldridge, M.J. and Jennings, N.R., eds., *Intelligent Agents III: Agent Theories, Architectures and languages*, Lecture Notes in Artificial Intelligence 1193, Springer, pp. 1-20.

Levy, R., Erol, K. and Mitchell, H. (1996) A Study of Infrastructure Requirements and Software Platforms for Autonomous Agents, In *Proceedings of iCSE'96*.

Malkoun, M.T. and Kendall, E.A. (1997) CLAIMS: Cooperative Layered Agents for Integrating Manufacturing Systems, In *Proceedings of PAAM'97*, London, UK.

McCabe, F.G. and Clark, K.L. (1995) April – Agent PRocess Interaction Language, In Wooldridge, M. and Jennings, N., eds., *Intelligent Agents: ECAI-94 Workshop on Agent Theories, Architectures, and Languages*, Amsterdam, the Netherlands, Lecture Notes in Artificial Intelligence 890, Springer, pp. 324-340.

Moulin, B. and Cloutier, L. (1994) Collaborative work based on multi-agent architectures: a methodological perspective, In Aminzadeh, F. and Jamshidi, M., eds., *Soft Computing: Fuzzy Logic, Neural Networks and Distributed Artificial Intelligence*, Prentice-Hall, pp. 261-296.

Moulin, B. and Brassard, M. (1996) A scenario-based design method and an environment for the development of multi-agent systems, In Zhang, C. and Lukose, D., eds., *Distributed artificial intelligence: architecture and modelling*, Lecture Notes in Artificial Intelligence 1087, Springer, pp. 216-231.

Ovalle, A. and Garbay, C. (1992) Towards a method for multi-agent system design, In *Proceedings of Expert Systems '92*, Cambridge University Press, pp. 93-106.

Searle, J.R. (1969) *Speech Acts*, Cambridge University Press.

Schreiber, A.T., Wielinga, B.J., and Akkermans, J.M., van der Velde, W. and de Hoog, R. (1994) CommonKADS – A Comprehensive Methodology for KBS Development. *IEEE Expert*, 9(6).

Shapiro, S., Lespérance, Y. and Levesque, H.J. (1998) Specifying Communicative Multi-Agent Systems with ConGolog, In Wobcke, W., Pagnucco, M. and Zhang, C., eds., *Agents and Multi-Agent Systems: Formalisms, Methodologies, and Applications*, Springer-Verlag, Berlin, pp. 1-14.

Shoham, Y. (1990) *Agent-Oriented Programming*, Technical Report No. STAN-CS-90-1335, Computer Science Department, Stanford University.

Steiner, D.D. (1996) IMAGINE: An Integrated Environment for Constructing Distributed Artificial Intelligence Systems, In O'Hare, G.M.P. and Jennings, N.R., eds., *Foundations of DAI*, Wiley Interscience, pp. 345-364.

Thomas, S.R. (1993) PLACA, an Agent Oriented Programming Language, PhD thesis, Computer Science Department, Stanford University, Stanford, CA.

Wooldridge, M. and Jennings, N. (1995) Intelligent Agents: Theory and Practice, *Knowledge Engineering Review*, 10(2):115-152.

17

Building Agent-Based Concurrent Design and Manufacturing Systems

17.1 INTRODUCTION

We have introduced the fundamental concepts of agent technology and agent-based concurrent design and manufacturing systems in Part 1 (Chapters 2 and 3), discussed the key issues in developing such systems in Part 2 (Chapters 4 through 12), described a number of example systems in Part 3 (Chapters 13, 14 and 15), and presented tools, frameworks, languages, standards, and methodologies in Chapter 16. In this Chapter, we provide some guidelines for practitioners to follow during prototype design and implementation. These guidelines will be useful for researchers and professional engineers who wish to develop industrial agent-based concurrent design and manufacturing systems.

It should be noted that research prototype systems are not developed to optimally solve actual problems but rather to test new concepts and ideas. As such they can turn out to be unnecessarily complex. Thus, industrial users should be careful when trying to utilize research approaches and mechanisms.

17.2 SELECTING OR DEVELOPING AN AGENT ARCHITECTURE

The design of the agent architecture is a key step in developing a multi-agent system. We have earlier introduced various agent architectures described in the literature, including those for agent-based concurrent design and manufacturing systems. Although researchers have made considerable efforts to develop so-called generic agent architectures, it is often difficult to apply these generic agent architectures directly to industrial applications. This section provides some guidelines for an appropriate agent architecture for a specific prototype or application.

In Chapter 6 (Section 6.5), we briefly compared different architectures from an agent behavior point of view. For the practitioner beginning his or her first prototype implementation, we suggest starting with a simple reactive agent architecture. This approach is also very useful for encapsulating an existing engineering tool or database.

However, for real industrial applications, the layered hybrid agent architectures incorporating features from reactive, deliberative and collaborative agent architectures are more appropriate. In the following paragraphs, we outline how to design agent architecture from an agent's internal organization point of view:

1) Classify the different types of agents needed for a specific application. Some agents may be developed by encapsulating existing engineering tools and databases to provide specific services; user interface agents will need implementing for interaction with managers and others users; some middle agents may be needed to facilitate communication and cooperation among agents; and so on. The various types of agents need different components and mechanisms for their internal organizations, which will result in different internal architectures.

2) Define the desired characteristics for these different types of agents. Each agent type should have a list of desired characteristics. For example, for a database agent (or an agent encapsulating an existing database), important desired characteristics would include being communicative, reactive, semi-autonomous, and persistent. For a user interface agent, important characteristics would include being communicative, collaborative, pro-active, and adaptive.

3) Design the necessary modules to implement the desired characteristics for each type of agent. These modules will include interfaces with the environment (e.g., communication, perception, and execution, etc.), knowledge models, and mechanisms for planning, reasoning, and control.

4) Organize these modules using the modular approach or the layered approach as described in Chapter 6 (Section 6.4.2).

During the agent architecture design, an agent development tool or framework may be selected to assist the design process, particularly for the design of the knowledge models and the mechanisms for planning, reasoning and control. We explore this further in Section 17.8.

17.3 SELECTING AN APPROACH FOR AGENT ORGANIZATION

Each specific application's characteristics govern many of the choices in the development effort. The system architecture must supply the underlying communication infrastructure for facilitating data exchange, as well as a shared language and communication protocol, and a control infrastructure for ensuring that agents can execute at appropriate times.

Obviously, the application's infrastructure will constrain the choice of agents, and vice versa. For example, an application involving humans and highly sophisticated software agents that plan and coordinate their own tasks and manage their own information flow requires an infrastructure that primarily supports the flow of messages among the participants. In contrast, if the agents are encapsulated legacy tools with no innate concept of goal-directed or cooperative behavior – that is, they are isolated programs that operate only when directly called – the infrastructure then needs to provide considerable functionality. This might include planning and scheduling tasks, invoking agents at appropriate times, and merging partial solutions returned by the various agents. This functionality could either be embedded in a sophisticated (infrastructure) architecture or provided by separate agents acting as controllers in a simple system architecture.

In Chapter 7, we have introduced various system architectures for agent organization (Section 7.4), and provided a comparison of these approaches (Section 7.5). This section provides some guidelines for practitioners on appropriate approaches for agent organization.

For anyone first developing a prototype for personal learning purposes, we suggest the autonomous agent approach using a small number of agents. This allows a practitioner to feel and understand the key differences between a pure agent-based system and a traditional centralized system. However, when the prototype becomes complex, with more agents and more difficult problems to be resolved, some degree of global system or subsystem control will be needed. Facilitators, mediators, or the like will be needed to facilitate communication and cooperation among the diverse agents, and shared media may be needed for information or knowledge sharing within a group of agents (such shared media may be temporary or permanent).

The federated approach tends to be widely used in implementing practical industrial agent-based concurrent design and manufacturing systems. Different types of middle agents may be employed according to the specific application. More than one type of middle agent can be implemented in one system.

When the problems to be resolved become more complex, particularly when other related technologies (such as Internet technology, CSCW and Groupware technology as discussed in Chapter 12 – Section 12.10) are to be integrated, a hybrid approach such as used in the MetaMorph II project (see Chapter 15 – Section 15.5) is more appropriate.

17.4 SELECTING OR DEVELOPING LANGUAGES AND PROTOCOLS FOR INTER-AGENT COMMUNICATION

At the time of writing, KQML (Knowledge Query and Manipulation Language) (see Chapter 16 – Section 16.4) is the most widely used language/protocol for communication in multi-agent systems. However, it has the disadvantage of being unnecessarily complex for many applications and non-standard (each research group adds or changes some performatives according to their needs).

Until accepted international standards become available for agent communication, KQML should be considered one of the candidates for the system being implemented. However, for a prototype implementation or a simple application, only a subset of KQML performatives may be needed. Or, a small number of performatives formed by combining several KQML performatives may be all that is needed. From the authors' first-hand experience, the following three performatives are fundamental:

- *Request:* to ask for provision of a service, or for bids;

- *Reply:* to answer requests;
- *Inform:* to notify agents without expecting a response.

In addition, two more performatives, i.e., *call-for-bids*, and *bid*, will be found useful, particularly for distributed task allocation using the Contract Net protocol.

The FIPA (Foundation for Intelligent Physical Agents) Consortium (see Chapter 16) has developed the agent communication language ACL. A detailed comparison of KQML and ACL is given in the annex of the relevant FIPA specification. KQML and ACL have many similarities as well as differences, which are listed in detail in this annex. Because the FIPA specifications are becoming increasingly used and accepted by industry, it is suggested that the FIPA communication standards should also be considered for any system being implemented.

17.5 SELECTING OR DEVELOPING PROTOCOLS AND MECHANISMS FOR COOPERATION, COORDINATION AND NEGOTIATION

The contract net (Smith, 1980) and its variants constitute the most widely used set of protocols and mechanisms for cooperation, coordination and negotiation. The contract net was introduced in Chapter 3 and further discussed in Chapters 8, 9 and 10. A practitioner can use the basic contract net protocol for initial prototype implementation, and then a modified or extended version for subsequent implementation.

17.6 SELECTING STANDARDS FOR INFORMATION EXCHANGE

We will not discuss comprehensive standards to be used when developing agent-based systems, because there are no such internationally accepted standards at this time. However, we note that FIPA is an international consortium with this objective, whose specifications are considered as being effectively standards by the members of FIPA consortium. The Holonic Manufacturing Systems consortium also has the objective of developing such standards.

In this section, we consider the selection of standards for data and knowledge exchange in agent-based concurrent design and manufacturing systems. The Knowledge Interchange Format (KIF) is a computer-oriented language for the interchange of knowledge among disparate agents (see Chapter 16 – Section 16.4 for details). It has widely been accepted by the agent research community, and has also been considered as a recommended standard for knowledge exchange in multi-agent systems by the FIPA Consortium. KIF is a good candidate as a common content format for knowledge exchange among agents, particularly when KQML is selected as the communication language for the same system.

In agent-based concurrent design and manufacturing systems, much of the information exchange among agents is product information exchange among various

CAD/CAM/CAE/CAPP tools, databases and manufacturing execution systems. Such information exchange occurs in large or even huge volume. The tools provided by different vendors were initially developed without considering interoperability with other tools. Also, these tools may be located in different manufacturing sites and in different organizations. Exchange of data between these tools can be based on STEP (ISO 10303) (Bjork and Wix, 1991), an International Standard for the computer-interpretable representation and exchange of product data. STEP provides a mechanism capable of describing product data throughout the life cycle of a product, independent of any particular system. It is therefore a good candidate for a standard for product information exchange in agent-based concurrent design and manufacturing systems. More information about STEP is available through the following web site:

http://www.nist.gov/sc4/www/stepdocs.htm

When applications are related to business transactions, such as those in electronic commerce, supply chain management or virtual enterprises, the EDI (Electronic Data Interchange) standards may be adopted. A well-known EDI standard is UN/EDIFACT (United Nations Electronic Data Interchange for Administration, Commerce and Transport)[1].

For some situations, the EDI protocols are too complex, particularly for practitioners implementing prototypes or simple applications. Most researchers and many developers are now turning instead to the emerging XML (eXtensible Markup Language) standards. Although XML is still evolving and many tools for it are still in development, it is already being supported by leading companies like Microsoft (http://www.microsoft.com/xml/) and IBM (http://www.ibm.com/xml/). It will undoubtedly become a widely accepted international standard. It is therefore strongly recommended that XML be used for implementing agent-based business applications. One of the main advantages of using XML is easy implementation in the Internet environment.

17.7 AGENT-ORIENTED ANALYSIS AND DESIGN

Like traditional software development, the development of an agent-based system should have the four phases of Analysis, Design, Implementation and Test. This section discusses only the first two, namely, agent-oriented analysis and design. The other two phases are discussed in subsequent sections.

At the time of writing, there is not yet any widely accepted method for agent-oriented analysis and design, although several research groups are working on formal methodologies for these areas. As noted in Chapter 16, most existing agent-oriented analysis and design methods, still under development in research laboratories, are

---

[1] For more information, visit the website of the United Nations Trade Data Interchange Directory at: http://www.unece.org/trade/untdid/.

various extensions of object-oriented analysis and design methods or are extensions of knowledge engineering methodologies. Until these methodologies for agent-oriented analysis and design are more widely accepted, it is suggested that existing object-oriented analysis and design methods and tools be used. This is usually possible since most industrial agent systems are still being implemented using object-oriented languages. Even where implementation is not directly through an object-oriented language (but for example, through a commercial agent development environment), many of these methods and tools can still be used because agents in the general sense are still objects.

The methodology for Object-Oriented Analysis and Design (OOAD) is becoming increasingly adopted and is of proven value in the modeling and design of object systems. The more recent approaches to OOAD use UML as the standard for the modeling language. OOAD can also be used with advantage for agent systems as its concern is with entities that send messages to each other to accomplish tasks (often in collaboration).

It is in the transformation of the OOAD artifacts (models, diagrams) to computer program code that there is a distinct difference between object systems and agent systems. In object-oriented programming languages (such as Smalltalk, C++, Java), "messages" are implemented by one object (sender) invoking a method of the other object (receiver). Although no message is actually sent, the "receiving" object carries out the activity desired, just as if this had been initiated via message sending. The effect is equivalent. In a true agent-based system, the messages are *actually* sent and the agents need to have communication interfaces (modules) as described earlier in Chapter 6. However, many agent systems are implemented using an object-oriented programming language and their messages can be implemented as described above for object systems.

For object systems, architectural and design features that experience has shown to be valuable are often formalized as "design patterns" (Gamma et al., 1995). These guidelines for good design of object systems are generally applicable to agent systems implemented using an object-oriented programming language, and even useful (perhaps with some adaptation) when agent systems are implemented using an agent development environment or agent-oriented programming language.

Similar guidelines to good practice in agent system development have also been formalized, as "agent design patterns" (Aridor and Lange, 1998; Kendall et al., 1995; Kendall et al, 1998; Shu and Norrie, 1999). A thorough knowledge of both object and agent design patterns is advantageous when developing multi-agent systems.

17.8 SELECTING FRAMEWORKS, TOOLS AND LANGUAGES

It is important to select an effective development tool or framework and/or the primary programming language (though more than one programming language is usually needed) before implementing an agent-based system. There are different ways to then start the prototype implementation. We suggest two alternatives to help the practitioner to start more easily:

- A traditional object-oriented programming language may be used when implementing a simple proof-of-concept prototype. In this case, the agents are usually modeled as objects, with communication among agents being by simple message passing. To test agent communication using local level network protocols, agents may be separated into groups each of which is implemented as one process, with communication among these groups being based, for example, on the TCP/IP protocol. This is a practical way to learn this new technology, which is particularly suitable for students working on a course project for a related course.

- As noted in Chapter 16, commercial frameworks or tools are already available for the development of multi-agent systems. Some research groups have also developed, or are developing, agent development tools for use by their own group or other collaborative groups. Using a framework or tool, particularly a commercial one, is a faster way to implement a prototype. It is also useful for a pilot system that is to be further developed into a real industrial application. In the authors' opinion, however, this is not a good way for students to learn about agent technology, especially when a commercial package is used. Note, though, that some research tools developed by universities were initially developed for educational purposes and are often suitable for student use.

- The selection of an effective commercial package (usually including methodology, framework and tools) is a difficult task. It is not appropriate for us to recommend such packages. Selection of a package depends on the application domain and the software, hardware, and management environments. A safe procedure would be to test different packages, with the assistance of their developers.

17.9 SIMULATION AND IMPLEMENTATION

We have previously discussed key issues in implementing agent-based systems (particularly agent-based concurrent design and manufacturing system). The implementation of a real agent-based system, however, even just a prototype, is a difficult task in our experience, although commercially available tools or frameworks (see Chapter 16) can make this task easier.

For the audiences of this book, primarily professional engineers (rather than professional software engineers) or graduate students, a practical approach would be to first implement the desired application as a simulation. This is common practice in the agent research community. The majority of agents in the application would be implemented as objects grouped within a number of separate programs (or processes). Only a few agents (such as the middle agents described in Chapter 3) would be implemented as real agents and programmed in separate processes. These could even be located in different computers. Many existing agent-building tools, including those listed in the previous chapter (Chapter 16) provide environments for such implementation and simulation of multi-agent systems.

17.10 TESTING, DEBUGGING AND EVALUATION

The testing and debugging of a multi-agent system requires significant effort, because even when individual agents in a multi-agent system are "correct", the overall system behavior might be different from expectation. Formal methodologies and well-developed tools for testing and debugging would be very desirable. However, as stated earlier, no widely accepted methodology or tools are available in this area. Of those that are available, most are research prototypes and are still under development.

We note here, four agent-building tools that provide functionality for testing and debugging a multi-agent system: BT Laboratories' ZEUS (Collis et al., 1998); JAFMAF at the University of Cincinnati (Chauhan, 1997); MAGE by the University of Kansas (Soh et al., 1999); and AgentBuilder (http://www.agentbuilder.com/) from Reticular Systems, Inc..

JAFMAF uses Petri Net based analysis tools to determine conversation coherency in a multi-agent system. MAGE provides functionality for testing, debugging, and analyzing a multi-agent system through its graphical user interface (GUI). The ZEUS Toolkit provides similar functionality through its visualization tool-set.

The AgentBuilder Toolkit, which is a commercial software package introduced in the previous chapter, includes tools for managing the agent-based software development process, analyzing the domain of agent operations, designing and developing networks of communicating agents, defining the behaviors of individual agents, and debugging and testing agent software.

The evaluation of a multi-agent system is very difficult because of the nature of such systems, and there are no comprehensive methodologies or tools available at the time of writing. An interesting Web page (http://www.isi.edu/soar/galk/Eval/), however, that has some useful concepts, focuses on RoboCup Multi-Agent Systems Evaluation.

REFERENCES

Aridor, Y. and Lange, D.B. (1998) Agent Design Patterns: Elements of Agent Application Design, In *Proceedings of the Second International Conference on Autonomous Agents*, Minneapolis, MN, pp. 108-115.

Bjork, B. and Wix, J. (1991) An Introduction to STEP. Technical Report, VTT Technical Research Centre of Finland and Wix McLelland Ltd., England, 1991.

Chauhan, D. (1997) JAFMAS: A Java-based Agent Framework for Multiagent Systems Development and Implementation, Master Thesis, ECECS Department, University of Cincinnati (http://www.ececs.uc.edu/~abaker/JAFMAS/)

Collis, J., Ndumu, D., Nwana, H. and Lee, L. (1998) The ZEUS Agent Building Toolkit, *BT Technology Journal*, 16(3):60-68. (http://www.labs.bt.com/projects/agents/research/collab2.htm)

Gamma, E., Helm, R., Johnson, R. and Vlissides, J. (1995) *Design Patterns: Elements of Reusable Object-Oriented Software*, Addison-Wesley.

Kendall, E.A., Pathak, C.V., Krishna, P.V.M. and Suresh, C.B. (1995) The Layered Agent Pattern Language, In *Pattern Languages of Programming (PLOP '97)*, pp. 53-63.

Kendall, E.A., Krishna, P.V.M., Pathak, C.V. and Suresh, C.B. (1998) Patterns of Intelligent and Mobile Agents, In *Proceedings of Autonomous Agents '98 (Agents '98)*, pp. 115-122.

Shu, S. and Norrie, D.H. (1999) Patterns for Adaptive Multi-Agent Systems in Intelligent Manufacturing, In *Proceedings of the Second International Workshop on Intelligent Manufacturing Systems (IMS '99)*, Leuven, Blegium, pp. 67-74.

Smith, R.G. (1980) The Contract Net Protocol: High-Level Communication and Control in a Distributed Problem Solver, *IEEE Transaction on Computers*, 29(12):1104-1111.

Soh, L.-K., Sevay, H. and Tsatsoulis, C. (1999) MAGE: Multi-Agent Graphical Environment, In *Proceedings of the 1999 AAAI Symposium Spring Series: Intelligent Agents in Cyberspace*, Stanford University, CA (http://www.ittc.ukans.edu/mage/)

Index